W0079594

Concepts of Chemical Engineering 4 Chemists

Concepts of Chemical Engineering 4 Chemists

Stefaan J.R. Simons
Department of Chemical Engineering,
University College London, London, UK

RSCPublishing

ISBN-13: 978-0-85404-951-6

A catalogue record for this book is available from the British Library

© The Royal Society of Chemistry 2007

All rights reserved

Apart from fair dealing for the purposes of research for non-commercial purposes or for private study, criticism or review, as permitted under the Copyright, Designs and Patents Act 1988 and the Copyright and Related Rights Regulations 2003, this publication may not be reproduced, stored or transmitted, in any form or by any means, without the prior permission in writing of The Royal Society of Chemistry, or in the case of reproduction in accordance with the terms of licences issued by the Copyright Licensing Agency in the UK, or in accordance with the terms of the licences issued by the appropriate Reproduction Rights Organization outside the UK. Enquiries concerning reproduction outside the terms stated here should be sent to The Royal Society of Chemistry at the address printed on this page.

Whilst every effort has been made to contact the owners of copyright material, we apologise to any copyright holders whose rights we may have unwittingly infringed.

Published by The Royal Society of Chemistry,
Thomas Graham House, Science Park, Milton Road,
Cambridge CB4 0WF, UK

Registered Charity Number 207890

For further information see our web site at www.rsc.org

Preface

This book is meant as a handbook and resource guide for chemists (and other scientists) who either find themselves working alongside chemical engineers or who are undertaking chemical engineering-type projects and who wish to communicate with their colleagues and understand chemical engineering principles. The book has arisen out of the short course, Concepts of Chemical Engineering for Chemists, held annually at UCL since 1999 and the forerunner to the Royal Society of Chemistry's "4 Chemists" series of professional training courses, of which it is now part. It can be used as accompanying material for the course, or as a stand-alone reference book. The course itself is designed to provide basic information on the main aspects of chemical engineering in a relatively simple, but practical, manner. Hence, while this book tries to emulate this, it also includes worked examples, plus extensive reference lists and bibliographies in order that the reader can research elsewhere for more detail and for aspects that are not covered in the book.

This book aims to give chemists an insight into the world of chemical engineering, outlining the basic concepts and explaining the terminology of, and systems approach to, process design. It can be said that chemists create new molecules and compounds and chemical engineers manufacture these into useful products on a commercial scale, but, of course, the two disciplines do not work in isolation; chemistry and chemical engineering are intertwined. One only has to look at the history of chemical engineering and its origins in chemistry (or, more correctly, applied chemistry) to appreciate the close relationship between the two and their shared foundation in molecular behaviour. The reader is referred to Darton *et al.*'s collection of visionary essays on chemical engineering's role in society[1] and the Whitesides report, "Beyond the Molecular Frontier",[2] which reflect on the importance of chemists and chemical engineering working effectively together to tackle the enormous challenges facing the world today.

To work effectively together, chemists and chemical engineers need to know how to communicate. This was the premise for the short course mentioned above and is the basis on which this book is written. Hence, the book does not focus on the derivation of mathematical formulae, but rather on the use of the governing principles in process design. Before I describe what the book contains, it may be useful here to take a brief look at what a typical chemical engineering degree course involves.

WHAT CHEMICAL ENGINEERING INVOLVES

Chemical Engineers are responsible for the design and operation of processes and of products and their application. Since they must consider processes in their entirety, from raw materials to finished products, they use a "systems" approach, enabling the prediction of the behaviour of both the process as a whole and of the individual plant items. Once a plant has been designed, chemical engineers are involved in its construction, commissioning and subsequent operation. Hence, in terms of education, a chemical engineer will undertake courses in:

- Mathematics (the emphasis is on engineering, after all).
- Science (notably chemistry, biology, material science).
- Process analysis (defining the mass, momentum and energy balances for the entire process).
- Thermodynamics (determining the fundamental parameters on which the process can be analysed).

Once the process has been defined, attention can then be paid to the "unit operations", the reactors, separators, *etc.* that make up the process route, or flowsheet (Figure 1). This involves the study of:

- Reaction engineering (the manipulation of molecular behaviour to determine the reaction routes to a specified product).
- Transport processes (the physical manipulations that underlie the process).
- Separation processes (the manner in which products are separated and purified).

Interwoven with all these topics are the crucial areas of safety (see Figure 2), risk analysis, plant and equipment design, process control and process economics. A practising chemical engineer will often find that the data he/she requires is either unreliable or incomplete[3] and, hence, he/she must make sound judgements based on mathematics, physics and chemistry to determine appropriate simplifying assumptions, while at

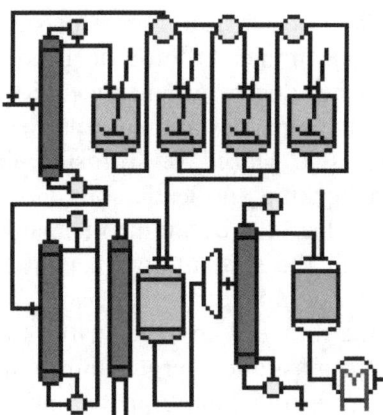

Figure 1 *Chemical engineers develop process flowsheets from mass and energy balances based on the conservation principle and using their knowledge of unit operations and thermodynamics*

Figure 2 *The explosion at the Buncefield oil depot in 2005 is a graphic example of how safety issues must always be paramount in chemical engineering (photo courtesy of Dr. Dave Otway, Department of Chemistry, University College Cork, taken by him from Ryanair flight FR903 STN-CORK at 11.40 am 11-12-05 10 min after take-off)*

the same time satisfying safety, environmental, operational and legal constraints.

Chemical Engineers are profoundly aware of their ethical and social responsibilities, encompassed in the notion of "sustainable development" which is becoming an increasing component of chemical engineering degree programmes. Often a chemical engineering student will supplement his or her degree programme with courses on environmental practice, law, management and entrepreneurship. In addition, a great deal of emphasis is placed on transferable skills training, in communication, teamworking and leadership.

The culmination of a degree programme in chemical engineering is the design project, in which the students work in teams to carry out the preliminary design of a complete process plant. This exercise involves the use of much of the material covered in the topic areas mentioned above, starting from the mass and energy balances and ending with a full economic appraisal, environmental impact and risk assessment (or, increasingly, a sustainability analysis on the socio-economic as well as environmental impacts) and safety analysis. Very often the teams will include MSc students with first degrees in chemistry, underlining the close relationship between the two disciplines and the need for mutual understanding in the development of effective and appropriate plant designs.

Sustainable development is defined as "development that meets the needs of the present without compromising the ability of future generation to meet their own needs". As stated in Ref. 4, technological change must be at the heart of sustainable development. In order to achieve a sustainable future, the basic principles that should guide technological (and societal) development are that consumption of resources should be minimised while that of non-renewable materials should cease, with preference given to renewable materials and energy sources. It is chemical engineers, together with chemists and other scientific disciplines, who will lead this technological revolution.[5] Although beyond the scope of this book, these issues must be at the forefront of all process and product developments and the concepts covered here are the bases on which to found this progress.

WHAT THIS BOOK CONTAINS

The book begins in Chapter 1 with the cornerstone of any process design, the development of mass and energy balances, based on the simple conservation principle. From such balances a chemical engineer will then go on to add more detail in order to come up with a fully optimised process flowsheet, providing the most efficient, safe and economic route to the production of the specified chemicals within the constraints of thermodynamics, material properties and environmental regulations. All chemical processes are dynamic in nature, involving the flow of material (fluid flow or momentum transfer) and the transfer of heat and mass across physical and chemical interfaces. The central part of any chemical process is the reactor, in which chemicals are brought together to produce the precursors to the eventual products. A chemical engineer must be able to predict and manipulate chemical reaction kinetics to be able to design such reactors. Hence, the concepts of chemical reaction engineering in relation to reactor design are considered in Chapter 2. The equally important principles behind momentum, heat and mass transfer are then covered in the following three chapters. Moving often huge quantities of material around a plant, through pipelines and in and out of vessels, requires a knowledge of fluid mechanics and its use in the design of the appropriate pumps and measurement equipment (Chapter 3). Maintaining rates of reactions and product quality requires the transfer of heat to and from chemicals, often through physical boundaries (*e.g.* pipe and vessel walls), and heat exchangers are a common means in which this transfer is achieved. The formulation of heat-transfer rate equations is described in Chapter 4 and their use demonstrated in the design of shell and tube heat exchangers. Mass transfer typically occurs in the separation of chemical components to remove impurities from the desired product. Distillation is a common separation technique employed in chemical plants and is described in Chapter 5.

Often, chemical plant design is informed by laboratory and pilot-scale experimentation. While the initial chapters in this book will inform the reader of the most important design parameters that need to be measured and determined from such experiments, how to then ensure that these parameters perform in the same way in a large-scale plant is the subject of Chapter 6. Although it is desirable to conduct the experimental work in the system for which the results are required, this is not always easy. The system of interest may be hazardous or expensive to build and run, while the fluids involved may be corrosive or toxic. In this case scale models are used, which overcome the above problems and allow extensive experimentation. In the majority of cases the model will be smaller in size than the actual, desired plant, but sometimes, due to the nature of the materials to be handled, the fluids involved may also be different. Scale-up is only possible if the model and plant are physically similar and, hence, the procedure is based on dimensionless groups. How to develop and use these groups is described in Chapter 6.

Many chemicals at some stage in a process (*i.e.* whether as raw materials, intermediates, products and by-products) are in powder form. Often, the handling of powders is mistakenly assumed to be relatively straight forward, leading to disastrous consequences (*e.g.* clogging of storage vessels, dust explosions, *etc.*). The characterisation and handling of powders is described in Chapter 7, with particular emphasis given to fluidisation as a common process operation. Particle science and technology is a rapidly maturing field but, surprisingly, there is still much reliance on empiricism in chemical engineering practice.

Throughout the design of a chemical plant, issues relating to safety, economics and environmental impact must be considered. By doing so, the risks associated with the plant can be minimised before actual construction. The same principle applies whatever the scale of the process. The field of process control (Chapter 8) considers all these issues and is, indeed, informed by the type of hazard analyses described in Chapter 10. The objectives of an effective control system are the safe and economic operation of a process plant within the constraints of environmental regulations, stakeholder requirements and what is physically possible. Processes require control in the first place because they are dynamic systems, so the concepts covered in the earlier chapters of this book are central to process control (*i.e.* control models are based on mass, energy and momentum balances derived with respect to time). Chapter 8 focuses on the key aspects of control systems.

The economic assessment of a proposed plant, known as project appraisal, is necessary at the design stage in order to determine its viability, the capital requirements and the expected return on investment. Such an analysis can kill a project at any stage in the design. Chapter 9 discusses how the planning for profitable operation is undertaken.

Last, but not least, safety is considered in terms of the analysis of the risks associated with potential hazards identified by detailed consideration of the proposed process flowsheet. Safety is the number one concern for chemical engineers and the reader should not confuse the fact that it is the focus of the final chapter in this book with its order of importance. However, in order to carry out a hazard study and risk assessment, one must understand the concepts on which a process flowsheet is developed, and these are covered in the preceding chapters. The procedure describe in Chapter 10 is recognised as best practice in the process industry sector.

The chapter authors of this book have tried to keep mathematical derivations to a minimum and have assumed that their readers have background knowledge of thermodynamics, since the latter is ubiquitous to both chemistry as well as chemical engineering. The authors are all either full time members of academic staff at UCL or, in the cases of Robert Thornton and Ken Sutherland, are retired industrialists who have extensive experience in their respective fields and have taught undergraduate courses on crucial aspects of chemical engineering practice. My thanks are due to them for their hard work in not only preparing the contents of this book but also in the development and delivery of the lectures and practical sessions of the "Concepts of Chemical Engineering for Chemists" course. Without their efforts, there would be neither.

Stef Simons
UCL

REFERENCES

1. R. Darton, D. Wood and R. Prince (eds), *Chemical Engineering: Visions of the World*, Elsevier Science, Amsterdam, The Netherlands, 2003.
2. Committee on Challenges for the Chemical Sciences in the 21st Century, *Beyond the Molecular Frontier: Challenges for Chemistry and Chemical Engineering*, National Research Council, National Academy of Sciences, Washington D.C., USA, 2003.
3. C.A. Heaton (ed), *An Introduction to Industrial Chemistry*, Leonard Hill, Glasgow, 1984.
4. K. Mulder (ed), *Sustainable Development for Engineers*, Greenleaf Publishing, Sheffiled, UK, 2006.
5. A. Azapagic, S. Perdan and R. Clift (eds), *Sustainable Development in Practice: Case Studies for Engineers and Scientists*, Wiley, Chichester, 2004.

Contents

Chapter 6 Scale-Up in Chemical Engineering 171
Tim Elson

CHAPTER 1

Process Analysis – The Importance of Mass and Energy Balances

ERIC S. FRAGA

1.1 INTRODUCTION

Process engineering includes the generation, study and analysis of process designs. All processes must obey some fundamental laws of conservation. We can group these into conservation of matter and conservation of energy.[†] Given a set of operations, if we draw a box around this set, the amount of mass going in must equal the amount going out; the same applies to the energy. Mass and energy balance operations are fundamental operations in the analysis of any process. This chapter describes some of the basic principles of mass and energy balances.

1.1.1 Nomenclature and Units of Measurement

In carrying out any analysis, it is important to ensure that all units of measurement used are consistent. For example, mass may be given in kg (kilogrammes), in lb (pounds) or in any other units. If two quantities are given in different units, one quantity must be converted to the same unit as the other quantity. Any book on chemical engineering (or physics and chemistry) will have conversion tables for standard units.

There are seven fundamental quantities that are typically used to describe chemical processes, mass, length, volume, force, pressure, energy and power, although some of these can be described in terms

---

[†] These two laws are separate in non-nuclear processes. For nuclear processes, we of course have the well-known equation, $E = mc^2$, which relates mass and energy. For this lecture, we will consider only non-nuclear processes but the same fundamental principles apply to all processes.

Table 1   *Some of the quantities encountered in process analysis with typical notation and units of measure*

Quantity	Notation	Dimension	Units
Time	t	T	s
Mass	m	M	kg
Mass flow	$\dot{m}$	$M\,T^{-1}$	$kg\,s^{-1}$
Mole	n	M	mol
Molar flow	$\dot{n}$	$M\,T^{-1}$	$mol\,s^{-1}$
Pressure	P	$M\,(T^2L^2)^{-1}$	bar
Energy	H,Q,W	$ML^2\,T^2$	Joule

of others in the list. For example, volume is length raised to the power 3; power is energy per unit time, pressure is force per area or force per length squared, and so on.

Chemical engineering uses some standard notation for many of the quantities we will encounter in process analysis. These are summarised in Table 1 where T is time, M is mass and L is length.

In describing processes, the variables that describe the condition of a process fall into two categories:

(i) *extensive* variables, which depend on (are proportional to) the size of the system, such as mass and volume, and

(ii) *intensive* variables, which do not depend on the size of the system, such as temperature, pressure, density and specific volume, and mass and mole fractions of individual system components.

The number of intensive variables that can be specified independently for a system at equilibrium is known as the *degrees of freedom* of the system.

Finally, it is important to note that the precision of quantities is often not arbitrary. Measuring tools have limits on the precision of measurement. Such measures will have a particular number of significant figures. Calculations with measurements may not result in an increase in the number of significant figures. There are two rules to follow to determine the number of significant figures in the result of calculations:

(i) When two or more quantities are combined by multiplication and/or division, the number of significant figures in the result should equal the lowest number of significant figures of any of the multiplicands or divisors. In the following example, one multiplicand has three significant figures and the other, four. Therefore, the result must have no more than three significant figures regardless of the number of figures that are generated by the calculation:

$$3.57 \times 4.286 = 15.30102 \Rightarrow 15.3$$

(ii) When two or more numbers are either added or subtracted, the positions of the last significant figure of each number relative to the decimal point should be compared. Of these positions, the one farthest to the left is the position of the last permissible significant figure of the sum or difference. It is important to make sure that all the numbers are represented with the same exponent if scientific notation is used:

$$
\begin{aligned}
1.53 \times 10^3 - 2.56 &= (1.53 \times 10^3) - (0.00256 \times 10^3) \\
&= (1.53 - 0.00256) \times 10^3 \\
&= 1.52744 \times 10^3 \\
&\Rightarrow 1.53 \times 10^3
\end{aligned}
$$

1.2 MASS BALANCES

Chemical processes may be classified as batch, continuous or semi-batch and as either steady-state or transient. Although the procedure required for performing mass, or material, balances depends on the type of process, most of the concepts translate directly to all types.

The general rule for mass balance in a *system box* (a box drawn around the complete process or the part of the process of interest) is:

$$\text{input} + \text{generation} - \text{output} - \text{consumption} = \text{accumulation} \qquad (1)$$

where,

- (i) *input* is the material entering through the system box. This will include feed and makeup streams;
- (ii) *generation* is the material produced within the system, such as the reaction products in a reactor;
- (iii) *output* is the material that leaves through the system boundaries. These will typically be the product streams of the process;
- (iv) *consumption* is the material consumed within the system, such as the reactants in a reactor;
- (v) *accumulation* is the amount of material that builds up within the system.

In a steady-state continuous process, the accumulation should always be zero, which leads to a more simple mass balance equation:

$$\text{input} + \text{generation} = \text{output} + \text{consumption} \qquad (2)$$

In the case of systems with no reaction, where mass is neither generated nor consumed, the result is even simpler:

$$\text{input} = \text{output} \qquad (3)$$

1.2.1 Process Analysis Procedure

The analysis of the mass balance of a process typically follows a number of steps:

(i) Draw and label a diagram for the process, clearly indicating the information given by the problem definition and the values that have been requested.
(ii) Choose a basis of calculation if required. If no extensive variables (*e.g.* amount or flow rate of a stream) have been defined, a basis of calculation is required and this must be an extensive variable.
(iii) Write down appropriate equations until zero degrees of freedom are achieved. In other words, write down enough equations so that the number of equations equals the number of unknown variables and such that all the unknown variables are referred to in the equations. Possible sources of equations include the following:

  (a) Mass balances. For a system with n species, n mass balance equations may be written down. These mass balance equations may be drawn from a total mass balance and from individual species mass balances.
  (b) Process specifications and conditions such as, for example, the separation achieved by a distillation unit or the conversion in a reactor.
  (c) Definitions such as the relationship between density, mass and volume or the relationship between mole fraction and total mass.

(iv) Identify the order in which the equations should be solved.
(v) Solve the equations for the unknown values.

These steps are illustrated by the following example in Section 1.2.2.

1.2.2   Example 1: Mass Balance on a Continuous Distillation Process

Suppose that a 1000 kmol h^{-1} feed stream, consisting of 30.0% by mole *n*-pentane and the remainder *n*-hexane, is to be separated into 95.0% pure pentane and 95.0% pure hexane streams using a distillation column. Determine the flow rates of the output streams through the use of mass balances, assuming steady-state operation. We will assume three digits of significance for this example.

1.2.2.1   *Solution.*

(i) The first step is to draw and label a flowsheet diagram indicating the process steps and all the streams. Figure 1 shows the layout of the distillation unit labelled with both the known variables and the variables we wish to determine. The system box for this example is also indicated. The streams we wish to consider are those that intersect this system box and consist of the feed stream and the two output streams. All other streams can be ignored in solving this example. The notation we have used is that F is the feed stream, D the distillate or tops product stream (which will be primarily pentane, the lighter of the two species), B the bottoms product stream (primarily hexane), p refers to pentane and h to hexane. We are interested in finding the values for $\dot{n}_D$ and $\dot{n}_B$, as indicated by boxed question marks in the figure. There are three streams and

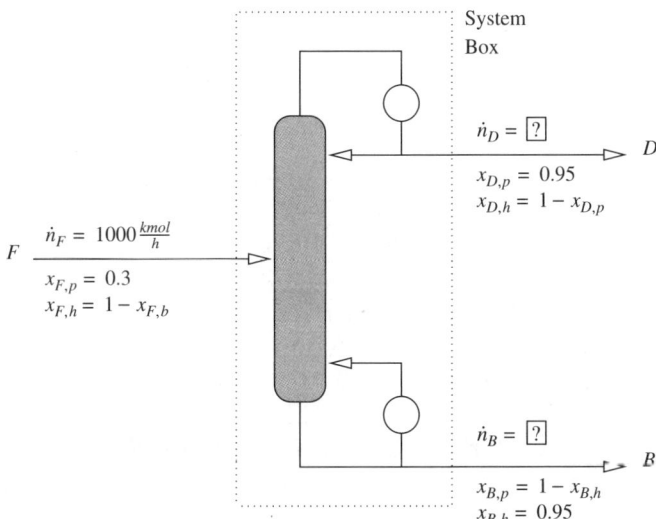

Figure 1   *Distillation process mass balance problem*

two unknowns. Strictly speaking some of the mole fractions are also unknown, but we have directly incorporated the equations we will use to determine values for $x_{F,p}$, $x_{D,h}$ and $x_{B,p}$ into the diagram, noting that by definition the mole fractions of the components in a stream add up to one. Note that we have assumed that the pentane-rich stream is the distillate output of the unit and that the hexane-rich stream is the bottoms product. This is based on their relative volatilities; pentane has a lower boiling point than hexane.

(ii) As this is the simplest case described above (steady-state, continuous and no reaction), we can use the simplest mass balance equation:

$$\text{input} = \text{output}$$

(iii) As indicated above, we have two unknowns. Therefore, we need to generate two independent equations that will allow us to solve for these unknowns. For mass balance problems, the general rule is that we can define n_c equations if there are n_c components in the streams involved in the mass balance problem defined by the system box chosen. The one exception is that if the system box includes only a splitter, there is only one independent equation that can be defined. This is because the splitter does not change the compositions of the streams involved, only the amounts or flows. For this example we have two components, and so we can define two mass balance equations. The choice of equations is a total mass balance and two individual component balances:

$$\dot{n}_F = \dot{n}_D + \dot{n}_B \tag{4}$$

$$x_{F,h}\dot{n}_F = x_{D,h}\dot{n}_D + x_{B,h}\dot{n}_B \tag{5}$$

$$x_{F,p}\dot{n}_F = x_{D,p}\dot{n}_D + x_{B,p}\dot{n}_B \tag{6}$$

(iv) We can choose any two of these three equations to solve our problem. For this example, we will choose the total mass balance, Equation (4), and the pentane mass balance, Equation (6). Together with the mole fraction definitions already labelled on the diagram, we are left with 0 degrees of freedom. The degrees of freedom are defined as the difference between the number of unknowns and the number of equations relating these unknowns. For the example, we have five unknowns – $\dot{n}_D$, $\dot{n}_B$, $x_{F,p}$, $x_{D,h}$ and $x_{B,p}$ – and five equations, three equations from the mole fractions in each stream adding up to one and two equations from the mass balances (Equations (4) and (6)). Therefore, we can solve these equations simultaneously to find the values of the unknowns.

(v) Before solving the equations, it is worth planning ahead to determine which order to solve the equations in. The full set of equations are given here:

$$\dot{n}_F = \dot{n}_D + \dot{n}_B \tag{7}$$

$$x_{F,p}\dot{n}_F = x_{D,p}\dot{n}_D + x_{B,p}\dot{n}_B \tag{8}$$

$$x_{F,p} = 1 - x_{F,h} \tag{9}$$

$$x_{D,h} = 1 - x_{D,p} \tag{10}$$

$$x_{B,p} = 1 - x_{B,h} \tag{11}$$

We can solve for $x_{F,p}$ immediately using Equation (9), as we know $x_{F,b}$. Likewise, we can solve for both $x_{D,h}$, with Equation (10), and $x_{B,p}$, with Equation (11), given the specifications on the output streams ($x_{D,p}$ and $x_{B,h}$). Finally, we solve the two mass balance equations, Equations (7) and (8), simultaneously.

(vi) The actual solution can now proceed:

$$
\begin{aligned}
x_{F,h} &= 1 - x_{F,p} \\
&= 1 - 0.300 = 0.700 \\
x_{D,h} &= 1 - x_{D,p} \\
&= 1 - 0.950 = 0.050 \\
x_{B,p} &= 1 - x_{B,h} \\
&= 1 - 0.950 = 0.050 \\
\dot{n}_D &= \dot{n}_F - \dot{n}_B \\
&= 1000 \text{ kmol h}^{-1} - \dot{n}_B
\end{aligned} \tag{12}
$$

$$x_{B,p}\dot{n}_B = x_{F,p}\dot{n}_F - x_{D,p}\dot{n}_D$$

$$= 0.300 \times 1000 \text{ kmol h}^{-1} - 0.950 \times (1000 \text{ kmol h}^{-1} - \dot{n}_B)$$

$$\Downarrow$$

$$0.050\,\dot{n}_B = 300 \text{ kmol h}^{-1} - 950 \text{ kmol h}^{-1} + 0.950\dot{n}_B \tag{13}$$

We now rearrange this last equation, Equation (13), to have $\dot{n}_B$ on the left hand side alone:

$$(0.050 - 0.950)\,\dot{n}_B = \frac{-650}{-0.900}\ \text{kmol h}^{-1}$$

$$\Downarrow$$

$$\dot{n}_B = \frac{-650}{-0.900}\ \text{kmol h}^{-1}$$

$$= 722\ \text{kmol h}^{-1}$$

This value can now be plugged back into the first mass balance equation, which we previously rearranged with $\dot{n}_D$ on the left hand side, Equation (12):

$$\dot{n}_D = 1000\ \text{kmol h}^{-1} - \dot{n}_B$$

$$= 1000\ \text{kmol h}^{-1} - 722\ \text{kmol h}^{-1}$$

$$= 278\ \text{kmol h}^{-1}$$

(vii) The example has been solved. We can now use the unused mass balance equation, in this case being the hexane component mass balance, Equation (5), to provide a check for consistency:

$$x_{F,h}\dot{n}_F = x_{D,h}\dot{n}_D + x_{B,h}\dot{n}_B$$

$$\Downarrow$$

$$0.700 \times 1000\ \text{kmol h}^{-1} = 0.050 \times 278\ \text{kmol h}^{-1} + 0.950 \times 722\ \text{kmol h}^{-1}$$

$$\Downarrow$$

$$700\ \text{kmol h}^{-1} = 13.9\ \text{kmol h}^{-1} + 686\ \text{kmol h}^{-1}$$

$$\Downarrow$$

$$700\ \text{kmol h}^{-1} = 700\ \text{kmol h}^{-1}$$

So the results are at least consistent, which gives some confidence in their correctness.

For processes involving reactions, the mass balance equation, Equation (3), used in the first example is not sufficient. Equations (1) or (2) must be used. The presence of reactions means that the *generation* and *consumption* terms in these equations are non-zero. The key difference between a simple separation process and a process involving reactions is the need to define these extra terms in terms of the amounts of the components in the system. Extra equations are required to satisfy the extra degrees of freedom introduced by having these terms present.

Two concepts are often used to describe the behaviour of a process involving reactions: *conversion* and *selectivity*. Conversion is defined with respect to a particular reactant, and it describes the extent of the reaction that takes place relative to the amount that could take place. If we consider the *limiting reactant*, the reactant that would be consumed first, based on the stoichiometry of the reaction, the definition of conversion is straight forward:

$$\text{conversion} \equiv \frac{\text{amount of reactant consumed}}{\text{amount of reactant fed}} \tag{14}$$

Selectivity is a concept that applies to processes with multiple simultaneous reactions. It is used to quantify the relative rates of the individual reactions. However, any discussion about multiple reactions and the analysis of these is beyond the scope of this chapter. Refer to the further reading material identified at the end of this chapter for more information.

1.2.3   Example 2: Mass Balance on a Process with Reaction

Suppose an initially empty tank is filled with 1000 mol of ethane and the remainder with air. A spark is used to ignite this mixture and the following combustion reaction takes place:

$$2C_2H_6 + 7O_2 \rightarrow 4CO_2 + 6H_2O \tag{15}$$

Assume that the amount of air provides twice the stoichiometric requirement of oxygen for this reaction, and that air is composed of 79% nitrogen and the remainder oxygen. Suppose that the reaction reaches a 90% conversion. What is the composition of the mixture in the tank at the end of the reaction?

1.2.3.1   Solution.

(i) As in the first example, the first step is to draw and label a diagram, as shown in Figure 2. This example is a batch problem, and so arrows indicate material flow in the sense of loading and discharging the tank. The subscripts for mass amounts are the stream index (1 for the original contents of the tank and 2 for the

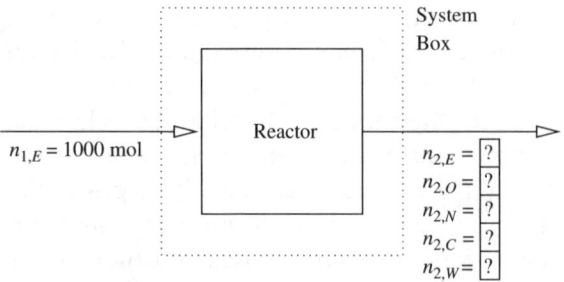

Figure 2 *Reaction process mass balance problem*

final contents after combustion) and the species involved. The key is E for ethanol, O for oxygen, N for nitrogen, C for carbon dioxide and W for water.

(ii) As in the first example, the boxed question marks highlight the variables that need to be determined. As there are five highlighted variables, we will require at least five equations. As there are five species involved, five mass balance equations can be defined. In this case, we should use Equation (1) but with *accumulation* set to zero to indicate that the tank is empty to start with and also empty at the end. This means that the mass balance equation we should use is identical in form to Equation (2):

$$\text{input} + \text{generation} = \text{output} + \text{consumption}$$

Our five mass balance equations, therefore, are as follows:

$$\left.\begin{array}{ll} n_{1,E} + n_{g,E} = n_{2,E} + n_{c,E} & \text{(Ethane balance)} \\ n_{1,O} + n_{g,O} = n_{2,O} + n_{c,O} & \text{(Oxygen balance)} \\ n_{1,N} + n_{g,N} = n_{2,N} + n_{c,N} & \text{(Nitrogen balance)} \\ n_{1,C} + n_{g,C} = n_{2,C} + n_{c,C} & \text{(Carbon dioxide balance)} \\ n_{1,W} + n_{g,W} = n_{2,W} + n_{c,W} & \text{(Water balance)} \end{array}\right\} \quad (16)$$

where the subscript g is used to indicate the amount generated and the subscript c indicates the amount consumed. At this point, all the variables except for $n_{1,E}$ are unknown. We have 19 unknown variables. As we have just defined 5 equations, we have 14 degrees of freedom remaining. To solve this problem, therefore, we need to define at least 14 more equations.

(iii) We can write down new equations relating the unknown and known variables by making use of the stoichiometric coefficients given by Equation (15). It is helpful to write all of these in terms of one of the consumption or generation terms. In this case, given that the key process specification is the conversion of ethane, it helps to write the equations in terms of the amount of ethane consumed:

$$\left.\begin{array}{ll} n_{g,E} = 0 & \text{(No ethane is generated)} \\[4pt] n_{g,O} = 0 & \text{(No oxygen is generated)} \\[4pt] n_{c,O} = \dfrac{7}{2} n_{c,E} & \\[8pt] n_{g,N} = 0 & \text{(No nitrogen is generated)} \\[4pt] n_{c,N} = 0 & \text{(No nitrogen is consumed)} \\[4pt] n_{g,C} = \dfrac{4}{2} n_{c,E} & \\[8pt] n_{c,C} = 0 & \text{(No carbon dioxide is consumed)} \\[4pt] n_{g,W} = \dfrac{6}{2} n_{c,E} & \\[8pt] n_{c,W} = 0 & \text{(No water is consumed)} \end{array}\right\} \quad (17)$$

This set of nine equations reduces the degrees of freedom to 5 as no new variables have been introduced.

(iv) Further equations can be defined on the basis of the specifications of the feed and the conversion of the reaction:

$$\left.\begin{array}{ll} n_{1,O} = 2 \times \dfrac{7}{2} n_{1,E} & \text{(Twice as much as required)} \\[10pt] n_{1,N} = 0.79 \times \dfrac{n_{1,O}}{0.21} & \text{(Nitrogen is remainder of air)} \\[10pt] n_{1,C} = 0 & \text{(No carbon dioxide in the feed)} \\[4pt] n_{1,W} = 0 & \text{(No water in the feed)} \\[4pt] 0.90 = \dfrac{n_{c,E}}{n_{1,E}} & \text{(Conversion of ethane)} \end{array}\right\} \quad (18)$$

This set of five equations also introduces no new unknown variables. The result is that we have 19 equations and 19 unknowns giving zero degrees of freedom.

(v) The system of 19 equations, comprising the three sets of equations above, Equations (16), (17) and (18), can be solved as follows. Given $n_{1,E} = 1000$ mol, the initial amount of ethane, we evaluate the equations in set Equation (18):

$$n_{1,\mathrm{O}} = 2 \times \frac{7}{2} n_{1,\mathrm{E}} = 7 \times 1000 \text{ mol} = 7000 \text{ mol}$$

$$n_{1,\mathrm{N}} = 0.79 \times \frac{n_{1,\mathrm{O}}}{0.21} = 0.79 \frac{7000 \text{ mol}}{0.21} \approx 26333 \text{ mol}$$

$$n_{c,\mathrm{E}} = 0.90 \times n_{1,\mathrm{E}} = 0.90 \times 1000 \text{ mol} = 900 \text{ mol}$$

Now determine the amounts generated and consumed for each species using the set of Equation (17):

$$n_{c,\mathrm{O}} = \frac{7}{2} n_{c,\mathrm{E}} = \frac{7}{2} \times 900 \text{ mol} = 3150 \text{ mol}$$

$$n_{g,\mathrm{C}} = \frac{4}{2} n_{c,\mathrm{E}} = \frac{4}{2} \times 900 \text{ mol} = 1800 \text{ mol}$$

$$n_{g,\mathrm{W}} = \frac{6}{2} n_{c,\mathrm{E}} = \frac{6}{2} \times 900 \text{ mol} = 2700 \text{ mol}$$

Finally, we use the mass balance equation, Equation (16), to determine the amount of each species in the output:

$$n_{2,\mathrm{E}} = n_{1,\mathrm{E}} + n_{g,\mathrm{E}} - n_{c,\mathrm{E}} = 1000 \text{ mol} + 0 - 900 \text{ mol} = 100 \text{ mol}$$

$$n_{2,\mathrm{O}} = n_{1,\mathrm{O}} + n_{g,\mathrm{O}} - n_{c,\mathrm{O}} = 7000 \text{ mol} + 0 - 3150 \text{ mol} = 3850 \text{ mol}$$

$$n_{2,\mathrm{N}} = n_{1,\mathrm{N}} + n_{g,\mathrm{N}} - n_{c,\mathrm{N}} = 26333 \text{ mol} + 0 - 0 = 26333 \text{ mol}$$

$$n_{2,\mathrm{C}} = n_{1,\mathrm{C}} + n_{g,\mathrm{C}} - n_{c,\mathrm{C}} = 0 + 1800 \text{ mol} - 0 = 1800 \text{ mol}$$

$$n_{2,\mathrm{W}} = n_{1,\mathrm{W}} + n_{g,\mathrm{W}} - n_{c,\mathrm{W}} = 0 + 2700 \text{ mol} - 0 = 2700 \text{ mol}$$

These two examples have illustrated some of the key steps in solving mass balance problems. In particular, the number of mass balance equations that are available for ensuring you have zero degrees of freedom and the use of the system box to limit yourself to the streams that are of interest. These steps are also key points for the analysis of energy balances in processes, the topic of Section 1.3 in this chapter.

1.3 ENERGY BALANCES

Energy balances can be treated in much the same way as material balances. The only fundamental difference is that there are three types of energy (for non-nuclear processes):

   (i) *Kinetic*: Energy due to the translational motion of the system as a whole relative to some frame of reference (the earth's surface, for instance) or to the rotation of the system about some axis.

  (ii) *Potential*: Energy due to the position of the system in a potential field. In chemical engineering, the potential field will typically be gravitational.

 (iii) *Internal*: All energy possessed by the system other than kinetic or potential. For example, the energy due to the motion of molecules relative to the centre of mass of the system and to the motion and interactions of the atomic and subatomic constituents of the molecules.

Energy may be transferred between a system and its surroundings in two ways:

   (i) As *heat*, or energy that flows as a result of a temperature difference between a system and its surroundings. The direction of flow is always from the higher temperature to the lower. Heat is defined as *positive* when it is transferred to the system from its surroundings.

  (ii) As *work*, or energy that flows in response to any driving force other than a temperature difference. For example, if a gas in a cylinder expands and moves a piston against a restraining force, the gas does work on the piston. Energy is transferred as work from the gas to its surroundings, including the piston. Positive work means work done by the system on its surroundings, although this convention is sometimes not followed and one should be careful to note the convention used by other people.

In this chapter, we will deal solely with heat.

Energy and work have units of force time distance, such as a Joule, which is a Newton metre. Energy is sometimes measured as the amount of heat that must be transferred to a specified mass of water to raise the temperature of the water at a specified temperature interval at a specified pressure (*e.g.* 1 kcal corresponds to raising 1 kilogramme of water from 15 to 16°C).

As in material balances, energy must be conserved. This conservation law is also known as the first law of thermodynamics. The full energy balance equation is

$$\Delta U + \Delta E_k + \Delta E_p = Q - W \tag{19}$$

where ΔU is the difference in internal energy of all the streams coming out of a system in relation to those coming in, ΔE_k the change in kinetic energy, ΔE_p the change in potential energy, Q the amount of heat put into the system and W the amount of work done by the system. For continuous processes, we wish to consider rates of energy, and so the equivalent equation is

$$\Delta \dot{U} + \Delta \dot{E}_k + \Delta \dot{E}_p = \Delta \dot{Q} - \Delta \dot{W}$$

In the following discussion, there exists an analogous equation for energy rates for each equation based on energy amounts.

For systems in which kinetic and potential energies are assumed not to change or where the changes in these energies are assumed to be negligible, the energy balance equation can be reduced to

$$\Delta U = Q - W$$

In this chapter, we will introduce some of the basic properties required to perform energy balances on a process. As internal energy, U, is typically difficult to measure or estimate, we will concentrate instead on changes in *enthalpy*. Specific enthalpy (enthalpy per unit mass), denoted by H, is defined as

$$\Delta \hat{H} = U - P\hat{V}$$

where P is the system pressure and $\hat{V}$ the specific volume of the material. The actual enthalpy, H, will be given by $m\hat{H}$. The energy balance equation we will use, therefore, will be

$$\Delta H = Q - W$$

and we will describe how to estimate changes in enthalpy due to changes in temperature and phase for streams.

In working with changes of energy, it is often useful to choose a reference state, a state in which one of the quantities is assumed to be zero. Enthalpy data, in fact, are typically given with reference to a particular state. It is often difficult, if not impossible, to actually determine the absolute values of internal energies. Finally, it is worth noting that internal energy is nearly independent of pressure for solids and liquids at a fixed temperature.

In working with materials, there are two types of heat:

  (i) *sensible* heat, which signifies heat that must be transferred to raise or lower the temperature of a substance, assuming no change in phase (solid, liquid, gas), and

 (ii) *latent* heat, which is the heat necessary to change from one phase to another.

The calculation of sensible heat is based on the *heat capacity* (at constant pressure) of the substance, $C_P(T)$, which is in units of heat (energy) per unit mass. Heat capacity information is typically in the form of coefficients for a polynomial expression

$$C_p(T) = a + bT + cT^2 + dT^3$$

and the values of the coefficients are given in physical property tables found in most chemical engineering reference books. To determine the change in enthalpy in heating a substance from one temperature, T_1, to another temperature, T_2, we integrate the polynomial over this temperature range

$$\int_{T_1}^{T_2} Cp(T)dT = \left[aT + b\frac{T^2}{2} + c\frac{T^3}{3} + d\frac{T^4}{4} \right]_{T_1}^{T_2} \tag{20}$$

and evaluate the right hand side of this equation as the difference of the polynomial evaluated at T_2 and at T_1. The relation between the heat capacity (integrated over a temperature interval) and the change in enthalpy is exact for ideal gases, exact for non-ideal gases only if the pressure is indeed constant, and a close approximation for solids and liquids.

For latent heat, we look up the corresponding entry in the tables for either the *latent heat of vapourisation* (or simply the heat of vapourisation) or the *heat of fusion*, depending on the type of phase change encountered (liquid to vapour and solid to liquid, respectively). These quantities are in units of energy per unit mass and are given for a specific reference state (often the 1 atm boiling point or melting point of the substance).

The calculation of the change in enthalpy from one temperature to another for a given substance will often be a multi-step process. The main principle is to identify a path of pressure and temperature changes that goes from state 1 to state 2, passing through states at which we have reference data (*e.g.* the heat of vapourisation) available. Any sections of the path that do not involve phase changes will simply require the calculation of sensible heat using the heat capacity equation above. Phase changes will then require the use of the appropriate latent heat quantity.

For process analysis including energy balances, the same procedure defined for mass balances is followed. The only change is that the source of equations, described in item 3 of the procedure, now includes the energy balance equation as well as the definition of different energy terms. Again, this procedure will be illustrated by example.

1.3.1 Example 3: Energy Balance on a Distillation Column

We again consider the distillation unit introduced in Example 1, updated with temperature information for each of the streams, including some that were previously considered to be internal to the system box. The temperatures have been estimated using a physical property estimation system. There are a number of such computer-based tools and most simulation software systems will include property estimation methods. Figure 3 shows these temperatures as well as the results obtained in the mass balance step above. As more streams are included in this diagram, we have new unknowns related to flow rates. Specifically, we now have the vapour stream, V, from the top of the column to the condenser and the liquid reflux stream, L, from the condenser back into the column. The relationship between the liquid reflux stream back into the column and the actual distillate product stream (D) is given by the *reflux ratio*:

$$R = \frac{\dot{n}_L}{\dot{n}_D} \tag{21}$$

For this particular configuration, the reflux ratio required to achieve the separation is $R = 1.6$.

Neglecting the effect of pressure on enthalpy, estimate the rate at which heat must be supplied.

1.3.1.1 Solution.

(i) Figure 3 is the flowsheet diagram for this problem with all the streams labelled, both with known quantities and with an indication of what we require to estimate to solve this problem. The specific answer to the question posed is the value of $\dot{Q}_r$, the

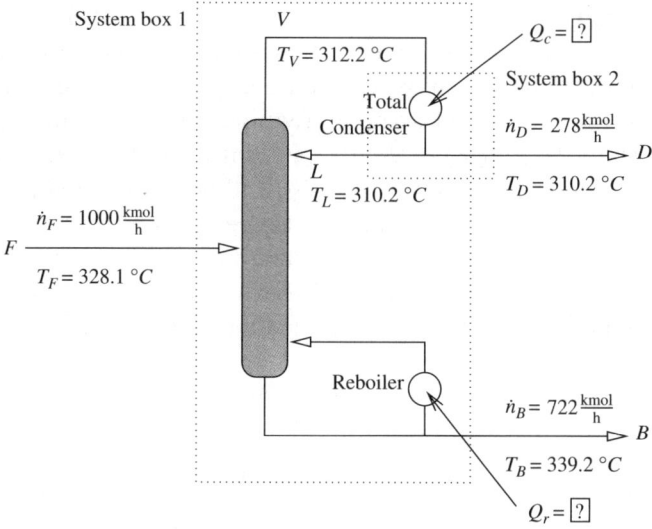

Figure 3 *Distillation unit with temperature and flow data*

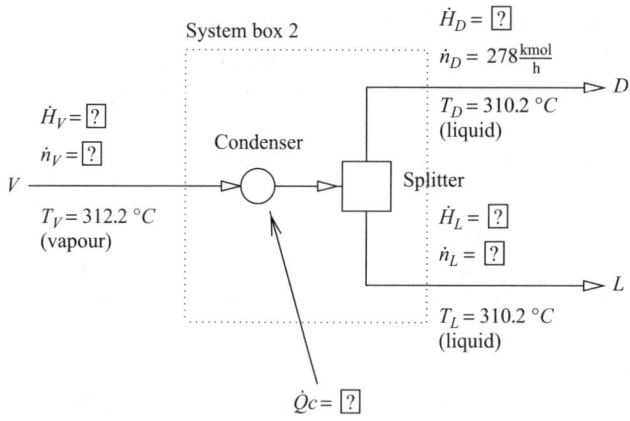

Figure 4 *System box 2 for Example 3*

rate of heat supplied to the reboiler. To determine this amount, we will need to determine the change in enthalpy of the output streams relative to the feed stream and the amount of cooling done in the condenser, $\dot{Q}_c$. The figure shows two system boxes, one around the whole unit, as was used in solving Example 1, and one around the condenser. The latter system box will be useful for determining $\dot{Q}_c$.

(ii) Figure 4 represents the contents of system box 2 in more detail. There are three streams: an input vapour stream that originates at the top of the column and two output streams: D, the distillate product of the unit and L, the liquid reflux going back into the column. The system box includes a splitter and a condenser. The variables identified as being required are the enthalpies of each stream, $\dot{H}_V$, $\dot{H}_D$ and $\dot{H}_L$, the amount of cooling required, Q_c and the flow rates of the vapour, $\dot{n}_V$, and liquid $\dot{n}_L$ streams. Material and energy balances must be addressed simultaneously. As the processing step within this system box is a splitter, we can only use one material balance, and so we choose to use a total material balance. The balance equations therefore are

$$\dot{n}_V = \dot{n}_D + \dot{n}_L \tag{22}$$

$$\Delta\dot{H} = \dot{Q}_c$$

Two equations with six unknowns give us four degrees of freedom, and so we cannot solve for the unknowns yet.

(iii) The remaining equations come from two sources: the process specification, specifically the reflux ratio, Equation (21), and one equation for the enthalpy of each stream by estimating sensible and latent heats for each of them relative to the reference temperature. For this problem, we will use 0°C=273.15 K as the reference temperature. Together with the balance equation, Equation (22), this gives us a total of six equations in six unknowns, which can be solved.

(iv) The estimate of the specific enthalpy of each stream requires data, available from a number of sources as discussed above, for the heat capacity equation and the latent heat of vapourisation. The data for this problem are summarised in Table 2. The coefficients for the heat capacity equation are based on units of kJ mol^{-1}°C^{-1}.

Table 2 *Data for calculation of enthalpies for sensible heat and the latent heat of vapourisation*

		Heat capacity coefficients $(J\ (mol^{-1}\,°C^{-1}))$ $a + bT + cT^2 + dT^3$				
Compound	*Phase*	*a*	*b* × 10^1	*c* × 10^4	*d* × 10^8	$\Delta\hat{H}_V$ *(J mol^{-1})*
n-Pentane	Liquid	155.40	4.368	0	0	27,634
	Gas	114.80	3.409	−1.899	4.226	
n-Hexane	Liquid	216.30	0	0	0	31,089
	Gas	137.44	4.095	−2.393	5.766	

(v) For each stream, we will have a contribution for each compound present. For instance, the enthalpy of the distillate stream will be composed of the enthalpy contribution of pentane and hexane. Each can be calculated individually:

$$\dot{H}_D = x_{D,p}\dot{n}_D(\hat{H}_{D,p} + \hat{H}_{v,p}) + x_{D,h}\dot{n}_D(\hat{H}_{D,h} + \hat{H}_{v,h})$$

where the subscripts D,p, for instance, indicate the distillate stream and pentane component and v,h indicate the latent heat of vapourisation for hexane. The specific enthalpy for each species will consist of a contribution from the gas phase and a contribution from the liquid phase, and so

$$\hat{H}_{D,p} = \hat{H}_{D,p,l} + \hat{H}_{D,p,g}$$

$$\hat{H}_{D,h} = \hat{H}_{D,h,l} + \hat{H}_{D,p,g}$$

where each of these is calculated using Equation (20) using the data from Table 2.

(vi) Table 3 shows the specific enthalpies for the two species before and after the condenser (noting that the specific enthalpies for the species in the liquid reflux stream will be the same as in the distillate output stream). The difference between the enthalpies is primarily due to the heats of vapourisation, but there is a small contribution from the drop in temperature across the condenser.

(vii) Using the specific enthalpies in Table 3 to get the actual stream enthalpies, allows us to then solve for the remaining three unknowns, with the following results: $\dot{n}_V = 722$ kmol h^{-1}, $\dot{n}_L = 444$ kmol h^{-1} and $\dot{Q}_c = -2.25 \times 10^7$ kJ h^{-1}.

Therefore, the condenser is expected to remove heat (as expected).

(viii) To determine the amount of heating required in the reboiler, we now look at the first system box. This box has three streams, the feed stream and the two output streams, and one heat input, $\dot{Q}_r$.

Table 3 *Specific enthalpies for species before and after the condenser*

Species	Specific enthalpies (J mol^{-1})	
	$\hat{H}_{V,i}$	$\dot{H}_{D,i}$
Pentane	38,521	7377
Hexane	38,601	8013

We have only one equation, the energy balance around the column:

$$\Delta \dot{H} = \dot{Q}_c + \dot{Q}_r \tag{23}$$

with unknowns Q_r and the enthalpies of the feed and bottoms streams (we already have the distillate stream enthalpy from the previous step). We calculate the feed and bottoms stream enthalpies in the same way as step 6 and then solve Equation (23) for Q_r to get $Q_r = 2.20 \times 10^7$ kJ h^{-1}, a positive value indicating that heat is an input.

1.4 SUMMARY

This chapter has introduced the concepts of mass and energy balances. These are essential steps in the analysis of any process. Simple examples have been used to illustrate the different steps, including not only formulating mass and energy balances but also simultaneously solving mass and energy balances together (the analysis of the condenser in the distillation unit).

The key to doing process analysis is the identification of the equations that may be used to achieve zero degrees of freedom. These equations will come from a number of sources, including the balance equations themselves (Equations (1) and (19)), process specifications (such as the purity of output streams and the reflux ratio), physical relations (such as the definition of enthalpy for liquid and vapour streams) and other constraints imposed by the problem. Once a full set of equations has been developed, the equations can be solved, usually with little difficulty, and the desired results obtained.

RECOMMENDED READING

1. J. Coulson, J. F. Richardson, J. R. Backhurst and J. H. Harker, *Coulson & Richardson's Chemical Engineering Volume 1 Fluid Flow, Heat Transfer and Mass Transfer*, 5th Edition, Butterworth-Heinemann, 1997.
2. R.M. Felder and R.W. Rousseau, *Elementary principles of chemical processes*, 3rd Edition, John Wiley & Sons, New York, 2000.
3. C.A. Heaton, (Editor), *An Introduction to Industrial Chemistry*, Leonard Hill, Glasgow, 1984.
4. R.H. Perry, D.W. Green and J.O. Maloney, *Perry's Chemical Engineers' Handbook*, 7th Edition, McGraw-Hill, 1997.

CHAPTER 2

Introduction to Chemical Reaction Engineering

GEORGE MANOS

2.1 INTRODUCTION

The major objective of chemical reaction engineering is the analysis and design of equipment for carrying out desirable chemical reactions, *i.e.* chemical reactors.

Apart from choosing operating conditions and type of reactor, including provisions for heat exchange, the design of a reactor involves calculating its size to accomplish the desired extent of a reaction. This requires knowledge of the rate of the chemical reaction, *i.e.* how fast the chemical reaction occurs. In other words, it requires knowledge of the kinetics of the chemical reaction. Very often, physical processes, such as mass and energy transfer, play an important role and influence the overall rate of a chemical reaction. The principles governing the physical processes are as important as those governing the chemical kinetics.

The essential feature in chemical reactor analysis and design is the formulation of conservation equations for mass and energy for the chosen type of reactor. These equations can be either algebraic or differential and their solution gives the extent of the reaction.

This chapter gives an introduction to the subject of chemical reaction engineering. The first part introduces basic definitions and concepts of chemical reaction engineering and chemical kinetics and the importance of mass and heat transfer to the overall chemical reaction rate. In the second part, the basic concepts of chemical reactor design are covered, including steady-state models and their use in the development of reactor design equations. Worked examples are given in the text to demonstrate these concepts.

2.1.1 Classification of Reactors

There are different ways to group chemical reactors. First of all they can be classified according to their operation, either *batch* or *continuous*, with *semibatch* or *semicontinuous* being a category between them. A batch reactor has no stream continuously flowing into or out of it and is charged with the reactants in a pure form or in a solution, heated to the reaction temperature and left for the reaction to occur for a specific time. At the end of the reaction the reactor is emptied, cleaned, maintained (if necessary) and then filled with reactants again. Batch reactors operate in such cycles and to calculate production rates the overall time, including filling and emptying of the reactor, must be considered. A semibatch reactor has either an inlet or an outlet stream only. An example is a hydrogenation reactor using a liquid reactant to produce a liquid product. Both liquid reactant and liquid product stay during the whole reaction time in the reactor and only hydrogen flows into it. In continuous reactors an inlet and an outlet stream flow continuously. Continuous reactors normally operate at steady state.

Another classification of chemical reactors is according to the phases being present, either *single phase* or *multiphase* reactors. Examples of multiphase reactors are gas–liquid, liquid–liquid, gas–solid or liquid–solid catalytic reactors. In the last category, all reactants and products are in the same phase, but the reaction is catalysed by a solid catalyst. Another group is gas–liquid–solid reactors, where one reactant is in the gas phase, another in the liquid phase and the reaction is catalysed by a solid catalyst. In multiphase reactors, in order for the reaction to occur, components have to diffuse from one phase to another. These mass transfer processes influence and determine, in combination with the chemical kinetics, the overall reaction rate, *i.e.* how fast the chemical reaction takes place. This interaction between mass transfer and chemical kinetics is very important in chemical reaction engineering. Since chemical reactions either produce or consume heat, heat removal is also very important. Heat transfer processes determine the reaction temperature and, hence, influence the reaction rate.

Finally, chemical reactors can be classified according to the mode of heat removal. We can have either *isothermal* or *non-isothermal* reactors, a sub-category of which is *adiabatic reactors*. Often it is desirable to use the heat released by an exothermic reaction for an endothermic reaction, in order to achieve higher heat integration.

2.2 CHEMICAL REACTION KINETICS

2.2.1 Definitions

A chemical reaction can generally be written as

$$\sum_{i=1}^{N} v_i A_i = 0$$

where A_i $(I = 1, 2, \ldots, N)$ are the chemical species participating in the chemical reaction and v_i are their stoichiometric coefficients. Stoichiometric coefficients of products are positive and of reactants, negative.

For example, the simple full oxidation of methane

$$CH_4 + 2O_2 \rightarrow CO_2 + 2H_2O \tag{1}$$

can be written as

$$CO_2 + 2H_2O - CH_4 - 2O_2 = 0 \tag{2}$$

In this example, the chemical species and respective stoichiometric coefficients are $A_1 = CO_2$, $A_2 = H_2O$, $A_3 = CH_4$, $A_4 = O_2$ and $v_1 = 1$, $v_2 = 2$, $v_3 = -1$, $v_4 = -2$.

2.2.1.1 Extent of Reaction. The extent of a reaction (ξ) is defined as the change in the number of moles of any reaction compound (reactant or product) due to the chemical reaction divided by its stoichiometric coefficient.

Let us consider the full oxidation of methane again. As the various components react according to the reaction stoichiometry, at any time the change in the number of moles of the reaction components has the following relationship:

$$\Delta n_{H_2O} = 2\Delta n_{CO_2} = -\Delta n_{O_2} = -2\Delta n_{CH_4} \tag{3}$$

Note: The change in the number of moles (Δn_i) during a reaction is negative for a reactant and positive for a product.

Therefore, the extent of reaction will have the same value, whatever compound we choose for its calculation

$$\xi = \frac{\Delta n_{CO_2}}{1} = \frac{\Delta n_{H_2O}}{2} = \frac{\Delta n_{CH_4}}{-1} = \frac{\Delta n_{O_2}}{-2} \tag{4}$$

In a complex reaction network of R reactions occurring simultaneously, the extent of the reaction ξ_j is

$$\xi_j = \frac{\Delta n_{ij}}{v_{ij}} \tag{5}$$

where Δn_{ij} is the change in the number of moles of the reaction component i due to the reaction j and v_{ij} is the stoichiometric coefficient of the reaction component i in the reaction j.

The total change in the number of moles of i is thus

$$\Delta n_i = \sum_{j=1}^{R} \Delta n_{ij} = \sum_{j=1}^{R} v_{ij}\xi_j \tag{6}$$

2.2.1.2 Conversion of a Reactant.

The conversion of a reactant A (X_A) is the fraction of reactant A converted/transformed to products.

$$X_A = \frac{n_{A0} - n_A}{n_{A0}} = \frac{|\Delta n_A|}{n_{A0}} = \frac{|\sum_{j=1}^{R} v_{Aj}\xi_j|}{n_{A0}} \tag{7}$$

where n_{A0} is the initial number of moles of A.

The conversion is defined with reference to a reactant and could be different for different reactants. As a matter of fact, in a general reaction the conversions of reactants A and B will only be equal when the initial (or feed) ratio of moles of A to B is equal to the stoichiometry of the reaction a/b, where a and b are the stoichiometric coefficients of A and B, respectively.

For example, if during a reaction at some time from the initial 50 mol of reactant A only 20 mol have been left unreacted, i.e. 30 mol have reacted to products, the conversion at that time would be

$$X_A = \frac{50 - 20}{50} = 0.60 = 60\%$$

The conversion can be defined for a continuous reactor as well. For example, 50 mol min^{-1} of reactant A enter a continuous reactor. At the outlet of the reactor only 20 mol min^{-1} of A exit the reactor. The conversion is then

$$X_A = \frac{50 \, \text{mol min}^{-1} - 20 \, \text{mol min}^{-1}}{50 \, \text{mol min}^{-1}} = 0.60 = 60\%$$

2.2.1.3 Yield. The yield of a specific product K (Y_K) is the amount of K produced expressed as a fraction of the maximum amount of K (according to the stoichiometry) that could be produced. This is the amount of reactant A converted to the specific product:

$$Y_K = \frac{n_K - n_{K0}}{n_{A0}} \frac{|v_A|}{|v_K|} \tag{8}$$

For example, let us consider the following reaction scheme for decomposition/isomerisation reactions

$$A \rightarrow 2K, \quad A \rightarrow B$$

From 100 initial moles of A, 30 are converted to produce 60 mol of K and 50 are converted to B. The total number of moles of A converted (to either K or B) is 80. The (total) conversion of A is 80%, but only 30% is converted to K. Hence, the yield of K is

$$Y_K = \frac{60}{100}\frac{1}{2} = 0.30 = 30\% \tag{9}$$

2.2.1.4 Selectivity. We now introduce the term selectivity for reaction networks. Selectivity can be defined in two ways.

First, the selectivity to a product P is the fraction of the converted reactant A (not the initial A) that is converted to the specific product K.

$$S_P = \frac{n_P - n_{P0}}{n_{A0} - n_A} \frac{|v_A|}{|v_P|} = \frac{Y_P}{X_A} \tag{10}$$

In the previous example the selectivity to P is

$$S_P = \frac{60}{80}\frac{1}{2} = 0.375 = 37.5\%$$

i.e. 37.5% of the converted A, not the initial A, is reacted to P.

The yield to a product is equal to the conversion multiplied by the selectivity

$$Y_K = S_K \times X_A \tag{11}$$

Secondly, selectivity can be defined with reference to another product. Overall or integrated selectivity is the ratio of the amount of one product (B) produced to the amount of another (C). For example, for reactions in series or in parallel

$$A \rightarrow B \rightarrow C \quad \text{or} \quad A \rightarrow B, \ A \rightarrow C$$

$$S_{B/C} = \frac{n_B - n_{B0}}{n_C - n_{C0}} \frac{|v_C|}{|v_B|} = \frac{Y_B}{Y_C} \tag{12}$$

Point selectivity can be defined as the ratio of the reaction rates of B and C at a specific point in a reactor at a specific time. We introduce the term 'reaction rate' in the next section.

$$S_{B/C} = \frac{r_B}{r_C} \tag{13}$$

2.2.1.5 Reaction Rate. The reaction rate (r) is the change of the extent of the reaction per unit time per unit volume. Consider a point somewhere in a reactor (continuous or batch) and some volume dV around it. Imagine that you are in this volume and can measure how many moles of a reaction species i (dn_{ij}) have been produced[†] during some time dt via reaction j.

The change of the extent of this reaction is

$$d\xi_j = \frac{dn_{ij}}{v_{ij}} \tag{14}$$

and the rate of reaction j is

$$r_j = \frac{d\xi_j}{dV dt} = \frac{1}{v_{ij}} \frac{dn_{ij}}{dV dt} \tag{15}$$

This is the general definition of reaction rate. Only in special cases is this equal to dC/dt (C = concentration).

In a reaction network of R reactions, the total change in the number of moles of a reaction component i is due to many reactions and given by

$$dn_i = \sum_{j=1}^{R} dn_{ij} = \left(\sum_{j=1}^{R} r_j v_{ij} \right) dV dt \tag{16}$$

The above definition of rate refers to a chemical reaction. If we would like to use a similar term for a component, its reaction rate, or better, the

---

[†] Reactant consumption is considered as a negative production.

rate of change in the number of moles of i is

$$R_i = \frac{dn_i}{dV dt} = \sum_{j=1}^{R} v_{ij} r_j \tag{17}$$

We use capital R for a component in order to distinguish it from small case r which symbolises the rate of a chemical reaction.

2.2.2 Chemical Reaction Thermodynamics

Consider the reversible reaction

$$aA + bB \Leftrightarrow cC + dD$$

At equilibrium, the rate of reaction from left to right is equal to the rate of the reaction from right to left, so that the net overall reaction rate is zero, *i.e.* nothing changes. The same amount of A and B react to C and D as C and D react to give A and B.

The equilibrium conversion is the conversion of a reversible reaction at its equilibrium. This is the maximum conversion that can be achieved. Reversible reactions cannot progress further than the equilibrium.

The equilibrium constant (K) of a gaseous reaction will be defined here for the case of gaseous reaction components behaving like ideal gases

$$K = K_P = \prod_{i=1}^{N} P_i^{v_i} \tag{18}$$

where P_i is the partial pressure of i. K and K_P have no units. The reason for this is that the more rigorous definition for ideal gases is

$$K_P = \prod_{i=1}^{N} \left(\frac{P_i}{P_0}\right)^{v_i} \tag{19}$$

where P_0 is a standard pressure. Normally $P_0 = 1$ bar so that arithmetically the first equation is valid. It is widely used because of its simplicity.

For the gaseous reversible reaction of the example above, Equation (19) can be rewritten as

$$K_P = \frac{P_C^c P_D^d}{P_A^a P_B^b} \tag{20}$$

K or K_P can be calculated from the Gibbs free energy of the reaction, ΔG°_R

$$\Delta G^{\circ}_{RT} = -RT \ln K \tag{21}$$

where R is the universal gas constant.

2.2.2.1 Temperature Dependence of K_P and Equilibrium Conversion. The temperature dependence of K_P is given by the equation

$$\frac{d \ln K_P}{dT} = \frac{\Delta H^{\circ}_{RT}}{RT^2} \tag{22}$$

or the integrated form

$$\ln\left(\frac{K_{P2}}{K_{P1}}\right) = \frac{\Delta H^{\circ}_R}{R}\left(\frac{1}{T_1} - \frac{1}{T_2}\right) \tag{23}$$

with the assumption that the reaction enthalpy ΔH°_R is independent from the temperature; *i.e.* it is almost constant between T_1 and T_2.

For exothermic reactions (*e.g.* oxidation, hydrogenation, polymerisation reactions, *etc.*) the equilibrium constant decreases with temperature. Consequently, the equilibrium (maximum) conversion decreases with temperature.

Note: Chemical reactions – even exothermic reactions – become faster at higher temperatures. Only their equilibrium conversion decreases with temperature.

For endothermic reactions, *e.g.* dehydrogenation, cracking reactions, *etc.*, the equilibrium constant and the equilibrium conversion increase with temperature.

2.2.2.2 Pressure Dependence of Equilibrium Conversion. For ideal gases, K_P is independent of pressure. However, the equilibrium conversion depends upon pressure:

$$K_P = \prod_{i=1}^{N} P_i^{v_i} = \left(\prod_{i=1}^{N} y_i^{v_i}\right) P_{\text{tot}}^{\left(\sum_{i=1}^{N} v_i\right)} \tag{24}$$

since $P_i = y_i P_{\text{tot}}$.

By introducing $K_y = \prod_{i=1}^{N} y_i^{v_i}$ we can write

$$K_P = K_y P_{\text{tot}}^{\left(\sum_{i=1}^{N} v_i\right)} \tag{25}$$

The higher K_y becomes, the higher is the equilibrium conversion (more products than reactants). To examine the influence of pressure, we have to consider three cases:

(a) More reactants than products $\sum_{i=1}^{N} v_i < 0$ (*e.g.* polymerisation, hydrogenation, hydration, alkylation, *etc.*). The equilibrium conversion increases with pressure; the higher the pressure, the more products at equilibrium.

(b) The same amount of products as reactants $\sum_{i=1}^{N} v_i = 0$ (*e.g.* isomerisation). No pressure effect.

(c) More products than reactants $\sum_{i=1}^{N} v_i > 0$ (*e.g.* cracking/decomposition, dehydrogenation, dehydration, dealkylation, *etc.*). The equilibrium conversion decreases with pressure.

2.2.2.3 Reaction Networks. In the case of a network of simultaneous reactions we use the following methodology:

(i) We calculate the equilibrium constant of each reaction according to the equation

$$\ln K_j = -\frac{\Delta G_R}{RT} \text{ or } K_j = \exp\left\{-\frac{\Delta G_R}{RT}\right\} \qquad (26)$$

(ii) We calculate K_{yj} for each reaction from the equation

$$K_j = K_{Pj} = K_{yj} P_{\text{tot}}^{\left(\sum_{i=1}^{N} v_i\right)} \qquad (27)$$

where

$$K_{yj} = \prod_{i=1}^{N} y_i^{v_{ij}} \qquad (28)$$

(iii) It is advisable to express the mole-fractions y_i as functions of the equilibrium extents of the different reactions ($\xi_{j,\text{eq}}$); in other words, use the extents of the reactions at equilibrium as independent variables.

(iv) Solve the system of the R algebraic Equations (28) to calculate all ($\xi_{j,\text{eq}}$) and then the mole-fractions of the reaction components at equilibrium.

2.2.3 Kinetics

2.2.3.1 Fundamentals – Effect of Concentration.   The simplest relation between reaction rate and concentrations of the reactants is a power law.
   For the reaction

$$aA + bB \rightarrow cC + dD$$

the rate of reaction can be written as

$$r = kC_A^m C_B^n \tag{29}$$

where k is the kinetic constant dependent upon reaction conditions (temperature and pressure), m is the order of the reaction with respect to A, n is the order of the reaction with respect to B and $(m + n)$ is the total reaction order.
   Generally, $m \neq a$, $n \neq b$. If $m = a$, $n = b$, then the reaction is called elementary. The reaction mechanism is the network of elementary steps (reactions) that add up to the reaction. The reaction mechanism describes how, and in what steps, the overall reaction occurs. An elementary reaction cannot be analysed in simpler steps; it occurs exactly as the reaction states.
   From the reaction mechanism, we can derive a reaction rate equation (kinetics) for the overall reaction. Below, we will discuss two such methods: rate limiting step and quasi-stationary state (pseudo-steady-state).

2.2.3.2 Rate Limiting Step.   According to this concept, the overall reaction rate is equal to the reaction rate of the slowest step in the reaction mechanism. If one of the elementary steps occurs at a much lower rate than the others, then that slowest step determines the overall reaction rate. The other steps are assumed to be so fast that they are already at equilibrium.
   We will apply this concept on the decomposition of N_2O_5 to explain the experimentally found first order of the reaction:

$$2N_2O_5 \rightarrow 4NO_2 + O_2$$

Reaction mechanism:

(1) $N_2O_5 \underset{k_1'}{\overset{k_1}{\rightleftharpoons}} NO_2 + NO_3$ $(2 \times /\text{occuring twice})$

   Two molecules of N_2O_5 have to be decomposed for the right amount of product to be produced (right stoichiometry). The formalism we use differs from a reaction equation with all the stoichiometric coefficients 2 (or -2), in that the equilibrium and/ or kinetic equation for this step would then be different.

(2) $NO_2 + NO_3 \xrightarrow{k_2} NO + O_2 + NO_2$ (slowest step)

(3) $NO + NO_3 \xrightarrow{k_3} 2NO_2$

The second reaction step is much slower than the others, so that the first reaction is in equilibrium.

The reaction of NO_2 and NO_3 to NO, O_2, NO_2 (second reaction) is very slow. What happens then is that a lot of NO_2 and NO_3 are produced from the fast first reaction but cannot react further. This will force these products to react back to N_2O_5 until an equilibrium is established. Now when a little of NO_2 and NO_3 react in reaction 2, immediately some more NO_2 and NO_3 will be produced by the fast first reaction, so that equilibrium will be established again.

Using the equilibrium constant for the first reaction step, we can write

$$K_1 = \frac{C_{NO_2} C_{NO_3}}{C_{N_2O_5}}$$

Note: At equilibrium, as the reaction rate from left to right is equal to the reaction rate from right to left. This means $K_1 = (k_1/k_1')$.

The overall reaction rate is equal to the reaction rate of the slowest step (2):

$$r = r_2 = k_2 C_{NO_2} C_{NO_3}$$

How fast the products are produced depends on how fast the second reaction occurs. It does not help if the first reaction is very fast. For the products of the overall reaction to be produced, we have to wait for the second reaction. On the other hand, when the second reaction occurs, NO and NO_3 react very fast in the third reaction. The third reaction does not build up a bottleneck.

Since from the equilibrium equation $C_{NO_2} C_{NO_3} = K_1 C_{N_2O_5}$, the reaction rate can be written as

$$r = k_2 K_1 C_{N_2O_5} \quad \text{or}$$

$$r = k C_{N_2O_5} \quad \text{where} \quad k = k_2 K_1 = k_2 \frac{k_1}{k_1'}$$

Thus, the kinetics are first order; the reaction rate is proportional to the concentration of the reactant. We have seen that even complicated reaction mechanisms can lead to simple effective kinetics.

2.2.3.3 Quasi-Stationary State.   The assumption in this method is that the concentration of intermediates is constant, after a short initial period. This means that the rate of formation of intermediates is equal to the rate of their disappearance/consumption.

   We consider the same reaction mechanism as before. In this mechanism NO_3 and NO are intermediate components and therefore

$$\frac{dC_{NO_3}}{dt} = 0 \quad \text{and} \quad \frac{dC_{NO}}{dt} = 0$$

For the rate of formation of the intermediates we can use the individual kinetics of the elementary reactions. NO_3 is formed and disappears in the first reversible reaction step and disappears in the second and third steps. Therefore,

$$\frac{dC_{NO_3}}{dt} = k_1 C_{N_2O_5} - k_1' C_{NO_2} C_{NO_3} - k_2 C_{NO_2} C_{NO_3}$$
$$- k_3 C_{NO} C_{NO_3} = 0$$

Applying the same procedure for the NO that is formed in the second reaction and that disappears in the third one, we get

$$\frac{dC_{NO}}{dt} = k_2 C_{NO_2} C_{NO_3} - k_3 C_{NO} C C_{NO_3} = 0$$

Solving this system of two equations for C_{NO} and C_{NO_3} we have

$$C_{NO} = \frac{k_2}{k_3} C_{NO_2} \quad \text{and} \quad C_{NO_3} = \frac{k_1 C_{N_2O_5}}{(2k_2 + k_1')C_{NO_2}}$$

Finally,

$$r = \frac{dC_{O_2}}{dt} = k_2 C_{NO_2} C_{NO_3} = \left(\frac{k_1 k_2}{2k_2 + k_1'} \right) C_{N_2O_5}$$

or

$$r = k_* C_{N_2O_5} \quad \text{with} \quad k_* = \frac{k_1 k_2}{2k_2 + k_1'}$$

This derived kinetic expression also explains the experimental observation of first-order kinetics in respect to N_2O_5, but the first-order kinetic constant is different from the one derived by the rate-limiting step method.

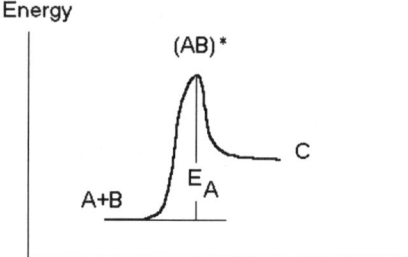

Figure 1 *Activation energy of a chemical reaction*

2.2.3.4 Temperature Dependence of Kinetic Constant. The dependence of k on temperature for an elementary reaction follows the Arrhenius equation

$$k = A \exp\left\{-\frac{E_A}{RT}\right\} \tag{30}$$

where A is known as the pre-exponential factor and E_A is the activation energy. The meaning of activation energy is illustrated in Figure 1.

A reaction $A+B \rightarrow C$ occurs via an intermediate activated complex $(AB)^*$.

$$A + B \rightarrow (AB)^* \rightarrow C$$

The activation energy actually represents the energy barrier the reactants have to overcome to form this activated intermediate.

The activation energy can be estimated from kinetic experimental data at different temperatures. The Arrhenius equation can also be written as

$$\ln k = \ln A - \frac{E_A}{R}\frac{1}{T} \tag{31}$$

If $\ln k$ is plotted against $1/T$, the graph is a straight line with slope equal to $(-E_A/R)$. By calculating the slope of the line, the activation energy E_A can be estimated.

2.2.4 Importance of Mass and Heat Transfer Processes

We will now examine how important the mass and heat transfer phenomena are.[‡] We will do so with the example of a solid catalysed reaction. In a catalytic reaction the reactants have to first form intermediate species with the active sites of the catalyst. The products are

---

[‡]The reader is referred to Chapters 4 and 5 for more detail on these subjects.

then formed much faster than in the absence of catalyst. What actually happens is that the activation energy of the route through the intermediate catalytic complex is much lower than the respective activation energy through the intermediate route in the absence of catalyst.

Let us examine the various steps of a catalytic reaction on a solid catalyst particle:

(i) Mass transfer of reactants from the bulk fluid phase through the boundary layer to the external surface of the catalyst particle.

A boundary layer is formed between the two phases (fluid and solid). This is a stagnant film that represents a layer of less movement of the fluid and hence builds up a zone with resistance to mass transfer. The mass transfer coefficient and generally the mass transfer rate depend on the fluid dynamics of the system. Higher fluid velocities significantly reduce the thickness of the film.

Normally mass transfer between two phases is modelled using a mass transfer coefficient (k_m). The flux (mol s^{-1} m$^{-2}_{ext.surf}$) of a component i is given by the formula

$$\hat{N}_i = k_m(C_{if} - C_{is}) \tag{32}$$

where C_{if} is the concentration of the component i in the fluid phase and C_{is} is the concentration on the external surface of the catalytic particle.

The above rate is expressed per unit of external surface. To express the rate per gramme of catalyst the flux has to be multiplied by the catalyst specific area (m$^2_{ext.surf}$ g$^{-1}_{cat}$).

(ii) Diffusion of the reactants into the pores of the particle where the majority of the active catalytic sites reside.

(iii) Adsorption of the reactants on the active sites.

(iv) Surface reaction to produce the products adsorbed on the active sites.

(v) Desorption of the products from the active sites.

(vi) Pore diffusion of the products to the external surface of the catalyst particle.

(vii) Mass transfer of the products from the external surface of the catalytic particle through the stagnant film to the bulk fluid phase.

The overall reaction rate, *i.e.* how fast the products are formed, depends on how fast is each one of the above phenomena.

Let us first examine the case where the catalytic reaction is very slow compared to the mass transfer (both external – from the fluid through the film to the solid particle – and internal – pore diffusion). In this case, the concentration in the particle is the same as in the bulk fluid phase. As the molecules move into the particle and react, the relatively fast mass transfer can replace the reacted molecules with new ones, keeping the level of the concentration constant.

The picture is completely different when the reaction is very fast. Now the mass transfer cannot cope with the rate of reacting molecules. Mass transfer cannot supply the new molecules to the inner part of the catalytic particle as fast, with the result that the concentration in the particle is much lower than in the bulk fluid phase. If, from the mass transfer phenomena, pore diffusion is the slowest step, additionally a falling concentration profile is formed in the particle. In both cases, the reaction occurs with a concentration much lower than the one we believe occurs, namely the one in the bulk fluid phase.

The picture is similar for heat transfer. As we know, reaction rates depend strongly upon temperature. Since the chemical reaction occurs in the catalytic particle, the transfer rate of the reaction heat will determine the temperature level in the particle. A fast heat transfer, compared with the rate of heat generation by the reaction, will lead to the particle temperature being equal to that in the bulk fluid phase, the one at which we believe the reaction takes place. However, if the heat transfer is very slow, the two temperatures will be different.

A major difference to mass transfer is that now the real reaction rate can be higher than the apparent one. The concentration in the particle is always lower or equal to the concentration in the bulk fluid phase and therefore, from this perspective, the real rate is always lower or, at best, equal to the apparent rate. However, in strong exothermic reactions heat transfer cannot cope with the high rates of generation of reaction heat and the temperature in the catalyst particle can be much higher than in the fluid phase, resulting in a much higher reaction rate than the apparent one.

Regarding mass transfer, the slowest step is normally pore diffusion rather than external mass transfer, whilst for heat transfer the slowest step is the interphase heat transfer between the particle and the fluid phase rather than the internal heat transfer in the solid particle.

2.2.5 Kinetics of a Catalytic Reaction

We conclude the first part of this chapter with a kinetic example. We will derive, using the rate limiting step approach, the Langmuir–Hinselwood kinetics of a simple catalytic (isomerisation) reaction: A $\Leftrightarrow$ B.

This reaction occurs via the following steps: A is adsorbed onto an active catalytic site, the adsorbed species AS undergoes a surface reaction to form

BS and finally B is desorbed to free the active site that can now catalyse another cycle. We assume that the surface reaction is the slowest and therefore the limiting step. The kinetic expression we derive is valid only under this assumption and if the assumption changes, another expression needs to be derived. Since the surface reaction is the slowest step, we assume that all other reaction steps are so fast that they are at equilibrium.

Reaction mechanism:

- Adsorption of reactant A:

$$A(g) + S(s) \rightleftharpoons AS(s)$$

  K_A is the adsorption equilibrium constant of A.
- Surface reaction:

$$AS(s) \rightleftharpoons BS(s)$$

  Slow surface reaction (rate-limiting step). k_1, k_2 are the kinetic constants of the forward and backward reactions.
- Desorption of product B:

$$BS(s) \rightleftharpoons B(g) + S(s)$$

  K_B is the adsorption equilibrium constant of B.

In the above mechanism, (g) shows that the respective species is in the gas phase and (s) in the solid phase (catalyst). We symbolise with C the concentrations in the gas phase and θ the concentration in the solid phase, usually in units of mol g_{cat}^{-1}.

As the surface reaction is the rate limiting step, the overall reaction rate is equal to the rate of this step:

$$r = k_1 \theta_{AS} - k_2 \theta_{BS}$$

This equation is not in the right kinetic form because it contains concentrations in the solid θ_i, which cannot be measured. We have to find a way to replace them with known quantities. The final kinetic equation should contain only concentrations of A or B in the gas phase and kinetic constants.

The other steps are at equilibrium and thus we can write

$$K_A = \frac{\theta_{AS}}{C_A \theta_S} \Rightarrow \theta_{AS} = K_A C_A \theta_S$$

$$K_B = \frac{\theta_{AS}}{C_A \theta_S} \Rightarrow \theta_{BS} = K_B C_B \theta_S$$

We wrote the same equation for B because the adsorption equilibrium constant is not the equilibrium constant of the final reaction step but of the reverse reaction.

The reaction rate equation can then be written as

$$r = (k_1 K_A C_A - k_2 K_B C_B)\theta_S$$

θ_S is the concentration of free active sites, that is, active sites that are not occupied by A or B. Therefore, it is not allowed to be in the kinetic expression. It is not a constant and we must replace it with a function of θ_{tot}, the total number of active sites, that is characteristic of the catalyst.

For θ_{tot} we can write

$$\theta_{AS} + \theta_{BS} + \theta_S = \theta_{tot} \Rightarrow K_A C_A \theta_S + K_B C_B \theta_S + 1 = \theta_{tot}$$

$$\Downarrow$$

$$\theta_S = \frac{1}{K_A C_A + K_B C_B + 1}\theta_{tot}$$

and $$r = \frac{(k_1 K_A C_A - k_2 K_B C_B)\theta_{tot}}{1 + K_A C_A + K_B C_B}$$

We can simplify the above equation by lumping θ_{tot} into the kinetic constants

$$k_1^* = k_1 \theta_{tot} \quad \text{and} \quad k_2^* = k_2 \theta_{tot}$$

The final Langmuir–Hinselwood kinetic expression is therefore

$$r = \frac{k_1^* K_A C_A - k_2^* K_B C_B}{1 + K_A C_A + K_B C_B}$$

2.3 CONCEPTS OF CHEMICAL REACTOR DESIGN

In order to determine design characteristics of chemical reactors, we have to formulate mole balance equations.

2.3.1 Mole Balances for Chemical Reactors

2.3.1.1 General Mole Balance Equation. The mole balance equation is applied for individual reaction mixture components. The first decision we have to make is the choice of the system, *i.e.* reactor space, in which we apply the mole balance. This system could be the whole reactor or a differential part of it.

The general mole balance equation for a reaction component i at a time t in words is

(Change in number of moles of i within the system between time t and $t + dt$)
= (Number of moles of i flowing into the system during the time interval dt)
  −(Number of moles of i flowing out of the system during the time interval dt)
  +(Number of moles of i formed by chemical reactions inside the system
      during the time interval dt)

If i is a reactant, its consumption can be considered as a negative formation. In terms of the right hand side, we should include all incoming and outgoing streams where the component i is present, as well as all reactions where i participates, being either a reactant or product.

The above general mole balance equation can be expressed in terms of rates as

(Rate of accumulation of moles of i within the system)
        = (Rate of molar flow of i into the system)
          −(Rate of molar flow of i out of the system)
          + (Rate of formation of i by chemical reactions inside the system)

In short

$$(\text{Accumulation}) = (\text{In}) - (\text{Out}) + (\text{Generation})$$

From the above mole balance equation we can develop the design equation for various reactor types. By solving the design equation we can then determine the time required for a batch reactor system or a reactor volume for a continuous flow system to reach a specific conversion of the reactant to products.

Before we develop the design equation, let us have a closer look at the terms of the balance equation.

(i) The rate of accumulation of component i in the system can be
    expressed as

$$\frac{dN_i}{dt} = \frac{d(C_i V_R)}{dt} \tag{33}$$

if the mole balance is applied to the whole reactor;
or it can be expressed as

$$\frac{dN_i}{dt} = \frac{d(C_i \, dV_R)}{dt} \tag{34}$$

if the mole balance is applied to a differential part of the reactor where C_i is the concentration of component i and V_R is the reactor volume.

(ii) The molar flow rate of component i can be expressed as

$$F_i = C_i \dot{V} \tag{35}$$

where $\dot{V}$ is the volumetric flow rate of the specific stream, inlet or outlet.

(iii) Since the reaction rate is defined as the number of moles reacting per unit time as well as per unit volume, to find the rate of formation of component i we should multiply the reaction rate by the volume of the system dV_R. Furthermore, we should take into account all reactions where i participates. Each reaction rate term should be multiplied by the stoichiometric coefficient of i in each reaction j

$$R_i = \sum_j v_{ij} r_j \tag{36}$$

In the following sections we will examine the following three reactor types and develop the mole balance/design equations for each:

- Perfectly mixed batch reactor
- Continuous stirred tank reactor (CSTR)
- Tubular plug flow reactor (PFR).

2.3.1.2 Batch Reactor. In a batch reactor there are no inlet or outlet streams: In = Out = 0. The total feed is charged into the reactor at the beginning and no withdrawal is made until the desired conversion level has been reached. Hence a reaction process occurring in a batch reactor is an unsteady one. All variables change with time. In addition, we assume that it is a perfectly mixed batch reactor, so that the concentrations of the reaction components, reactants or products are the same over the whole reactor volume. This assumption allows us to consider applying the mole balance equation across the whole reactor. With the term reactor we mean the space where the reaction(s) take place. For liquid phase reactions the reactor volume is smaller than the size of the physical reactor. It is the volume of the liquid phase, where the reaction(s) take(s) place.

Without inlet or outlet terms the mole balance becomes

$$\frac{dN_i}{dt} = \frac{d(C_i V_R)}{dt} = \left(\sum_j v_{ij} r_j \right) V_R \tag{37}$$

Only in the special case of constant reactor volume (V_R = const) can Equation (37) be simplified into the more familiar form in terms of the concentration derivative

$$\frac{dC_i}{dt} = \sum_j v_{ij} r_j \tag{38}$$

2.3.1.3 Continuous Stirred Tank Reactor.

A CSTR is a continuous flow reactor. It is a tank reactor, i.e. its height is comparable to its diameter, with powerful stirring. We assume that the stirring is so thorough that the reaction mixture is perfectly mixed. This means that each of the concentrations of the reaction mixture components has the same value everywhere inside the reactor. This assumption allows us again to formulate the mole balance equations in the whole reactor volume. In addition, the thorough stirring ensures that the temperature is the same everywhere in the reactor.

We are going to formulate the mole balance equation in a CSTR operating at steady state, where nothing changes with time. This means that there is no accumulation inside the reactor and that the inlet as well as outlet streams are at steady state. Following the general equation the mole balance for a component i in a CSTR is

$$0 = F_{i,in} - F_{i,out} + \left(\sum_j v_{ij} r_j \right) V_R \tag{39}$$

or in concentration terms

$$0 = C_{i,in} \dot{V}_{in} - C_{i,out} \dot{V}_{out} + \left(\sum_j v_{ij} r_j \right) V_R \tag{40}$$

where the subscripts 'in' and 'out' indicate, respectively, the inlet and the outlet characteristics (molar flow rates F_i or volumetric flow rate $\dot{V}$).

2.3.1.4 Plug Flow Reactor.

A PFR is a continuous flow reactor. It is an ideal tubular type reactor. The assumption we make is that the reaction mixture stream has the same velocity across the reactor cross-sectional area. In other words, the velocity profile across the reactor is a flat one. In a PFR there is no axial mixing along the reactor. The condition of plug flow is met in highly turbulent flows, as is usually the case in chemical reactors.

Since the concentration changes along the PFR, we cannot apply the mole balance equation to the whole reactor. We would not know what concentrations to assign to the reaction rate term. We are going instead to choose a differential slice of the reactor, between lengths z and $(z + dz)$. Instead of length we are going to use volume positions, namely V and $(V + dV)$. For this system the molar flow rates at the positions V and $(V + dV)$ are the inlet and outlet terms, accordingly, in the mole balance equation. For the steady-state condition (*i.e.* no accumulation) the mole balance becomes

$$0 = (F_i)_V - (F_i)_{V+dV} + \left(\sum_j v_{ij} r_j \right) dV \tag{41}$$

which can be written as

$$dF_i = \left(\sum_j v_{ij} r_j \right) dV \quad \text{or} \quad \frac{dF_i}{dV} = \sum_j v_{ij} r_j \tag{42}$$

Introducing concentrations, the above equation becomes

$$\frac{d(C_i \dot{V})}{dV} = \sum_j v_{ij} r_j \tag{43}$$

2.3.2 Reactor Design Equation

Integrating the mole balance equations gives us the design equations for the different reactor types:

- Batch reactor:

$$t = \int_{N_{i0}}^{N_i} \frac{dN_i}{\left(\sum_j v_{ij} r_j \right) V_R} \tag{44}$$

- CSTR:

$$V_R = \frac{F_{i,in} - F_{i,out}}{- \sum_j v_{ij} r_j} \tag{45}$$

- PFR:

$$V_R = \int_{F_{i,in}}^{F_{i,out}} \frac{dF_i}{\sum_j v_{ij} r_j} \tag{46}$$

In the next paragraph we are going to rewrite the above equations using conversion and introduce the term 'residence time'.

2.3.2.1 Design Equations Using Conversion. Recall from Section 2.2.1.2 the definition of conversion of a reactant A:

- Batch reactor:

$$X_A = \frac{N_{A0} - N_A}{N_{A0}} \text{ and } N_A = N_{A0}(1 - X_A) \tag{47}$$

- Continuous flow reactor:

$$X_A = \frac{F_{A,in} - F_{A,out}}{F_{A,in}} \text{ and } F_{A,out} = F_{A,in}(1 - X_A) \tag{48}$$

By substituting for N_i and F_i in the design equations, these are transformed into the following:

- Batch reactor:

$$t = C_{A0} \int_0^{X_A} \frac{dX_A}{-\sum_j \nu_{Aj} r_j} \tag{49}$$

- CSTR:

$$V_R = \frac{F_{A,in} X_A}{-\sum_j \nu_{Aj} r_j} \tag{50}$$

- PFR:

$$V_R = F_{A,in} \int_0^{X_A} \frac{dX_A}{-\sum_j \nu_{Aj} r_j} \tag{51}$$

The above equations are valid for the case of individual reactors. For the case of a series of continuous reactors, where the conversion at any point is defined with reference to the feed of the first reactor ($F_{A,0}$), the equations change to

- CSTR:

$$V_R = \frac{F_{A,0}(X_{A,out} - X_{A,in})}{-\sum_j \nu_{Aj} r_j} \tag{52}$$

- PFR:

$$V_R = F_{A,0} \int_{X_{A,in}}^{X_{A,out}} \frac{\mathrm{d}X_A}{-\sum_j \nu_{Aj} r_j} \tag{53}$$

These equations are particularly useful for graphical reactor design, when the plots of $1/(-r_j)$ versus X_A are given.

2.3.2.2 Design of a Batch Reactor. As seen above, for a batch reactor the design equation estimates only the time required for a reaction to proceed to a desired conversion level. If we want to design a batch reactor, *i.e.* estimate its volume, we have to consider other factors such as production rate, as demonstrated in the following example.

Example:
Chemical B is to be manufactured in a batch reactor at an average rate of 5 kmol h^{-1} through the liquid phase isomerisation reaction $A \rightarrow B$. At the end of each run, it takes 5 h to empty the reactor, carry out any repairs and maintenance work and refill the reactor. For a working year of 7000 h, determine the volume of the reactor and the number of batches.

Data:
Reaction rate constant of the first-order reaction: $k = 0.1 \text{ h}^{-1}$; concentration of A in solution: $C_{A0} = 1 \text{ kmol m}^{-3}$; final conversion of A: $X_A = 90\%$.

Solution:
Applying the design equation for a batch reactor with $r_A = kC_A = kC_{A0}(1 - X_A)$, $\nu_A = -1$, we estimate the reaction time needed to convert 90% of the reactant:

$$t = \int_0^{0.9} \frac{C_{A0} \mathrm{d}X_A}{kC_{A0}(1 - X_A)} = -\frac{\ln 0.1}{k} = 23 \text{ h}$$

What we have found so far is the reaction time, the time needed for 90% of A to be converted to B. We have to add to this the time needed for emptying the reactor, repairing/maintaining and refilling it, what we will call operation time (5 h), in order to find the total batch cycle time: 28 h.

Having calculated the total batch cycle, 28 h, we can estimate how much B we should produce with one batch during these 28 h. This is equal to the average production rate multiplied by the batch time: $N_B = 5\,\text{kmol h}^{-1} \times 28\,\text{h} = 140\,\text{kmol}$ of **B**. We know on the other hand that this amount is equal to the amount of A converted, 1:1 stoichiometry: $N_B = N_{A0}(1 - X_A) = (1 - X_A)C_{A0}V_R$. Hence:

$$V_R = \frac{N_B}{(1 - X_A)C_{A0}} = 156\,\text{m}^3$$

Obviously, the number of batch cycles per working year is 7000 h/28 h = 250 batches per year.

2.3.2.3   Residence Time of Continuous Flow Reactors.   The definition of the residence time of a continuous flow reactor is simply the ratio of its volume to its volumetric flow rate: $\tau = (V_R/\dot{V})$. It is the time that every reaction component element, *i.e.* every molecule, stays in the reactor.

In order to be able to express the equations in terms of residence time rather than reactor volume, the assumption is made that the volumetric flow rates of the inlet and outlet streams are the same. Dividing then both sides of the design Equations, (45, 46) and (50, 51), by the total volumetric flow rate, we get

- CSTR:

$$\tau = \frac{C_{i,\text{in}} - C_{i,\text{out}}}{-\sum_j v_{ij}r_j} \quad \text{or} \quad \tau = \frac{C_{A,\text{in}}X_A}{-\sum_j v_{Aj}r_j} \tag{54}$$

- PFR:

$$\tau = \int_{C_{i,\text{in}}}^{C_{i,\text{out}}} \frac{dC_i}{\sum_j v_{ij}r_j} \quad \text{or} \quad \tau = C_{A,\text{in}} \int_0^{X_A} \frac{dX_A}{-\sum_j v_{Aj}r_j} \tag{55}$$

It is obvious that, for continuous flow reactors, designing the reactor means estimating its residence time rather than volume. Continuous flow reactors of the same type with different volumes but the same residence time will give the same conversion.

2.3.3   Comparison between Continuous Stirred Tank Reactor and Plug Flow Reactor

In order to compare the performance of the two types of continuous flow reactors, let us consider an example of a first-order reaction of a reactant A into products ($v_A = -1$) with a reaction rate constant at the reaction temperature of $k = 0.5\,\text{min}^{-1}$. In order to estimate the residence time of a CSTR needed for a conversion, for example, equal to $X_A = 80\%$, we only have to apply Equation (54), where $r = kC_{A,out}$ and $C_{A,out} = (1 - X_A)C_{A,in}$.

Thus,

$$\tau_{CSTR} = \frac{X_A}{k(1 - X_A)} = 8 \text{ min}$$

Let us now see how much conversion can be achieved with a PFR of the same residence time.

From the mole balance for A in the PFR we have:

$$\frac{dC_A}{d\tau} = -kC_A \quad \text{or} \quad \frac{dX_A}{d\tau} = k(1 - X_A)$$

whose solution is $X_A = 1 - \exp(-k\tau_{CSTR}) = 98\%$

Obviously, in a PFR much more reactant is converted than in a CSTR of the same size. What is the reason for this? As the reaction fluid enters a CSTR, the reactant concentration jumps from its inlet value to the outlet value. In a CSTR all the reaction takes place at the same low concentration value across the whole reactor. Hence, the reaction rate has overall in the reactor a relatively low value achieving a relatively low conversion. In a PFR the reactant concentration decreases gradually (Figure 2). At and near the inlet it is relatively high, resulting in a high value of reaction rate at that part of the reactor. With the reaction being faster, the overall conversion is higher. This happens where we have positive kinetics, *i.e.* kinetics where the reaction rate depends positively

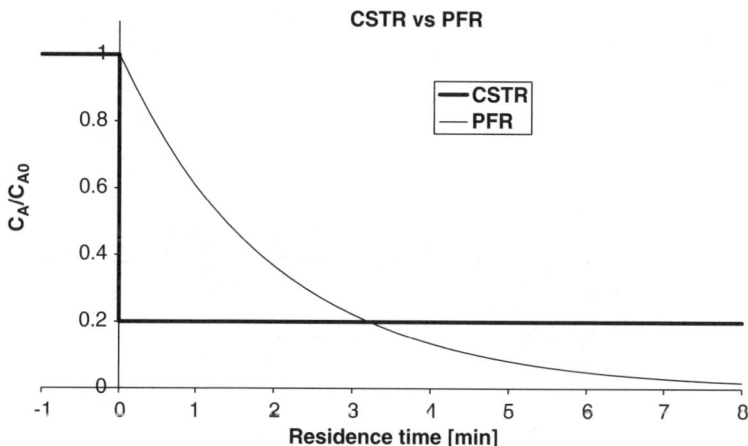

Figure 2 *Comparison of reactant exit concentration of a CSTR and PFR of the same residence time*

on the reactant concentration; the reaction rate increases as the concentration increases, as occurs with first-order kinetics and, generally, positive order kinetics.

The above example brings us to another point that seems a little odd. In a CSTR, how does the reactant concentration jump from its inlet value to its outlet value? In order to explain this apparent discontinuity, we have to consider what happens inside a CSTR rather than considering it as a 'black box'. We know that a CSTR is a thoroughly stirred reactor. This means that inside this reactor a high flow rate stream, many times more than the incoming flow rate, recirculates continuously. With such a high internal recirculation stream, the incoming stream looks like a 'drop in the ocean'. A drop in the ocean does not affect the concentration of the ocean water. A red drop is not going to change the shade of the pink water inside the CSTR. The internal concentration of a CSTR where the reaction mixture is vigorously recirculated is not going to take notice of the inlet concentration. Actually, we can use a model to represent a CSTR consisting of a tubular reactor, PFR, with a high recycle stream. Such a system is going to be considered in the next section.

2.3.4   Recycle Reactor

A recycle reactor, *i.e.* a tubular reactor with a recycle stream (Figure 3) could be a real system, not just a model for a CSTR.

The tubular reactor will be modelled like a PFR and for simplicity we are going to consider no change in volumetric flow rate due to reaction; in other words, the inlet volumetric flow rate and the outlet volumetric flow rate are considered the same, $\dot{V}$. Again we are going to consider first-order kinetics. We are going to represent the recycle volumetric flow by $(R\dot{V})$, where $R = $ (recycle volumetric flow/outlet volumetric flow) and is called "recycle ratio".

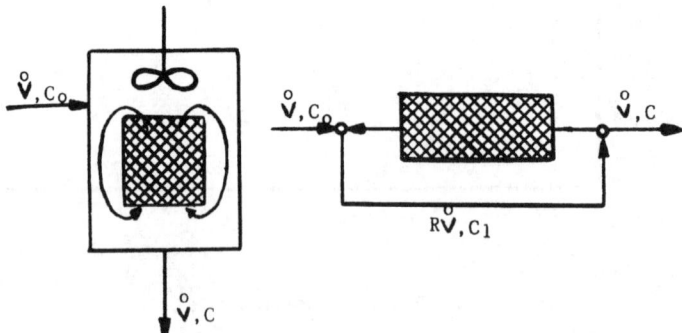

Figure 3   *Scheme of a recycle reactor and its model*

We know how to apply the mole balance equation for the PFR, but we should be careful. The inlet concentration for the PFR is not the same as the inlet concentration for the whole system, the recycle reactor, since the inlet stream is mixed with the recycle stream before the combined stream enters the PFR. It would probably be better to write down the balance equation around the mixing point:

$$\dot{V}C_{A0} + R\dot{V}C_A = (R+1)\dot{V}C_{A1} \quad \text{or} \quad C_{A1} = \frac{C_{A0} + RC_A}{R+1} \tag{56}$$

The mole balance for PFR, taking into account that the volumetric flow rate through the PFR is $(R+1)\dot{V}$, can be written as

$$(R+1)\dot{V}\frac{dC_A}{dV_R} = -kC_A \quad \text{or} \quad (R+1)\frac{dC_A}{d\tau} = -kC_A \tag{57}$$

where $\tau = (V_R/\dot{V})$ is the residence time of the whole recycle reactor system, not the residence time of the PFR alone, on its own.

Hence

$$\frac{dC_A}{C_A} = -\frac{k}{R+1}d\tau \quad \text{or} \quad \frac{C_A}{C_{A1}} = \exp\left(-\frac{k\tau}{R+1}\right) \tag{58}$$

Going back to the mixing Equation (57), we can replace C_{A1} from the above equation

$$C_A \exp\left(\frac{k\tau}{R+1}\right) = \frac{C_{A0} + RC_A}{(R+1)} \tag{58*}$$

Solving for C_A, we get the following equation:

$$C_A = \frac{C_{A0}}{(R+1)\left[\exp\left(\frac{k\tau}{R+1}\right) - R\right]} \quad \text{or}$$

$$\frac{C_A}{C_{A0}} = \frac{1}{(R+1)\left[\exp\left(\frac{k\tau}{R+1}\right) - R\right]} \tag{59}$$

When the recycle ratio R becomes very large (theoretically infinite) the above expression becomes

$$\frac{C_A}{C_{A,0}} = \frac{1}{1+k\tau}$$

which is the expression for a CSTR.

Obviously the other extreme case is when $R=0$, when the above formula becomes identical with that of the PFR: $(C_A/C_{A0}) = \exp(-k\tau)$.

2.3.5 CSTRs in Series

We consider now a system of n CSTRs in series (Figure 4). The outcoming stream from a CSTR is the incoming stream of the next one. For simplicity, we are going to assume the same volumetric flow rate and we are going to examine the reactor system for a reaction of first-order kinetics.

Let us write down the mole balance for reactant A for each reactor i:

$$0 = \dot{V}C_{i-1} - \dot{V}C_i - kC_iV_i \quad \text{or} \quad \frac{C_i}{C_{i-1}} = \frac{1}{1+k\tau_i} \tag{60}$$

where $\tau_i = (V_i/\dot{V})$ is the residence time of the i-th reactor.

Multiplying each of these equations for $i = 1, 2, \ldots, n$ results in a correlation between the outlet concentration of the system, *i.e.* outlet concentration of the n-th reactor, and inlet concentration of the system C_0, as well as an expression for the overall conversion:

$$\frac{C_n}{C_0} = \frac{1}{(1+k\tau_1)(1+k\tau_2)(1+k\tau_3)\ldots(1+k\tau_n)} \tag{61}$$

$$X_A = 1 - \frac{1}{(1+k\tau_1)(1+k\tau_2)(1+k\tau_3)\ldots(1+k\tau_n)} \tag{62}$$

In the case of equal volumes $V_1 = V_2 = V_3 = \cdots = V_n = V$ and, hence, equal residence times $\tau_1 = \tau_2 = \tau_3 = \cdots = \tau_n = \tau = (V/\dot{V})$, the above equations are simplified to

$$\frac{C_n}{C_0} = \frac{1}{(1+k\tau)^n} \tag{63}$$

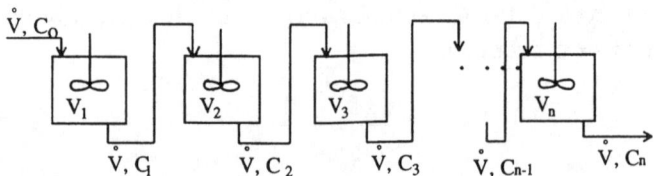

Figure 4 *CSTRs in series*

and

$$X_A = 1 - \frac{1}{(1 + k\tau)^n} \tag{64}$$

Let us now apply the above equation in order to calculate the conversion of a varying number of CSTRs in series with total volume $V_{tot} = 0.6 \text{ m}^3$ and volumetric flow rate $\dot{V} = 0.1 \text{ m}^3 \text{ h}^{-1}$ for a first-order reaction with reaction rate constant $k = 0.25 \text{ h}^{-1}$. The results are presented in Table 1.

It is obvious from the table that the conversion of the series of CSTRs increases with the number of CSTRs (total volume is constant), approaching an upper limit that is the value of the conversion of a PFR with the same volume as the total volume of the CSTR system. The PFR itself can be considered as an infinite number of differential homogeneous slices, each one of which behaves like a CSTR. The increase in conversion with the number of CSTRs is at first very rapid, levelling off later and hence, in practice, a number of CSTRs between 10 and 20 has an overall conversion similar to the equivalent total volume PFR. How

Table 1 *Conversion of a system of CSTRs in series (total volume = 0.6 m³, volumetric flow rate = 0.1 m³ h⁻¹) for a first-order irreversible reaction with kinetic constant equal to 0.25 h⁻¹*

Number of CSTRs	Volume of one reactor (m³)	Residence time (h)	Conversion
1	0.600	6.000	0.600
2	0.300	3.000	0.673
3	0.200	2.000	0.704
4	0.150	1.500	0.720
5	0.120	1.200	0.731
6	0.100	1.000	0.738
7	0.086	0.857	0.743
8	0.075	0.750	0.747
9	0.067	0.667	0.750
10	0.060	0.600	0.753
11	0.055	0.545	0.755
12	0.050	0.500	0.757
13	0.046	0.462	0.758
14	0.043	0.429	0.759
15	0.040	0.400	0.761
16	0.038	0.375	0.762
17	0.035	0.353	0.762
18	0.033	0.333	0.763
19	0.032	0.316	0.764
20	0.030	0.300	0.765
100	0.006	0.060	0.774
PFR	0.6	6.000	0.777

quickly the conversion of the CSTRs in series approaches that of the PFR depends upon the residence time as well as how fast the reaction is.

What we are going to consider now is to solve the same problem for arbitrary kinetics:

$$r_i = f(C_i) \tag{65}$$

Let us write the mole balance equation for the reactant in the i-th CSTR:

$$0 = \dot{V}C_{i-1} - \dot{V}C_i - r_iV_i$$

Solving for r_i we get

$$r_i = \left(\frac{\dot{V}}{V_i}\right)C_{i-1} - \left(\frac{\dot{V}}{V_i}\right)C_i \tag{66}$$

C_i is the solution of the system of Equations (65) and (66). Graphically this means that C_i is the intercept of the kinetics curve with the straight line of gradient $(\dot{V}/V_i)$ and the X-intercept is the entrance concentration of the i-th CSTR, *i.e.* the exit concentration of the $(i-1)$th CSTR. The estimation of the concentrations in the CSTRs one after the other is shown in Figure 5.

2.3.6 Multiple Reactions

So far we have considered reactor systems where only one reaction is occurring. However, this is seldom the case. Take as an example the partial oxidation reaction over a selective catalyst. Often this reaction is

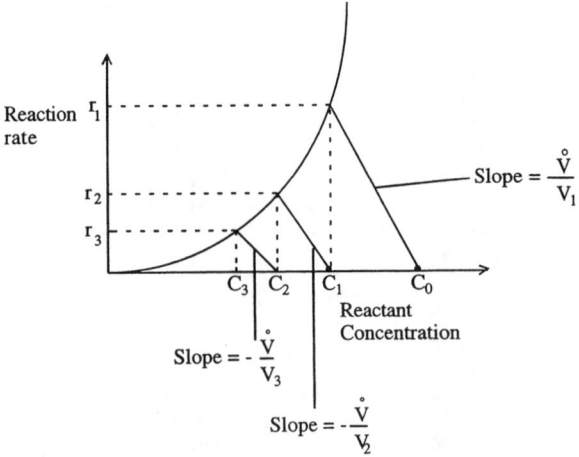

Figure 5   *Solution of the problem of CSTRs in series for arbitrary kinetics*

accompanied by full oxidation reactions of the reactant besides the desired partially oxidised product. Full oxidation of reactant decreases the amount of reactant converted to the desired product, while full oxidation of the desired product converts high value products to low value products or waste. Both side reactions result in the decrease of the overall yield to the desired product.

We are going to consider two basic types of multiple reactions

- Series: A → B → C

- Parallel: $\begin{array}{l} A \to B \\ A \to C \end{array}$

For a series reaction network the most important variable is either time in batch systems or residence time in continuous flow systems. For the reaction system A → B → C the concentration profiles with respect to time in a batch reactor (or residence time in a PFR) are given in Figure 6.

Obviously the reactant concentration decreases with time, exponentially if the reactions are first order, as in the example used in the plot. The concentration of intermediate B will increase in the beginning, as a lot of B is formed from reactant A, whose concentration is still high, but only a little is consumed to C as there is still little B to power this second reaction. As more and more B is formed however, while at the same time less and less A is left, the reaction rate of the second reaction will become greater than that of the first reaction, resulting in a maximum of the intermediate concentration.

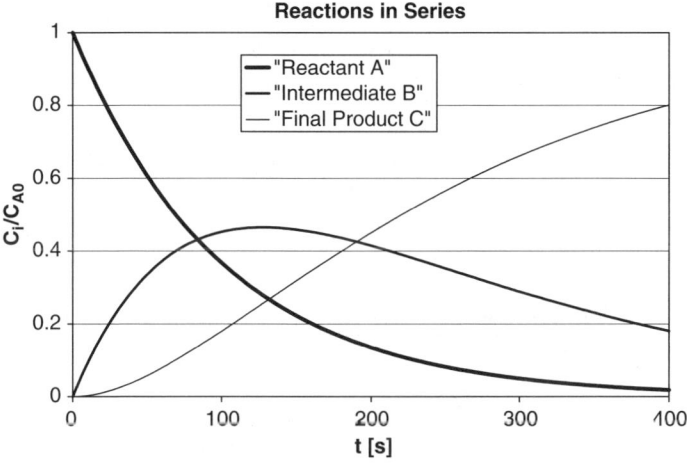

Figure 6 *Concentration profiles of reaction components in a batch reactor for a system of reactions in series*

The relative position of the maximum, as well as its value, depends on the ratio of the kinetic constants, *i.e.* which of the two reactions is faster and by how much.

In a parallel reaction network of first-order reactions, the selectivity does not depend upon reaction time or residence time, since both products are formed by the same reactant and with the same concentration. The concentration of one of the two products will be higher, but their ratio will be the same during reaction in a batch reactor or at any position in a PFR. The most important parameters for a parallel reaction system are the reaction conditions, such as concentrations and temperature, as well as reactor type. An example is given in the following section.

2.3.7   PFR with Continuous Uniform Feed of Reactant Along the Whole Reactor

Consider a product C being produced in a PFR by a second-order reaction:

$$A + B \rightarrow C, \quad r_C = k_1 C_A C_B \tag{67}$$

However, an undesired product A_2 is also formed by the second-order dimerisation reaction:

$$2\,A \rightarrow A_2, \quad r_{A_2} = k_2 C_A^2 \tag{68}$$

For the maximum selectivity of C compared to A_2 (S_{C/A_2}), we need to keep the concentration of A as low as possible. To minimise the concentration of A along the reactor, a different feeding arrangement is realised. Only B is fed into the inlet of the PFR, while A is added uniformly along the whole reactor length, see Figure 7.

In the following paragraph we are going to formulate the mole balance equations for A and B. In this reactor arrangement we have two feeds; direct feed into the reactor inlet of a solution of B with volumetric flow rate $\dot{V}_B$ and concentration C_{B0}, and sidewise feed of a solution of A with concentration C_{A0} and a uniform rate of feed of f_A,

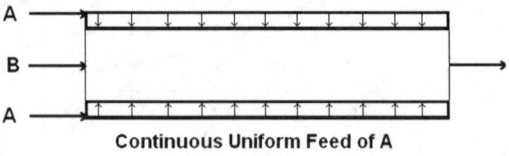

Continuous Uniform Feed of A

Figure 7   *Tubular reactor with continuous feed of one of the reactants along the whole reactor*

expressed as the volumetric flow rate of solution A per unit volume of reactor. The total volumetric flow rate increases along the reactor, since more solution A is added along the whole reactor, and is given by the following formula:

$$\dot{V} = \dot{V}_B + \int_0^V f_A \, dV = \dot{V}_B + f_A V \tag{69}$$

or

$$\frac{d\dot{V}}{dV} = f_A \tag{70}$$

The mole balance for A applied to a differential slice of the reactor is as follows. Remember that there are two inlet terms for A, one term expressing A carried forward with the main flow inside the reactor and a second term through the wall: $f_A dV C_{A0}$.

$$0 = (\dot{V} C_A)_V + f_A C_{A0} dV - (\dot{V} C_A)_{V+dV} - (k_1 C_A C_B + 2 k_2 C_A^2) dV \tag{71}$$

In differential form this equation is

$$\frac{d(\dot{V} C_A)}{dV} = f_A C_{A0} - (k_1 C_A C_B + 2 k_2 C_A^2)$$

$$\dot{V} \frac{dC_A}{dV} + f_A C_A = f_A C_{A0} - (k_1 C_A C_B + 2 k_2 C_A^2)$$

$$(V_B + f_A \dot{V}) \frac{dC_A}{dV} = -(k_1 C_A C_B + 2 k_2 C_A^2) + f_A (C_{A0} - C_A) \tag{72}$$

Mole balance for B:

$$0 = (\dot{V} C_B)_V - (\dot{V} C_B)_{V+dV} - k_1 C_A C_B dV$$

$$\dot{V} \frac{dC_B}{dV} + f_A C_A = -k_1 C_A C_B$$

$$(V_B + f_A \dot{V}) \frac{dC_B}{dV} = -k_1 C_A C_B - f_A C_A \tag{73}$$

In order to design this reactor, Equations (72) and (73) must be solved simultaneously. However, their solution is outside the scope of these notes.

RECOMMENDED READING

1. S. Fogler, *Elements of Chemical Reaction Engineering*, 3rd edn, Prentice Hall, Englewood Cliffs, 1998.
2. O. Levenspiel, *Chemical Reaction Engineering*, 3rd edn, Wiley, New York, 1998.
3. J.M. Smith, *Chemical Engineering Kinetics*, 3rd edn, McGraw Hill, New York, 1981.
4. C. Hill, *An Introduction to Chemical Engineering Kinetics and Reactor Design*, Wiley, New York, 1977.
5. G.F. Froment and K.B. Bischoff, *Chemical Reactor Analysis and Design*, 2nd edn, Wiley, New York, 1990.

CHAPTER 3

Concepts of Fluid Flow

TIM ELSON

3.1 INTRODUCTION

Fluid flow, also known as fluid mechanics, momentum transfer and momentum transport, is a wide-ranging subject, fundamental to many aspects of chemical engineering. It is impossible to cover the whole field here, so we will concentrate upon a few aspects of the subject.

The objective of this chapter is to introduce you to some of the fluid flow problems encountered in the process design of a typical pipe-flow system, illustrated in Figure 1, where a fluid is pumped from one vessel into another through a section of pipework.

In particular we will be looking at the following:

- modes of fluid flow;
- pipe flow;
- pressure drops in pipes and pipe fittings;
- flow meters;
- pumps.

Some other types of flow situation are also mentioned. Sample calculations are given in *Section 3.10*, illustrating the use of some of the principles presented here.

There are many standard texts of fluid flow, *e.g.* Coulson & Richardson,[1] Kay and Neddermann[2] and Massey.[3] Perry[4,5] is also a useful reference source of methods and data. Schaschke[6] presents a large number of useful worked examples in fluid mechanics. In many recent texts, fluid mechanics or momentum transfer has been treated in parallel with the two other transport or transfer processes, heat and mass transfer. The classic text here is Bird, Stewart and Lightfoot.[7]

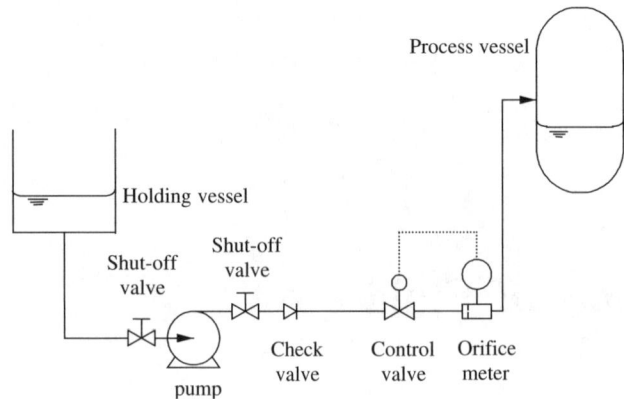

Figure 1 *Pipe-flow system*

3.2 DIMENSIONLESS GROUPS

Chemical engineering in general, and fluid flow in particular, utilises many dimensionless groups, which are discussed in more detail in Chapter 6 "*Scale-up in Chemical Engineering*". Since we will use a piping system as an example in this chapter, we will now consider the pertinent dimensionless groups for pipe flow.

3.2.1 Example: Pipe Flow

Neglecting end effects, *i.e.* $l \gg d$, the pressure drop per unit length of pipe due to friction $(p_1-p_2)/l = \Delta p_\mathrm{f}/l$, will be a function of the pipe diameter d, the surface roughness e, average flow velocity $\bar{u}$ across the cross-section, and fluid density ρ and viscosity μ (Figure 2).
  Thus

$$\frac{\Delta p_\mathrm{f}}{l} = f(d, e, \bar{u}, \rho, \mu) \tag{1}$$

  These six variables (considering $\Delta p_\mathrm{f}/l$ as a single variable), $m = 6$, may be expressed in terms of $n = 3$ fundamental dimensions, *i.e.* mass, length

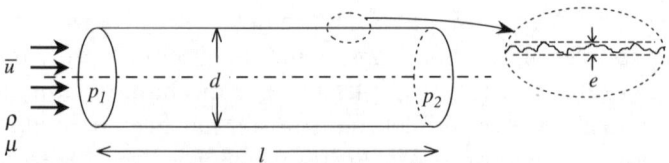

Figure 2 *Flow of a fluid in a pipe*

and time. Using Buckingham's Π theorem, these six variables may be rearranged into $m - n = 6 - 3 = 3$ new dimensionless variables:

$$\frac{\Delta p_f}{l} \frac{d}{\rho \bar{u}^2} = f\left(\frac{\rho \bar{u} d}{\mu}, \frac{e}{d}\right) \tag{2}$$

The procedure for obtaining these dimensionless groups is illustrated in Chapter 6 and in many chemical engineering textbooks.[1,3]

We will return to these dimensionless groups later in *Section 3.6*, where we will see experimental results of pressure-drop measurements in pipes shown in dimensionless form (Figure 7), and go on to use them to calculate pressure drops in the pipework, *Section 3.10*.

3.3 VISCOSITY

3.3.1 Newton's Law of Viscosity

Let us consider a fluid of viscosity μ held between two large flat parallel plates at distance Y apart, as illustrated in Figure 3.

The upper plate is kept in motion in the x-direction at a velocity U by applying a force F to the plate in the x-direction. The lower plate is kept stationary. The force per unit area required to maintain the upper plate (of area A) in motion is given by

$$\frac{F}{A} = \mu \frac{U}{Y} \tag{3}$$

or, in differential form

$$\tau = -\mu \frac{du}{dy} \tag{4}$$

where τ is the shear stress, the shear force per unit area, and du/dy is the velocity gradient, also called the *rate of shear* or *shear rate*. Equation (4) is known as *Newton's Law of Viscosity*.

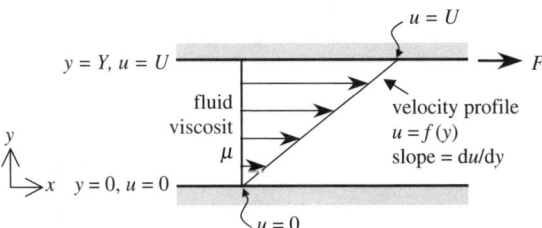

Figure 3 *Flow between parallel plates*

Viscosity, μ, is a property of the fluid and is a measure of its ability to transfer shear stresses. Note that μ [=] M L^{-1} T^{-1}, Pa s, N s m^{-2}.

3.3.2 Dynamic and Kinematic Viscosity

The viscosity μ referred to above is more correctly called the *dynamic viscosity*, and should not be confused with v, the *kinematic viscosity*. The kinematic viscosity is the ratio of dynamic viscosity to fluid density, $v - \mu/\rho$ [—] L^2 T^{-1}, m^2 s^{-1}.

3.3.3 Typical Values of Viscosity

3.3.3.1 Liquids. The viscosity of different liquids varies widely.[8] Their viscosity decreases with increasing temperature. Generally, liquid viscosity is very sensitive to temperature. For non-polar and non-associating liquids, viscosity follows Arrhenius type behaviour, $\mu = \mu_0 \exp(-E_\mu/RT)$.

Water at 20°C has a dynamic viscosity, μ, of 0.001 Pa s = 1 mPa s = 1 cP (centiPoise). The Poise (P) is the c.g.s. unit of dynamic viscosity. The kinematic viscosity, v, of water at 20°C is 10^{-6} m^2 s^{-1} = 1 cS (centiStoke). The Stoke (S) is the c.g.s. unit of kinematic viscosity. Viscosity data are often quoted in units of cP, P, cS and S in the literature.

Typical values of viscosities of liquids and gases are given in Table 1. It can be seen that viscosities of liquids vary widely over many orders of magnitude.

Table 1 *Typical values of viscosity of some gases and liquids at 20°C, 1 atm*

Substance	Dynamic viscosity (Pa s)	Kinematic viscosity (m^2 s^{-1})
Gases		
Hydrogen	0.0000088	0.00010
Air	0.000018	0.000015
Liquids		
Benzene	0.00064	0.00000073
Water	0.001	0.000001
Mercury	0.0016	0.00000012
Ethylene glycol	0.021	0.0000019
Olive oil	0.1	0.0001
Castor oil	0.6	0.0006
Glycerine	1.4	0.0011
Ball-point pen ink	~11	~0.01
Corn syrup	~100	~0.07
Bitumen	~1,000	~1

3.3.3.2 Non-Newtonian Liquids. Newtonian liquids are those that obey Newton's law of viscosity, *i.e.* viscosity does not vary with rate of shear $\mu \neq f(du/dy)$. Many liquids, especially those containing significant quantities of particles or high molecular weight polymers, have viscosities that vary with rate of shear, $\mu = f(du/dy)$ (at constant T and p), and so do not obey Newton's law. These liquids are called *non-Newtonian*. Common examples are toothpaste, drilling muds, non-drip paints, wallpaper paste, egg white, polymer solutions and melts, silicone rubbers, cement slurries and printing inks. Some non-Newtonian fluids, especially polymer solutions and melts, and silicone rubbers, have elastic as well as viscous properties and these are called *viscoelastic*. The behaviour of non-Newtonian fluids is beyond the scope of this chapter, but there are many textbooks on the subject, for example, Chhabra and Richardson.[9]

3.3.3.3 Gases. At moderate pressures, gas viscosity increases with increasing temperature, $\mu \propto T^{0.7-1.0}$, and is relatively independent of pressure.

Air at STP, $\mu \approx 1.7 \times 10^{-5}$ Pa s, $v \approx 1.4 \times 10^{-5}$ m^2 s^{-1}.

3.4 LAMINAR AND TURBULENT FLOW

There are two distinct modes of flow, laminar and turbulent. Fluid inertia tends to allow fluctuations to grow and give rise to turbulent eddies. Viscosity on the other hand, tends to damp out these fluctuations. A ratio of forces, inertial to viscous, is used to characterise the nature of the flow and is called the *Reynolds Number, Re*. For pipe flow this takes the form:

$$Re = \frac{\rho \bar{u} d}{\mu} = \frac{\bar{u} d}{\nu} \tag{5}$$

The Reynolds Number being a ratio of like quantities, forces in this case, is dimensionless. At low *Re*, flows tend to be laminar, and at high *Re*, flows tend to be turbulent. As we shall see later, the limiting values of *Re* for laminar and turbulent flow and the transition between these modes of flow is a function of the flow geometry.

Table 2 compares the characteristics of laminar and turbulent flow with particular reference to flow in a pipe. Many of the characteristics described also apply to other flow situations.

3.4.1 Boundary Layers

Although the bulk of the flow may be turbulent, the presence of an interface or solid boundary causes boundary layers to form. The

Table 2 *Comparison of laminar and turbulent flow in a pipe*

Laminar flow	*Turbulent flow*
• Laminar flows predominantly occur in higher viscosity fluids and flows in small diameter vessels	• Turbulent flow is the most common type of flow found industrially
• $Re_{pipe} \lesssim 2000$	• $Re_{pipe} \gtrsim 5000$
• Many flow problems can be solved analytically	• Flows are much less well defined than in Laminar flow, and turbulence is an empirically based subject
• Laminar flow is characterised by streamlines – stratified layers of fluid – with transport by molecular diffusion (due to viscosity) between them	• Superimposed upon an average flow in the x-direction is a random three-dimensional flow caused by turbulent eddies. Eddies dominate the transport processes
• A coloured dye trace injected into pipe shows a clearly defined thread extending over the whole length of the pipe	• The dye trace thread is broken down and becomes mixed with the fluid, eventually colouring the fluid over the whole width of the pipe
• Viscous forces dominate	• Inertial forces dominate
• Frictional losses $\propto \mu\, u$	• Frictional losses $\propto \mu\, u^2$
• Laminar velocity profile	• Turbulent velocity profile

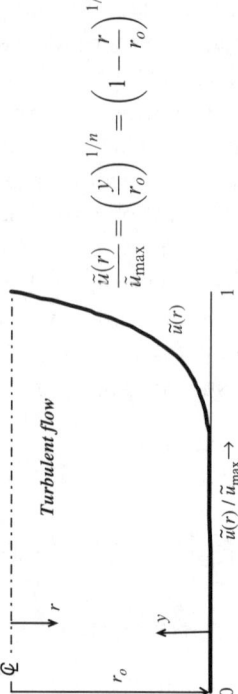

$$\frac{u(r)}{u_{max}} = 1 - \left(\frac{r}{r_0}\right)^2$$

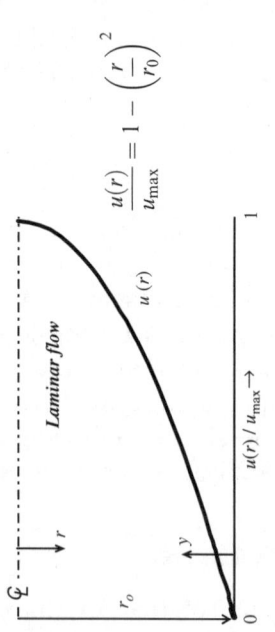

$$\frac{\tilde{u}(r)}{\tilde{u}_{max}} = \left(\frac{y}{r_o}\right)^{1/n} = \left(1 - \frac{r}{r_o}\right)^{1/n}$$

• Mean velocity in laminar flow:

$$\bar{u} = \frac{Q_v}{A} = 0.5\, u_{max}$$

where $\tilde{u}$ is the mean velocity at a point (r), $n = f(Re)$, experimentally $6 < n < 10$. At $Re = 10^5$, $n = 7$ "Prandtl 1/7th power law"

• Mean velocity in turbulent flow:

$$\bar{u} = \frac{Q_v}{A} \approx 0.82\, \tilde{u}_{max}$$

presence of a solid wall dampens turbulent fluctuations in the fluid layers next to the wall and close to the wall the flow becomes laminar in nature. At the wall, fluid is brought to a halt. This is known as the non-slip boundary condition.

3.4.1.1 Universal Velocity Profile. The Universal Velocity Profile is often used to describe the velocity near a wall. This profile is divided into three sections or layers and is shown in Figure 4. Here, dimensionless velocity, $u^+ = \tilde{u}\sqrt{\rho/\tau_w}$, is plotted against dimensionless distance from wall, $y^+ = y\sqrt{\tau_w\rho/\mu}$, where τ_w is the shear stress at the wall.

- *Laminar sub-layer*: $0 < y^+ < 5$, Newton's Law of Viscosity (4), describes the flow and gives $u^+ = y^+$; here laminar flow and transfer by molecular diffusion dominate.
- *Buffer zone*: $5 \leq y^+ \leq 30$, $u^+ = 5.0\ln y^+ - 3.05$; here both laminar and turbulent flow considerations are important; and
- *Turbulent core*: $30 \leq y^+$, $u^+ = 2.5\ln y^+ + 5.5$; here turbulent flow and transfer by eddy motion dominate.

Alternatively, a single laminar boundary layer, with $u^+ = y^+$ reaching up to $y^+ \approx 11.6$, is often used, beyond which there is the turbulent core.

3.5 BALANCE OR CONSERVATION EQUATIONS

3.5.1 General Form of the Conservation Equations

Three balance equations are used in solving fluid flow problems. Each is a conservation equation, conserving mass, momentum and energy

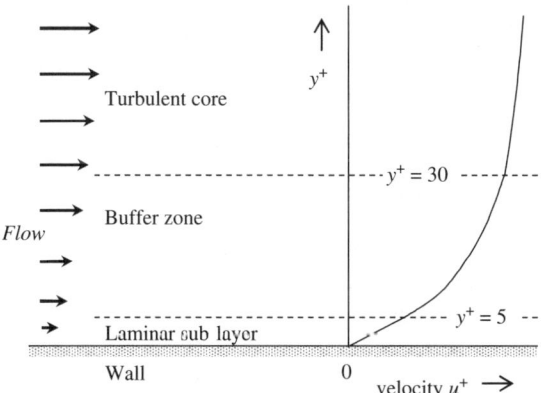

Figure 4 *Boundary layers and the universal velocity profile*

Table 3   *The conservation equations*

Equation	Conserved property	Conservation law
Continuity	Mass	Conservation of mass
Momentum	Momentum	Newton's 2nd law
Energy	Energy	1st law of thermodynamics

respectively. They are listed in Table 3, with the respective conserved property and conservation law.

The general form of the conservation equations is:

$$
\begin{array}{ccccccc}
\text{Rate of} & & \text{Rate of} & & \text{Rate of} & & \text{Rate of} \\
\text{input} & - & \text{output} & + & \text{generation} & = & \text{accumulation} \\
\text{of property} & & \text{of property} & & \text{of property} & & \text{of property}
\end{array} \tag{6}
$$

Note that the conservation equations are rate equations. In the examples below we consider only the steady-state forms of the conservation equations. Thus the accumulation terms will be zero. In most cases we will just present the final forms of the equations.

3.5.2   Control Volumes

These conservation equations are applied to control volumes. Figure 5 shows a generalised control volume, with one inlet (subscript 1) and one outlet (subscript 2), which we will use.

Here A represents the cross-sectional area of the inlet and outlet, z their elevations, u the fluid's mean velocity and ρ its density, m_T is the total mass within the control volume, q rate of heat input to the control volume per unit mass and w_x shaft work output from the control volume per unit mass. Note this is the *Engineers' convention* for shaft work. In the *Chemists' convention*, positive shaft work is an input to the control volume, *i.e.* the direction of positive shaft work is opposite.

3.5.3   Continuity Equation (Mass Balance)

Considering total mass and steady state:

$$
\begin{array}{ccccccc}
\text{Rate of} & & \text{Rate of} & & \text{Rate of} & & \text{Rate of} \\
\text{input} & - & \text{output} & + & \text{generation} & = & \text{accumulation} \\
\text{of mass} & & \text{of mass} & & \text{of mass} & & \text{of mass}
\end{array} \tag{7}
$$

$$
\begin{array}{ccccccc}
G_1 & - & G_2 & + & 0 & = & 0
\end{array}
$$

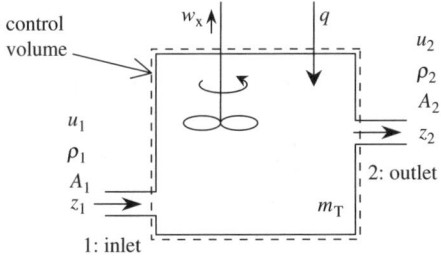

Figure 5 *Generalised control volume*

or

$$G_1 = G_2 \tag{8}$$

i.e.

$$\rho_1 u_1 A_1 = \rho_2 u_2 A_2 \tag{9}$$

If the fluid is incompressible, *e.g.* a liquid, $\rho_1 = \rho_2 = \rho$, then

$$u_1 A_1 = u_2 A_2, \quad \text{or} \quad Q_{v1} = Q_{v2} \tag{10}$$

Thus, at steady state the overall mass flow rate in equals the overall mass flow rate out and if the fluid is incompressible, the volumetric flow rate in equals the volumetric flow rate out.

If one is considering a chemical reaction and performing a component mass balance, the generation term of the components taking part in the reaction will not be zero, but related to the rate of reaction. However, if one considers total mass, the generation term will be zero unless the reaction is nuclear.

[See *Derivation of head loss in a sudden expansion* later in this section and the *Venturi Meter* in *Section 3.7* for examples of the use of the Continuity Equation.]

3.5.4 Steady-State Momentum (Force) Balance Equation

Rate of input of momentum		Rate of output of momentum		Rate of generation of momentum		Rate of accumulation of momentum
	−		+		=	
Gu_1	−	Gu_2	+	ΣF	=	0

$$\tag{11}$$

As momentum is a vector, the steady-state momentum equation above is a vector equation and, in general, needs to be applied in each of the three directions (x, y, z) in which momentum (*i.e.* velocity) has a component. The vector parameters in the momentum equations are shown here emboldened.

As the rate of momentum transfer is equal to a force, momentum balances are equivalent to force balances.

The generation term is equal to ΣF, the sum of forces acting on the control volume, which is comprised of the three following terms:

$p_1A_1 - p_2A_2$: the net force exerted on the fluid within the control volume by the pressure of the surrounding fluid at the inlet (point 1) and at the outlet (point 2); A is the cross-sectional area of the inlets and outlets perpendicular to the direction under consideration and so is shown as a vector;

F: the net force exerted on the fluid within the control volume by the surrounding/control volume walls, *e.g.* frictional forces;

$m_T g$: the body force exerted on the fluid within the control volume by gravitational acceleration.

Thus

$$G(\boldsymbol{u}_1 - \boldsymbol{u}_2) + (p_1\boldsymbol{A}_1 - p_2\boldsymbol{A}_2) + \boldsymbol{F} + m_T\boldsymbol{g} = 0 \qquad (12)$$

[See *Derivation of head loss in a sudden expansion* later in this section for an example of the use of the momentum equation.]

3.5.5   Steady-State Energy Balances

3.5.5.1   Steady Flow Energy Equation.

$$g(z_2 - z_1) + \left(\frac{u_2^2}{2} - \frac{u_1^2}{2}\right) + (h_2 - h_1) = q - w_x \qquad (13)$$

where h is the specific enthalpy of the fluid.

[Units: specific energy, energy per unit mass of fluid flowing [J kg^{-1}, m^2 s^{-2}]. To get rates [J s^{-1}, W] simply multiply each term by mass flow rate, G.]

The Steady Flow Energy Equation relates the changes in potential and kinetic energy, and enthalpy of streams flowing in and out of a control volume, to the rate of heat transfer to, and the rate of shaft work from, the control volume. Here the generation term is zero. If a chemical reaction were occurring within the control volume, the heat of reaction would need to be accounted for in a generation term. Note that if we are using the Chemists' convention for shaft work, the right-hand side of Equation (13) would become $q + w_x$.

3.5.5.2 Mechanical Energy Balance Equation. Assuming no heat transfer to or from the surroundings ($q = 0$) and that frictional losses end up being dissipated as heat, the Steady Flow Energy Equation for a constant density fluid becomes:

$$\frac{p_2 - p_1}{\rho} + \left(\frac{u_2^2}{2} - \frac{u_1^2}{2}\right) + g(z_2 - z_1) + g\Delta h_f + w_x = 0 \qquad (14)$$

and is known as the *Mechanical Energy Balance Equation.*

[Units: energy per unit mass of fluid flowing [J kg^{-1}, m^2 s^{-2}]. To get rates [J s^{-1}, W] multiply terms by mass flow rate, G.]

Here Δh_f is the *head loss due to friction* and is directly related to the pressure loss due to friction Δp_f by $\Delta p_f = \rho g \Delta h_f$.

The Mechanical Energy Balance Equation (14), is the key equation when calculating pressure changes through piping systems. See *Section 3.10* for examples of the use of the Mechanical Energy Balance Equation in such calculations.

3.5.5.3 Bernoulli's Equation. Bernoulli's Equation is obtained by integrating a momentum (force) balance along a flow streamline assuming steady state, with an incompressible ($\rho = $ constant) and inviscid ($\mu = 0$) fluid. As we integrate a momentum or force balance along a streamline, we get an energy balance: $\int F dx = $ force $\times$ distance moved $=$ work done $=$ energy exerted. Bernoulli's Equation is therefore an energy balance.

Bernoulli's Equation gives us the total energy in a flowing fluid and sums the energy due to fluid pressure, kinetic energy and potential energy.

$$\frac{p}{\rho g} \quad + \quad \frac{u^2}{2g} \quad + \quad z \quad = \quad \text{constant along a streamline}$$

$$\begin{array}{ccccccc}
\text{pressure} & + & \text{velocity} & + & \text{gravity or} & = & \text{total head of fluid} \\
\text{head} & & \text{head} & & \text{static head} & &
\end{array}$$

$$(15)$$

Each term on the left-hand side has the same dimensions, length (energy per unit weight of fluid) in this version of Bernoulli's Equation. Each term can be thought of as a *head* of fluid, and the sum of these terms is called the *total head* of fluid.

The use of the term head is simply an extension of the practice of expressing pressures, p, as heights or heads of fluid, h, where $p = \rho g h$, *e.g.* 1 atm $= 760$ mm Hg $= 406.8$ in H_2O, *etc.*

[Note: Equation (15) is just one of several versions of Bernoulli's Equation where the terms are expressed in different but self-consistent sets of units (energy per unit mass with (15) multiplied by ρ or energy per unit volume [pressure] with (15) multiplied by ρg).]

Bernoulli's Equation is normally applied between two points, 1 and 2, along a streamline.

$$\frac{p_1}{\rho g} + \frac{u_1^2}{2g} + z_1 = \frac{p_2}{\rho g} + \frac{u_2^2}{2g} + z_2 \tag{16}$$

Equation (16) is similar to (14), the Mechanical Energy Balance Equation, but with $g\Delta h_f$ and w_x both equal to zero.

[See *Venturi Meter* in *section 3.7* below for an example of the use of Bernoulli's Equation.]

3.5.6 Example of use of the Conservation Equations

3.5.6.1 Derivation of Head Loss in a Sudden Expansion. This example illustrates the use of all three conservation equations to derive an expression for the head loss through a sudden expansion in a pipe or as a pipe enters a large vessel, an *exit loss*, and is illustrated in Figure 6.

In Figure 6, section 1 is immediately downstream of the expansion where radial acceleration of fluid is small so pressure is uniform across the cross-section (and area is A_2 not A_1). Section 2 is sufficiently far downstream for flow to be parallel and pressure constant across the cross-section.

The Momentum Equation (12) is:

$$G(u_1 - u_2) + (p_1 A_1 - p_2 A_2) + F + m_T g = 0$$

Assume that the sudden expansion is horizontal, $z_1 = z_2$ and $g = 0$, and that there is negligible wall friction, $F = 0$, as length is short. In addition,

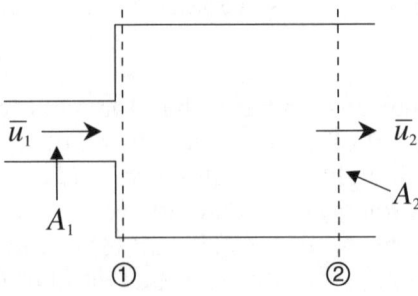

Figure 6 *Flow through a sudden expansion*

let us assume that the velocity is constant across each cross-section (*i.e.* turbulent flow, flat profile).

$$(p_1 - p_2)A_2 = G(\bar{u}_2 - \bar{u}_1) = \rho Q(\bar{u}_2 - \bar{u}_1) = \rho \bar{u}_2 A_2(\bar{u}_2 - \bar{u}_1)$$

giving

$$\frac{(p_1 - p_2)}{\rho} = \bar{u}_2(\bar{u}_2 - \bar{u}_1) \tag{17}$$

The Mechanical Energy Balance Equation (14) for a constant density fluid is:

$$\frac{p_2 - p_1}{\rho} + \left(\frac{\bar{u}_2^2}{2} - \frac{\bar{u}_1^2}{2}\right) + g(z_2 - z_1) + g\Delta h_f + w_x = 0$$

or, as $z_1 = z_2$ and $w_x = 0$,

$$\Delta h_f = \frac{1}{g}\left[\frac{p_1 - p_2}{\rho} + \frac{\bar{u}_1^2 - \bar{u}_2^2}{2}\right]$$

Using Equation (17)

$$\Delta h_f = \frac{1}{g}\left[\bar{u}_2(\bar{u}_2 - \bar{u}_1) + \frac{\bar{u}_1^2 - \bar{u}_2^2}{2}\right] = \frac{1}{2g}(\bar{u}_1 - \bar{u}_2)^2 \tag{18}$$

The Continuity Equation (9), assuming constant density, ρ, gives:

$$\bar{u}_1 A_1 = \bar{u}_2 A_2$$

and Equation (18) then becomes:

$$\Delta h_f = \frac{\bar{u}_1^2}{2g}\left(1 - \frac{A_1}{A_2}\right)^2 = \frac{\bar{u}_2^2}{2g}\left(\frac{A_2}{A_1} - 1\right)^2 = K\frac{\bar{u}_1^2}{2g} \tag{19}$$

where $K = (1-(A_1/A_2))^2$, and as $\bar{u}_1^2/2g$ is one velocity head. K is known as the number of velocity heads for the fitting loss.

If $A_2 \gg A_1$, *e.g.* flow from a pipe into a large vessel or reservoir, $K \approx 1$. Thus the exit loss from a pipe into a large vessel is 1 velocity head, *i.e.* all kinetic energy in the flowing fluid is lost when it comes to rest after entering a large vessel. This is the value reported later in Table 5 for an Exit Loss. Losses in fittings are described in more detail in the next section.

3.6   PIPE FLOW

With turbulent flows we normally have to resort to the correlation of experimental data. Friction between fluids and surfaces are normally characterised by friction factors. Different flow situations give rise to different friction factors.

3.6.1   Friction Factors

3.6.1.1   Fanning Friction Factor.   The Fanning friction factor for pipe flow, c_f, is defined by:

$$\tau_w = c_f \frac{1}{2} \rho u^2 \tag{20}$$

where τ_w is the pipe wall shear stress, the frictional shear force per unit area of pipe wall. Applying a momentum (force) balance to a pipe of diameter, d, and length, l, gives $\Delta p_f \times (\pi d^2/4) = \tau_w \times \pi d l$ and so the relation between Fanning friction factor and pipe pressure drop may be obtained:

$$\Delta p_f = 2c_f \frac{l}{d} \rho u^2 \quad \text{or} \quad c_f = \frac{\Delta p_f}{2\rho u^2} \frac{d}{l} \tag{21}$$

Recall that, using dimensional analysis for pipe flow, we saw (see *Section 3.2*):

$$\frac{\Delta p_f}{l} \frac{d}{\rho \bar{u}^2} = f\left(\frac{\rho \bar{u} d}{\mu}, \frac{e}{d}\right)$$

or,

$$c_f = f\left(Re, \frac{e}{d}\right) \tag{22}$$

3.6.1.2   Other Friction Factors.   Beware, there are other definitions of friction factor.

- *Blasius or Darcy* friction factor, λ or f: $\Delta h_f = \lambda(\text{or } f)(l/d)(\bar{u}^2/2g)$, giving $\lambda = f = 4c_f$
- *Coulson and Richardson* friction factor,[1] ϕ: $\tau_w = \phi \rho \bar{u}^2$, giving $\phi = c_f/2$

The form of the c_f–Re relationship depends upon the flow regime.

3.6.1.3 *Laminar Flow, $Re \leq 2000$.* A momentum balance on a differential fluid element yields the laminar velocity profile (see Table 2) and

$$\frac{\Delta p_f}{l} = \frac{32\mu\bar{u}}{d^2} \tag{23}$$

which is known as the Hagen–Poiseuille Equation, and indicates that for laminar flow,

$$\Delta p_f \propto \mu\bar{u}$$

Rearranging (23) yields

$$c_f = \frac{16}{Re} \tag{24}$$

Note: roughness has no effect on laminar flow in pipes.

3.6.1.4 *Transition, $2000 < Re < 5000$.* Flow fluctuates between laminar and turbulent. Pressure drops fluctuate and are uncertain. Avoid the transition region.

3.6.1.5 *Turbulent flow, $Re \geq 5000$.* In turbulent flow, one has to rely on correlation of experimental data

Smooth pipes, $e = 0$

Blasius:

$$c_f = 0.0791\, Re^{-1/4} \tag{25}$$

$$\text{for } 3000 < Re < 10^5$$

giving

$$\Delta p_f \propto \rho\, \bar{u}^{1.75}$$

Nikuradse:

$$\frac{1}{\sqrt{c_f}} = 4.0 \log_{10}(Re\sqrt{c_f}) - 0.40 \tag{26}$$

$$\text{for } 3000 < Re < 3 \times 10^6$$

Rough pipes, $e > 0$

Nikuradse:

$$\frac{1}{\sqrt{c_f}} = 4.0 \log_{10}\left(\frac{r_0}{e}\right) + 4.6 \tag{27}$$

$$\text{for } \left(\frac{r_0}{e}\right)(Re\sqrt{c_f}) > 0.005, \text{ fully rough pipes}$$

Table 4 *Typical values of pipe roughness[1,4,21,22]*

Pipe	e, absolute roughness (mm)
Drawn tubing	0.0015
Commercial steel & wrought iron	0.05
Cast iron	0.25
Concrete	0.3–3
Riveted steel	1–10

Moody:

$$c_f = 0.001375 \left[1 + \left(20,000 \frac{e}{d} + \frac{10^6}{Re} \right)^{1/3} \right] \qquad (28)$$

for $3000 < Re < 10^7$ and $e/d < 0.1$

giving $\Delta p_f \propto \rho \bar{u}^2$ for fully rough pipes

Typical values of pipe roughness, e, are given in Table 4.

Figure 7 shows the form of c_f as a function of Re and e/d as predicted by Equations (24)–(26) and (28).

With knowledge of c_f from the above correlations and using Equation (21) we can calculate frictional pressure drops in straight pipes. However, we also need to be able to evaluate frictional losses in pipe fittings.

3.6.2 Losses in Fittings

The flow in pipe fittings, *e.g.* bends and valves, is generally too complex to determine theoretically. For *turbulent flow*, these minor losses are approximately equal to the square of the flow velocity. Thus, we define a loss coefficient, K,

$$\Delta p_{loss} = K \frac{\rho \bar{u}^2}{2}, \quad \text{or} \quad \Delta h_{loss} = K \frac{\bar{u}^2}{2g} \qquad (29)$$

where $\bar{u}^2/2g$ is *one velocity head* (see *Section 3.5*). So K is known as the *number of velocity heads* for a fitting loss.

These minor fitting losses can also be expressed in terms of an equivalent length of straight horizontal pipe, l_e, giving the same frictional loss.

$$\Delta h_{loss} = 4 c_f \frac{l_e}{d} \frac{\bar{u}^2}{2g} \equiv K \frac{\bar{u}^2}{2g} \qquad (30)$$

Table 5 *Typical losses in fittings*[1,4]

Fitting	n, number of pipe diameters	K, number of velocity heads
45° elbows:		
Standard radius	15	0.3
Long radius	10	0.2
90° elbows:		
Standard radius	35	0.7
Long radius	22	0.45
Square or mitre	60	1.2
T-piece:		
Entry from leg	60	1.2
Entry into leg	90	1.8
Unions and couplings	Small, 2	Small, 0.04
Globe valve:		
Fully open	300	6
1/2 open	450	9
Gate valve:		
Fully open	7	0.15
3/4 open	40	1
1/2 open	200	4
1/4 open	800	16
Check valve:		
Swing	100	2
Disk	500	10
Ball	3500	70
Entrance loss	25	0.5
Exit loss	50	1
Orifice meter	$\Delta p_{loss} \sim 60\%$ of $(p_1 - p_2)$	
Venturi meter	$\Delta p_{loss} \sim 10\%$ of $(p_1 - p_2)$	

thus

$$K = 4c_f \frac{l_e}{d} = 4c_f n \tag{31}$$

where $n = l_e/d$, the *number of pipe diameters* of straight pipe that gives an equivalent frictional head or pressure loss.

Therefore, the total pressure or head loss due to friction in pipework due to the pipe and fittings is given by

$$\Delta p_{f_T} = 2c_f \frac{(l + l_e)}{d} \rho \bar{u}^2 = 2c_f \frac{l_T}{d} \rho \bar{u}^2$$

$$\text{or} \quad \Delta h_{f_T} = 4c_f \frac{(l + l_e)}{d} \frac{\bar{u}^2}{2g} = 4c_f \frac{l_T}{d} \frac{\bar{u}^2}{2g} \tag{32}$$

Table 5 shows typical losses in a range of fittings expressed as values of K and n.

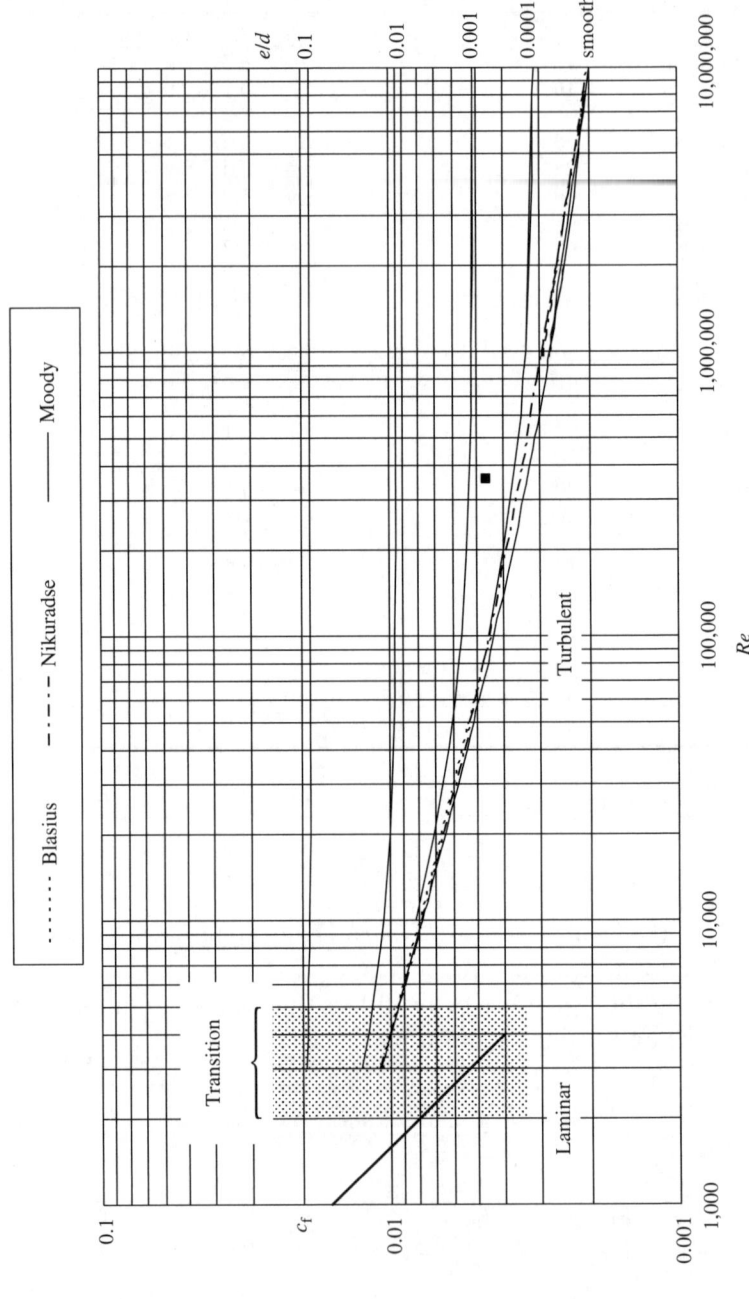

Figure 7 *Fanning friction factor chart for pipe flow.* (■) *piping system example (see Section 3.10)* $Re = 363,000$, $e/d = 0.0005$, $c_f = 0.00459$

3.6.3 Economic Velocities

Economic velocities for turbulent flow (the most common industrially), taking into account piping capital costs (piping, fittings, pumps, *etc.*) and running costs (pressure drops), are primarily a function of fluid density $[\Delta p_{f_T} \propto \rho \bar{u}^2]$. Details are given in the literature.[4] Order of magnitude economic velocities are:

- for low viscosity liquids, $\bar{u}_{\text{economic}} \approx 1\,\text{m}\,\text{s}^{-1}$
- for gases, $\bar{u}_{\text{economic}} \approx 10\,\text{m}\,\text{s}^{-1}$

3.7 FLOW MEASUREMENT

The details of design and construction of various flow meters are shown in many standard chemical engineering and fluid mechanics texts. Some reviews and advice in choosing flowmeters for particular duties are available in the literature.[10–12]

3.7.1 Variable Head Meters[13,14]

3.7.1.1 Venturi Meter. Figure 8 illustrates a Venturi meter. Fluid is accelerated through a gradual contraction, half angle approximately 20°, through a throat, and then decelerated though an even more gradual expansion, approximately 7°, for good pressure recovery.

The Continuity and Bernoulli Equations may be used to derive equations relating flow rate to measured pressure difference for the Venturi meter (and the orifice plate meter discussed below).

Applying Bernoulli's Equation (16), along a streamline between points 1 and 2 and assuming the meter is horizontal, $z_1 = z_2$, and that the velocities are constant across a cross-section, gives:

$$\frac{p_1 - p_2}{\rho g} = \frac{\bar{u}_2^2 - \bar{u}_1^2}{2g} \tag{33}$$

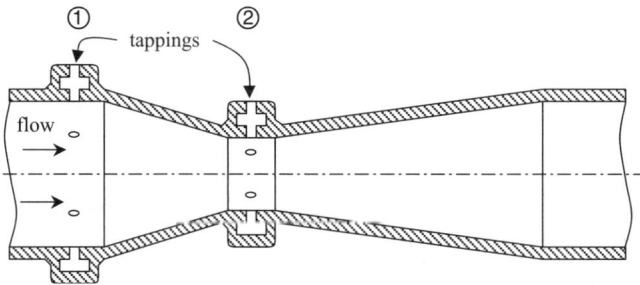

Figure 8 *Venturi meter*

Continuity, assuming constant density ρ, Equation (10), gives $\bar{u}_2 = \bar{u}_1(A_1/A_2)$. Substituting for $\bar{u}_2$ into Equation (33) and rearranging gives:

$$\bar{u}_1 = \left[\frac{2}{\rho}\frac{(p_1 - p_2)}{(A_1/A_2)^2 - 1}\right]^{1/2} \tag{34}$$

or

$$G = \rho\bar{u}_1 A_1 = A_1\left[\frac{2\rho(p_1 - p_2)}{(A_1/A_2)^2 - 1}\right]^{1/2} \tag{35}$$

Adding C_d, a *discharge coefficient*, to account for frictional losses (assumed zero in Bernoulli's Equation) and other non-idealities, we get Equation (36), the operating equation for the Venturi meter.

$$\bar{u}_1 = C_d\left[\frac{2}{\rho}\frac{(p_1 - p_2)}{(A_1/A_2)^2 - 1}\right]^{1/2} \tag{36}$$

or

$$G = C_d\rho\bar{u}_1 A_1 = C_d A_1\left[\frac{2\rho(p_1 - p_2)}{(A_1/A_2)^2 - 1}\right]^{1/2} \tag{37}$$

C_d for a well-designed Venturi meter is normally in the range 0.97–0.99. The appropriate British Standards[13,14] give correlations of C_d values for standard geometry meters.

3.7.1.2 Orifice Plate Meter.   Figure 9 illustrates an orifice plate meter. Here fluid is suddenly accelerated as it passes through an orifice in a plate, which is normally held between two flanges. Pressure is measured upstream (point 1) and downstream (point 2) of the orifice. There are a number of standard positions for the pressure tapings, "*D* & *D*/2", *i.e.* one pipe diameter upstream and one half diameter downstream of the orifice, or "flange tapings", or "corner tapings".

Equation (37) is also used for the orifice plate meter. The values of C_d are lower for orifice plate meters. They can vary significantly with flow rate and are a function of tapping location, as the measured pressure difference is a function of tapping position (see the pressure profile illustrated in Figure 9). For well-designed meters operating in their optimum range, C_d values typically range between 0.6 and 0.7. The appropriate British Standards[13,14] give correlations of C_d values for standard geometry meters.

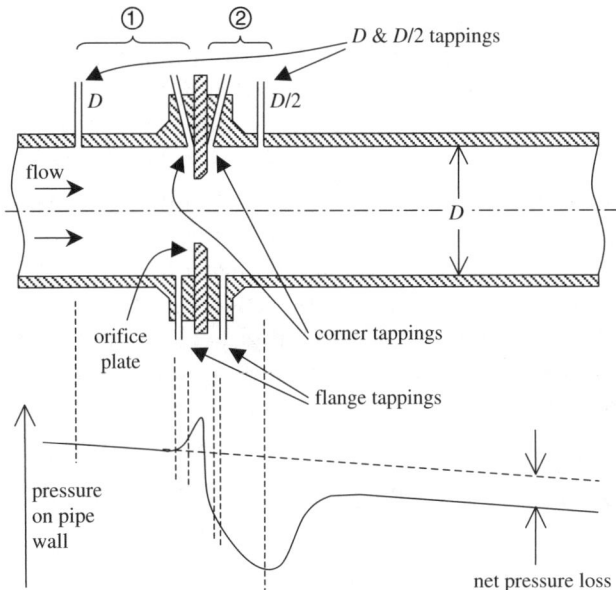

Figure 9 *Orifice plate meter showing various tapping locations and pressure profile*

3.7.2 Variable Area Meters

3.7.2.1 Rotameter. The rotameter consists of a vertical tapered tube, the inside diameter of which increases with height, and is normally made of glass. Inside the tube is a spherical or cylindrical "float". At a steady flow rate the float settles at a height where the weight of the float is balanced by the drag and buoyancy forces acting on it. As the flow rate increases, the float rises in the tapered tube (so giving a greater annular area for flow between the float and the tube walls) until the weight of the float is again balanced by the drag and buoyancy forces. The height of the float is related to the flow rate.

3.7.2.2 Variable Area Orifices. Variable area orifices consist of an orifice and a concentric obstruction. Fluid flows through the orifice around the concentric obstruction. As the flow rate increases, the area between the orifice and the obstruction changes, either by displacement of a spring loaded obstruction relative to a fixed orifice or by displacement of a spring loaded orifice relative to a fixed obstruction. The displacement is related to the flow rate.

3.7.3 Some Other Flowmeters

3.7.3.1 Vortex Shedding Flowmeter. At Reynolds numbers greater than ~ 1000, vortices are shed behind bluff bodies giving an oscillating wake. The rate of oscillation is proportional to the flow rate and is measured by the meter and the flow rate determined. See *Chapter 7 "An Introduction to Particle Systems"* for further information on oscillating wakes.

3.7.3.2 Magnetic Flowmeter. When a conductor moves though a magnetic field, an electromagnetic force (emf), is generated. The magnetic flowmeter uses this principle to measure flow rates in electrically conducting liquids. A magnetic field is applied across a section of non-conducting pipe and, if the liquid is electrically conducting, an emf will be generated that is proportional to liquid velocity. The emf is picked up by electrodes in the pipe wall and converted into a flow rate.

3.8 PUMPS AND PUMPING

3.8.1 Positive Displacement Pumps

In positive displacement pumps, a discrete "parcel" of fluid is taken from the pump inlet to the pump outlet where it is discharged. This continues even if there is substantial resistance to flow at the outlet. Positive displacement pumps may develop very high pressures and pressure relief devices should be fitted.

3.8.1.1 Reciprocating Pumps. The pump may deliver fluid on every other stroke of the reciprocating piston, known as "single-acting", or on each stroke known as "double-acting", see Figures 10(a) and 10(b), respectively. Flow through the pump is controlled by inlet and outlet valves. Delivery of fluid is pulsating.

3.8.1.2 Diaphragm Pump. The movement of the fluid is generated by a flexible diaphragm, driven by a reciprocating piston or by compressed air/gas (Figure 11). The diaphragm isolates the moving parts of the pump (except the inlet and outlet valves) from the fluid being pumped. Delivery of fluid is pulsating.

3.8.1.3 Other Positive Displacement Pumps. There is a wide range of other types of positive displacement pumps, including gear pumps (internal and external), lobe pumps and screw pumps.[5,15] These pumps have a wide range of uses and delivery of the fluid may be continuous or pulsating, depending upon the design of the pump.

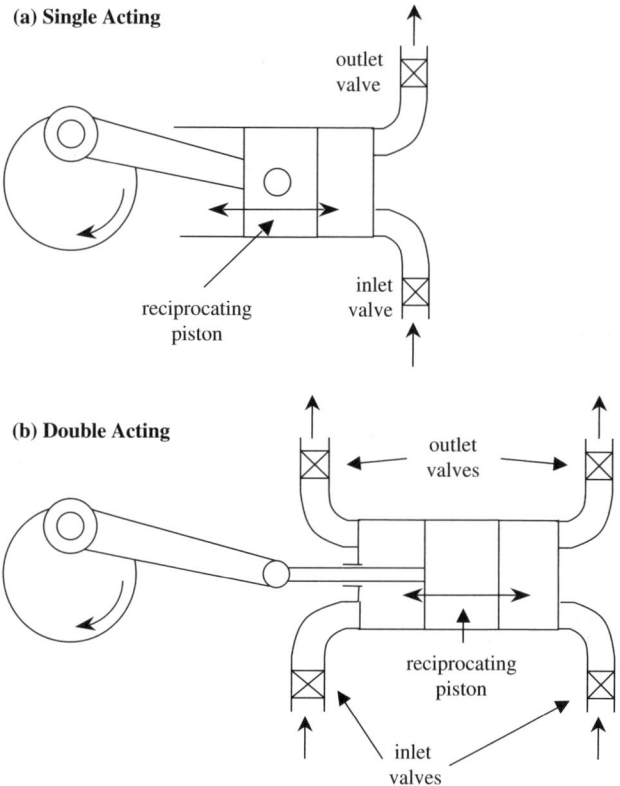

(a) Single Acting

outlet
valve

reciprocating
piston

inlet
valve

(b) Double Acting

outlet
valves

reciprocating
piston

inlet
valves

Figure 10 *Reciprocating piston pumps*

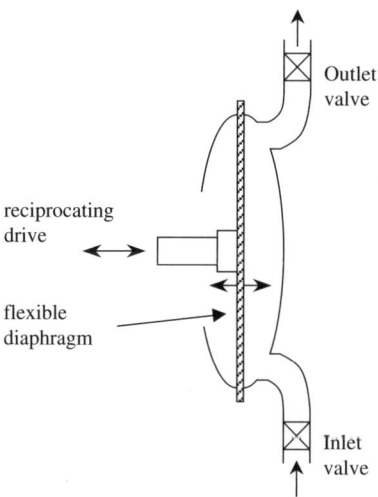

Outlet
valve

reciprocating
drive

flexible
diaphragm

Inlet
valve

Figure 11 *Diaphragm pump*

3.8.2   Non-Positive Displacement Pumps

3.8.2.1   Centrifugal Pump.   The most common pump in the process industries is the centrifugal pump, illustrated in Figure 12. It is used almost everywhere, except when very large heads are required, although multistage centrifugal pumps can develop pressures of 200 atm or more.

The impeller shaft is normally coupled directly to a motor drive and driven at high speed, up to ~ 4000 rpm. Fluid enters the pump and flows into the "eye" of the rotating impeller. In the impeller the fluid is guided by curved vanes and gains high velocity and hence high kinetic energy. The fluid then flows into the "volute casing" where the fluid is slowed and the kinetic energy gained in the impeller is converted into pressure head (*c.f. Bernoulli's* Equation (16), above).

There are many designs of impellers for different duties, *e.g.* for low viscosity fluids the impeller is normally shrouded; for solids suspensions and slurries the impeller may be open-faced to prevent blockage.

The delivery is continuous. Inlet and outlet valves are not required, but a non-return, or "check" valve is normally installed on the pump outlet line to prevent reverse flow when the pump is switched off. The outlet may be blocked (for short periods) without damage to the pump.

3.8.2.2   Characteristic Curves.   The pressure head versus flow relationship for a centrifugal pump, called the *characteristic curve*, and illustrated in Figure 13, depends very much upon the design of the impeller, its vanes and the volute casing.

Maximum head is attained at zero flow, when the maximum kinetic energy gained by the fluid within the impeller is converted into pressure. As the flow through the pump is increased, the fluid leaving the outlet takes with it kinetic energy. Therefore, less of the kinetic energy gained

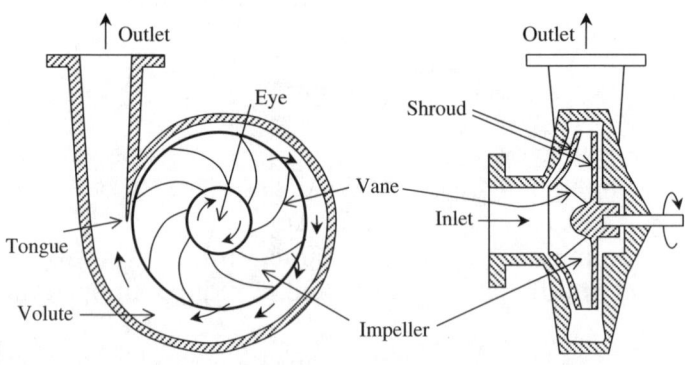

Figure 12   *Centrifugal pump (motor drive not shown)*

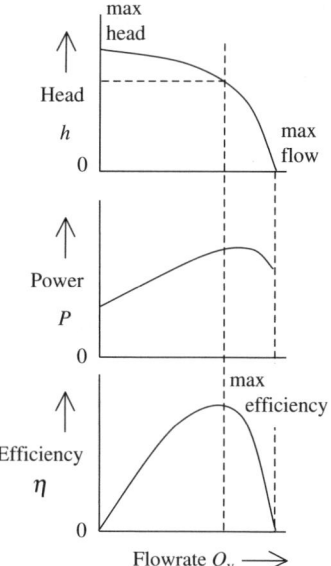

Figure 13 *Characteristic curves*

within the impeller gets converted into pressure and hence the head generated falls.

In a well-designed system, a centrifugal pump will normally operate at or close to its maximum efficiency point. Pump efficiencies, η, are typically in the range 60–80%, larger pumps having the higher efficiencies.

3.8.3 Matching Centrifugal Pumps to Flow Requirements

3.8.3.1 Pump Characteristic and System Head. The pump characteristic curve needs to be matched with the head loss through a piping system, which is known as the *system head*. The system head increases approximately in proportion to the square of the flow rate ($\Delta p \propto u^2$). An example of a system head calculation is given in Section 3.10.

The intersection of the characteristic curve (pump head) and the system head, illustrated in Figure 14, gives the *operating point* (flow rate) of the pump and pipe system. At this point the head generated by the pump is just balanced by the system head.

Different pumps with different characteristic curves will intersect the system head at different flow rates and give different operating points.

3.8.3.2 Control Valves. A flow system will normally require control and a control valve would normally be inserted into the pipeline.

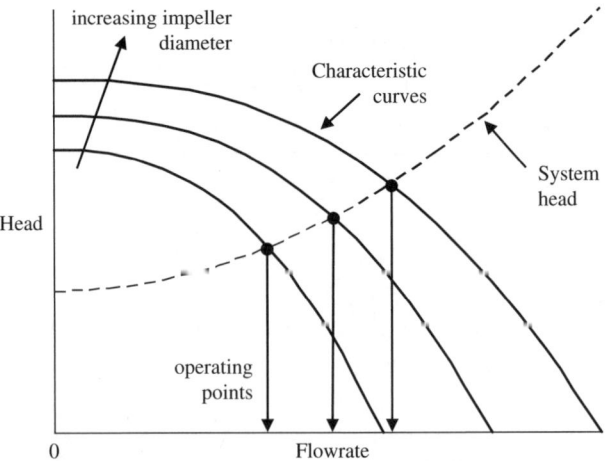

Figure 14   *Characteristic curves and system head*

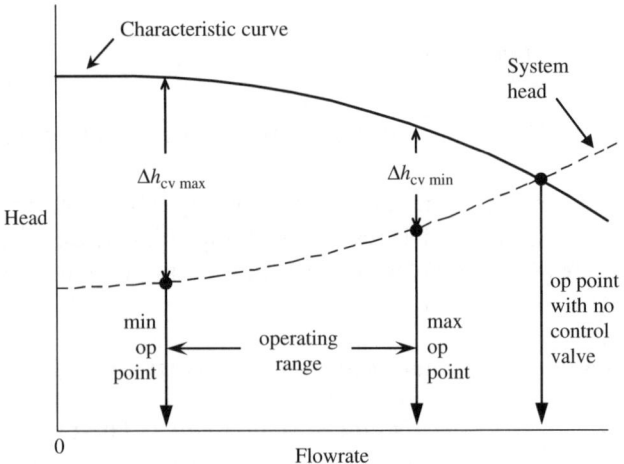

Figure 15   *Effect of control valves on operating point*

The required head loss (pressure drop), across the control valve, Δh_{cv}, at any desired operating point (op point) can easily be determined, see Figure 15. An operating range of head losses required across a control valve, $\Delta h_{cv\ min}$ to $\Delta h_{cv\ max}$, say, may also be determined.

3.8.3.3   Net Positive Suction Head – NPSH.   The pressure of liquid inside a pump must always exceed the liquid's vapour pressure, otherwise the liquid will tend to vaporise within the pump, causing cavitation. As the fluid passes on through the pump its pressure increases and any vapour bubbles collapse rapidly causing mechanical damage to the pump, which is often severe. Pumps may have their operating life reduced dramatically as a result.

The point of minimum pressure is not at the pump inlet, but will be somewhere within the impeller of the pump. Therefore, the absolute pressure (or head) of the liquid at the pump inlet, the suction side, must always be greater than the fluid vapour pressure by an amount, called the *Net Positive Suction Head* or *NPSH*. Manufacturers will quote the NPSH requirements for their pumps.

The Mechanical Energy Balance Equation (14), may be used to determine the fluid pressure at the inlet to the pump (see *Section 3.10*). This must exceed the liquid's vapour pressure by the NPSH requirement of the pump, otherwise the pipework should be redesigned or a different pump with a smaller NPSH requirement installed. This explains why pumps are normally located at the lowest part of the pipework (maximum head) and close to the source (minimum inlet pipework pressure drop).

3.8.4 Pump Power Requirements

Assuming ρ = constant and $A_1{=}A_2$, then $u_1 = u_2$, as Q_v = constant, the mechanical energy balance over a pump (see Figure 16) gives the work done on the fluid as: $-w_x = \int_{p_1}^{p_2} \mathrm{d}p/\rho = -\Delta p/\rho$ [J kg^{-1}].

$$\text{Theoretical power required} = Gw_x = \rho Q_v \frac{\Delta p}{\rho} = Q_v \Delta p \, [\text{J s}^{-1}, \text{W}] \qquad (38)$$

$$\text{Actual power required} = \frac{Gw_x}{\eta} = \frac{Q_v \Delta p}{\eta} [\text{J s}^{-1}, \text{W}] \qquad (39)$$

3.9 OTHER FLOWS

3.9.1 Equivalent Hydraulic Diameter

When the pipe or duct is non-circular or non-full, the equivalent hydraulic diameter, d_e, is often used in place of d, the pipe diameter,

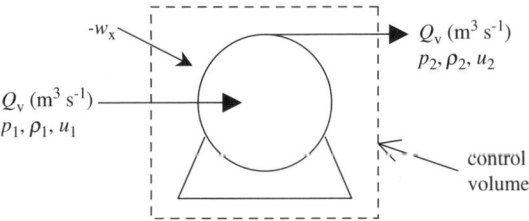

Figure 16 *Pump energy balance control volume*

in *turbulent flow* equations (although not for very narrow or slotted flow cross-sections).

$$d_e = 4 \times \frac{\text{flow cross-sectional area}}{\text{wetted perimeter}} \tag{40}$$

and the head loss due to friction becomes

$$\Delta h_f = 2c_f \frac{l}{d_e} \frac{\bar{u}^2}{g} \tag{41}$$

3.9.1.1 Example of Equivalent Hydraulic Diameters. Rectangular duct, sides a and b:

$$d_e = 4 \times \frac{ab}{2(a+b)} = \frac{2ab}{(a+b)} \tag{42}$$

Parallel plates, width b, distance a apart, $a \ll b$:

$$d_e = 4 \times \frac{ab}{2(a+b)} = 2a \tag{43}$$

Equilateral triangular duct, side a:

$$d_e = 4 \times \frac{(1/2)a^2 \sin 60°}{3a} = 0.577a \tag{44}$$

Open channel flow, non-full sloping duct, depth a width b:

$$d_e = 4 \times \frac{ab}{2a+b} = \frac{4ab}{2a+b} \tag{45}$$

With open channel flow the driving force is gravity; there is no net pressure force as the free surface is uniformly exposed to the atmosphere. Thus, in the Mechanical Energy Balance Equation (14), the $(p_2-p_1)/\rho$ and w_x terms are both zero and the $g(z_2-z_1)$ term quantifies the driving force and is given by $gl\sin\theta$, where l is the length of the duct which is at an angle θ to the horizontal.

3.9.2 Flow Around Bodies

3.9.2.1 Spheres and Cylinders. The drag force F exerted by a fluid on a body would be expected to be a function of the body size (area), relative velocity with fluid, fluid density and viscosity, and body shape (see Figure 17). Thus $F = f(A, u_\infty, \rho, \mu, \text{shape})$.

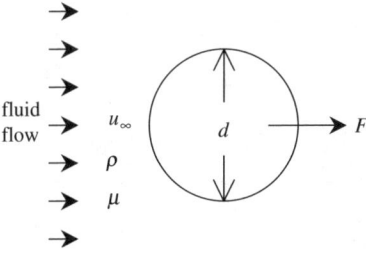

Figure 17 *Drag on a sphere or cylinder*

Dimensional analysis yields, $C_D = f(Re, \text{shape})$, where

- Drag coefficient:

$$C_D = \frac{F}{\frac{1}{2}\rho u_\infty^2 A_p} \equiv \frac{\text{drag force}}{\text{inertial forces}}$$

- Reynolds No:

$$Re = \frac{\rho u_\infty d}{\mu} \equiv \frac{\text{inertial forces}}{\text{viscous forces}}$$

A_p is the body's projected area in the direction of flow. For a sphere, $A_p = \pi d^2/4$, and for a cylinder $A_p = dL$ ($L =$ length of the cylinder).

C_D is a complex function of Re and geometry. The curves for cylinders and spheres show many similarities (see Figure 18), and may be divided into the following flow regimes:

- laminar: $Re \lesssim 0.5$;
- transition: $0.5 \lesssim Re \lesssim 1000$;
- turbulent, with laminar boundary layer: $1000 \lesssim Re \lesssim 20{,}000$;
- turbulent, with turbulent boundary layer: $Re \gtrsim 200{,}000$.

Further discussion on flow around and drag forces on bodies is available in the literature.[1,2,4]

3.9.2.2 Flow Through a Packed Bed of Spheres.

The Ergun Equation (46), gives the pressure drop through a packed bed of spheres:[2]

$$\frac{\Delta p}{l} = 150 \frac{(1-\varepsilon)^2}{\varepsilon^3} \frac{\mu u}{d_p^2} + 1.75 \frac{(1-\varepsilon)}{\varepsilon^3} \frac{\rho u^2}{d_p} \tag{46}$$

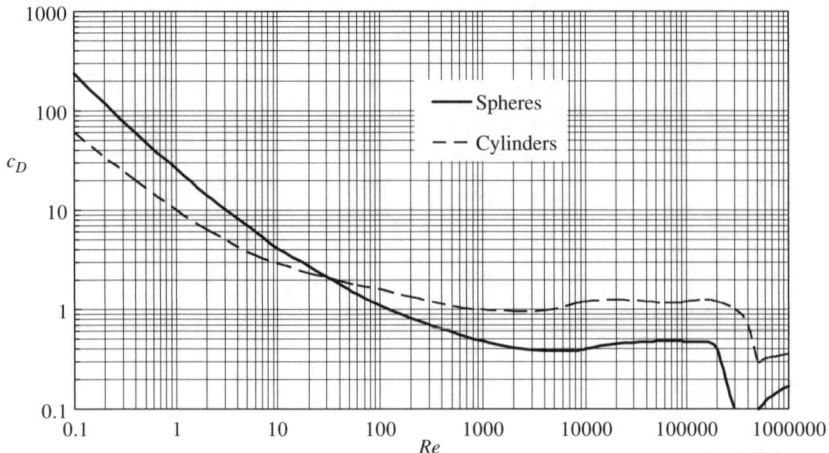

Figure 18   *Drag coefficients*

We see that $\Delta p/l$, the frictional pressure drop per unit depth of bed, is made up of two components. The first term on the right-hand-side accounts for viscous (laminar) frictional losses, $\propto \mu u$, and dominates at low Reynolds numbers. The second term on the right-hand-side accounts for the inertial (turbulent) frictional losses, $\propto \rho u^2$, and dominates at high Reynolds numbers. For further information about flow through packed beds, see *Chapter 7 "An Introduction to Particle Systems"*.

3.9.3   Stirred Tanks[16–20]

The power P to turn an impeller is given by $2\pi N \times M$, (angular velocity × torque). This power would be expected to be a function of impeller diameter and speed, and geometry. Gravitational acceleration should also be included in the analysis as there is normally a free liquid surface present and gravity affects its shape and the flow within the vessel.

Thus

$$P = f(D, N, \rho, \mu, g, \text{geometry}).$$

Assuming geometric similarity, dimensional analysis yields, $Po = f(Re, Fr)$, where

- Power No:

$$Po = \frac{P}{\rho N^3 D^5} \equiv \frac{\text{drag force on impeller}}{\text{inertial forces}}$$

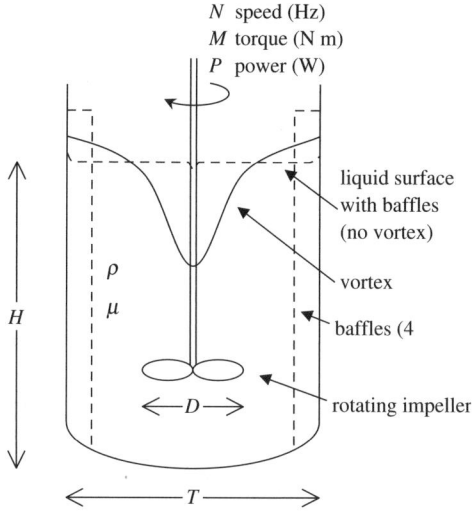

Figure 19 *Stirred Tank*

- Impeller Reynolds No:

$$Re_I = \frac{\rho N D^2}{\mu} \equiv \frac{\text{inertial forces}}{\text{viscous forces}}$$

- Froude No:

$$Fr = \frac{N^2 D}{g} \equiv \frac{\text{inertial forces}}{\text{gravitational forces}}$$

Po is a complex function of Re_I and geometry. Figure 20 shows the *Po* versus Re_I relationships for a number of impellers in tanks fitted with baffles, which suppress the formation of a vortex (see dashed lines in Figure 19), and hence where *Fr* has little effect. The curves may be divided into the following flow regimes:

- laminar: $Re_I \lesssim 10$;
- transition: $10 \lesssim Re_I \lesssim 5000$;
- turbulent: $Re_I \gtrsim 5000$.

3.10 EXAMPLE CALCULATIONS FOR THE PIPE-FLOW SYSTEM

This section illustrates the use of material presented earlier to determine the pressure drop in a pipe system, mating of a pump to the pipe system and the head loss over a control valve.

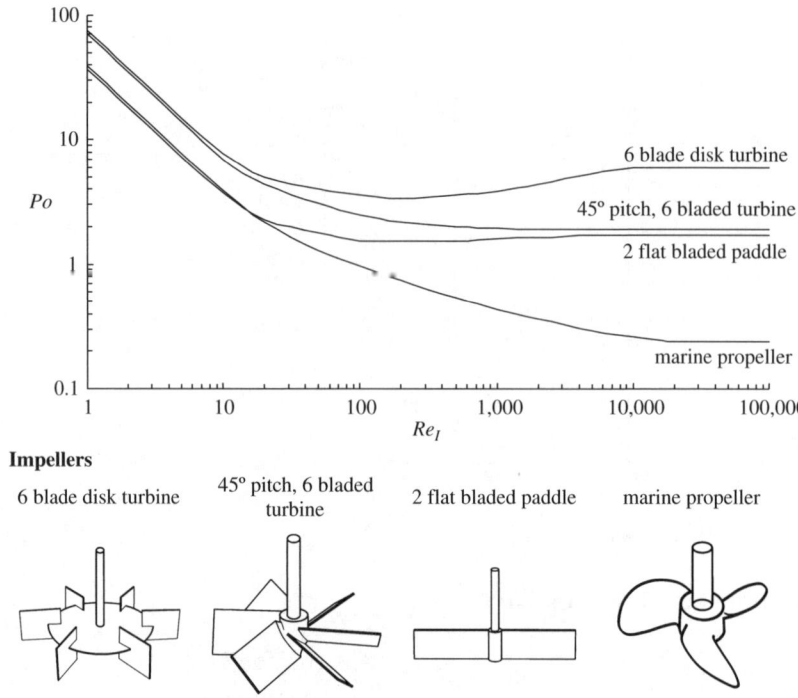

Figure 20   *Po versus Re₁ curves for baffled tanks for different impellers*

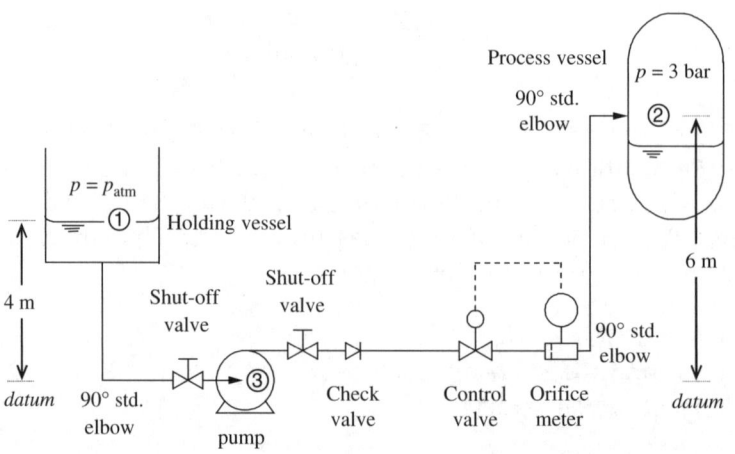

Figure 21   *Example pipe-flow system*

10 kg s⁻¹ of water at 80°C is pumped from a holding vessel through a pipe system to a process vessel, which is illustrated in Figure 21. The pipe is of made of commercial steel and has an internal diameter of 100 mm. The pipework has a total length of straight pipe of 35 m.

We are required to specify a suitable centrifugal pump to achieve the required flow.

3.10.1 Data

3.10.1.1 Datum and Reference Points.

Elevation datum	Level with the inlet to the pump, but can be any convenient level
Point ①, subscript 1	At the minimum level of liquid in the holding vessel
Point ②, subscript 2	Inside the pressure vessel, level with the pipe inlet
Point ③, subscript 3	At the inlet to the pump

3.10.1.2 Fluid Properties, Water at 80°C.

Density	$\rho = 972$ kg m^{-3}
Viscosity	$\mu = 0.351 \times 10^{-3}$ Pa s
Vapour pressure	$p_v = 0.4748 \times 10^5$ Pa

3.10.1.3 Pipe Properties.

Total length of straight pipe	$l = 35$ m
Internal diameter	$d = 0.1$ m
Roughness	$e = 0.05 \times 10^{-3}$ m (see Table 4)
Elevation of ①	$z_1 = 4$ m
Elevation of ②	$z_2 = 6$ m
Elevation of ③	$z_3 = 0$ m
Relative roughness	$e/d = 0.05 \times 10^{-3}/0.1 = 0.0005$
Pipe cross-sectional area	$A = \pi d^2/4 = \pi\, 0.1^2/4 = 7.85 \times 10^{-3}$ m^2

3.10.1.4 Flow Properties.

Mass flow rate	$G = 10$ kg s^{-1}
Volumetric flow rate	$Q_v = G/\rho = 10.3 \times 10^{-3}$ m^3 s^{-1}

(*continued*)

Mean velocity	$\bar{u} = Q_v/A = 1.31$ m s^{-1}
Reynolds number	$Re = \dfrac{\rho \bar{u} d}{\mu} = \dfrac{972 \times 1.31 \times 0.1}{0.000351} = 363{,}000$, *i.e.* turbulent

3.10.1.5 Pipe Fittings. Orifice meter and control valve will be considered separately, see Table 5.

No. fitting	Each fitting n	Subtotal n
1 × entrance loss from holding vessel	25	25
2 × shut off valves (open gate valves)	7	2 × 7
1 × check valve (swing)	100	100
3 × 90° standard radius bends	35	3 × 35
1 × exit loss into process vessel	50	50
		Total: $n = 294$

Equivalent length of straight pipe: $l_e = n \times d = 294 \times 0.1 = 29.4$ m.

3.10.2 Friction Losses

By using Equation (32), total frictional head loss:

$$\Delta h_{f_T} = 4c_f \frac{(l + l_e)}{d} \frac{\bar{u}^2}{2g} = 4c_f \frac{l_T}{d} \frac{\bar{u}^2}{2g}$$

Fanning friction factor, from Equation (28),

$$c_f = 0.001375 \left[1 + \left(20{,}000 \frac{e}{d} + \frac{10^6}{Re} \right)^{1/3} \right]$$

$$= 0.001375 \left[1 + \left(20{,}000 \times 0.0005 + \frac{10^6}{363000} \right)^{1/3} \right] = 0.00459$$

or look up c_f for appropriate values of Re and e/d in Figure (7), see also point ■.

Total equivalent length of straight pipe $l_T = l + l_e = 35 + 29.4 = 64.4$ m. Substituting

$$\Delta h_{f_T} = 4c_f \frac{l_T}{d} \frac{\bar{u}^2}{2g} = 4 \times 0.00459 \times \frac{64.4}{0.1} \times \frac{1.31^2}{2 \times 9.81} = 1.03 \text{ m H}_2\text{O}$$

3.10.3 Orifice Meter

Equation (37):

$$G = C_d A_1 \left[\frac{2\rho(p_1 - p_2)_{\text{orifice}}}{(A_1/A_2)^2 - 1} \right]^{1/2} = C_d A_1 \left[\frac{2\,\rho\Delta p_{\text{orifice}}}{(A_1/A_2)^2 - 1} \right]^{1/2}$$

typical discharge coefficient for orifice $C_d = 0.65$ (say)
pipe csa. $A_1 = \pi\, d^2/4 = 7.85 \times 10^{-3}$ m^2
orifice csa. $A_2 = 1/2\, A_1$ say $= 3.92 \times 10^{-3}$ m^2 (a typical ratio of areas).
 From Equation (37)

$$10 = 0.65 \times 7.85 \times 10^{-3} \left[\frac{2 \times 972 \times \Delta p_{\text{orifice}}}{(2)^2 - 1} \right]^{1/2}$$

$\therefore \Delta p_{\text{orifice}} = 5930$ Pa $\equiv 0.62$ m H$_2$O. This is the pressure difference across the flowmeter that would be measured and from which the flow rate would be determined for control purposes.

Overall frictional loss across orifice meter, see Table (5) $\Delta p_{\text{loss}} \approx$ 60% $\times \Delta p_{\text{orifice}} \approx 3560$ Pa $\therefore \Delta h_{\text{loss}} = \Delta p_{\text{loss}}/\rho\, g = 0.37$ m H$_2$O

3.10.4 Pump Shaft Work

Use the Mechanical Energy Balance Equation (14):

$$\frac{p_2 - p_1}{\rho} + \left(\frac{u_2^2}{2} - \frac{u_1^2}{2} \right) + g(z_2 - z_1) + g\Delta h_f + w_x = 0$$

and apply between two points. . .

Point 1: chosen at the liquid surface in the holding vessel ①
$u_1 \approx 0$, $p_1 = p_{\text{atm}} = 1.013$ bar, $z_1 = 4$ m.

Point 2: chosen inside the pressure vessel, level with the pipe inlet ②
$u_2 \approx 0$, $p_2 = 3$ bar, $z_2 = 6$ m.

 Substituting into Equation (14):

$$\frac{(3 - 1.013) \times 10^5}{972} + \left(\frac{0^2}{2} - \frac{0^2}{2} \right) + 9.81(6 - 4) + 9.81(1.03 + 0.37) + w_x = 0$$

$$204.4 \quad + \quad 0 \quad + \quad 19.6 \quad + \quad 13.7 \quad + w_x = 0$$

and solving yields theoretical pump power, $w_x = -237.7$ J kg^{-1}

$$P_{ideal} = -Gw_x = -10 \times -237.7 = 2.38 \text{ kW}$$

Actual pump power, assuming 70% efficiency, $P_{actual} = 2.38 \,/\, 0.7 = 3.4$ kW $= 4.6$ hp.

3.10.5   System Head

The required shaft work for the pump, w_x, when expressed as a head, is known as the system head, Δh_{system}. This is the head required from the pump to achieve the required flow through the pipe system.
   System head

$$\Delta h_{system} = -\frac{w_x}{g} = -\frac{-237.7}{9.81} = 24.2 \text{ m H}_2\text{O}$$

3.10.6   Pump Characteristics

Figure 22 shows characteristic curves and NPSH requirements for a family of different sized centrifugal pumps adapted from curves given by Perry.[23] Superimposed is the system head curve for the example piping system. The system head curve is generated by repeating the above calculations for Δh_{system} over the required range of flow rates.
   It can be seen that a pump with an impeller diameter of 138 mm is suitable for the required duty, in that it will deliver a head which comfortably exceeds the required system head of 24.2 m at the required flow rate of 0.0103 m^3 s^{-1}. A smaller diameter impeller, e.g. 125 mm, will not simultaneously deliver the required combination of head and flow rate and is therefore not suitable for this duty.

3.10.7   Control Valve

Note that at the required flow rate, the head produced by the pump exceeds the required system head. The excess head, ~ 6 m H$_2$O, will be taken up by the control valve, Δh_{cv}. Also by changing the head across the control valve, Δh_{cv}, the operating point of the system, i.e. the flow rate, will change. Figure 22 may be used to determine this effect.

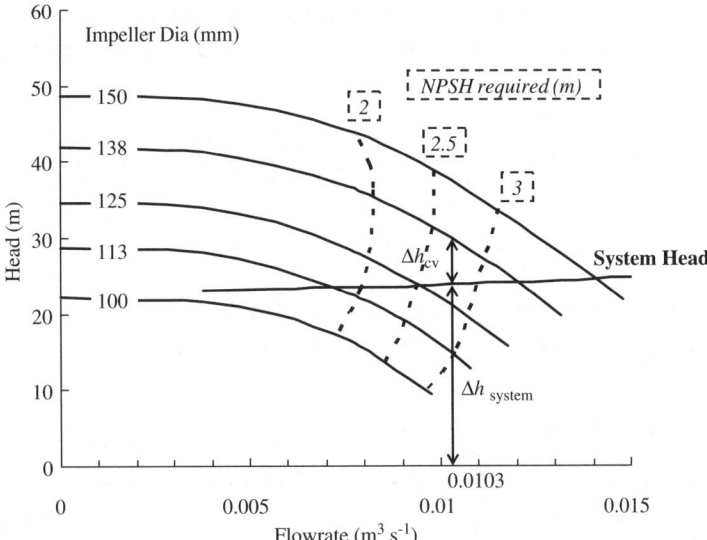

Figure 22 *Pump characteristics curves and system head. Adapted from Perry*[23]

3.10.8 Net Positive Suction Head

The absolute pressure (or head) of the liquid at the pump inlet, the suction side, must always be greater than the fluid vapour pressure by the NPSH. Let us use the Mechanical Energy Balance Equation (14), to find the pressure at the pump inlet.

Point 1: no change, at the liquid surface in the holding vessel ①: $u_1 \approx 0$, $p_1 = p_{atm} = 1.013$ bar, $z_1 = 4$ m.
Point 2: now chosen to be the inlet to the pump ③ where we need to determine the pressure: $u_3 = \bar{u}$, $z_3 = 0$, p_3 is the unknown pressure at pump inlet.

Note: Δh_f has to be recalculated based on the shorter length of pipe from the holding vessel to the pump inlet, 8 m, and the fewer fittings in this section of pipework. In this case $l_e = 6.7$ m, giving $\Delta h_f = 0.24$ m H_2O.

$$\frac{p_3 - p_1}{\rho} + \left(\frac{u_3^2}{2} - \frac{u_1^2}{2}\right) + g(z_3 - z_1) + g\Delta h_f + w_x = 0$$

$$\frac{p_3 - 1.013 \times 10^5}{972} + \left(\frac{1.31^2}{2} - \frac{0^2}{2}\right) + 9.81(0 - 4) + 9.81 \times 0.24 + 0 = 0$$

$$\frac{p_3}{972} - 104.2 + 0.86 - 39.2 + 2.3 + 0 = 0$$

$$p_3 = 1.36 \times 10^5 \, Pa = 1.36 \, bar$$

NPSH available is the difference between absolute pressure (p_3) of the liquid at the pump and the fluid vapour pressure (p_v), expressed as a head.

$$\text{NPSH}_{\text{available}} = \frac{p_3 - p_v}{\rho g} = \frac{(1.36 - 0.475) \times 10^5}{972 \times 9.81} = 9.4\,\text{m}\,\text{H}_2\text{O}$$

From Figure 22 it is seen that the NPSH required by the pump is less than 3 m.

Therefore, as the NPSH available exceeds that required by the selected pump, the pump is suitable. Our task is now complete.

NOMENCLATURE

Symbol	Description	Dimensions	SI unit
A	Area, cross-sectional area	L^2	m^2
a, b	Dimensions, lengths	L	m
C_D	Drag coefficient	—	—
C_d	Discharge coefficient	—	—
c_f	Fanning friction factor	—	—
d, D	Diameter, pipe diameter, impeller diameter	L	m
e	Absolute roughness	L	mm
F	Force	$M\,L\,T^{-2}$	N
f	Function	—	—
f	Friction factor	—	—
Fr	Froude number	—	
g	Gravitational acceleration	$L\,T^{-2}$	$m\,s^{-2}$
G	Mass flow rate	$M\,T^{-1}$	$kg\,s^{-1}$
h	Head of fluid	L	m *of fluid*
	Specific enthalpy	$L^2\,T^{-2}$	$J\,kg^{-1}$
K	Loss coefficient, number of velocity heads	—	
l, L	(Pipe) length, cylinder length	L	m
m	Number of variables	—	—
m_T	Total mass within control volume	M	kg
n	Number of fundamental dimensions, Number of pipe diameters	—	—
N	Rate of rotation	T^{-1}	Hz

(*continued*)

p	Pressure	$M\,L^{-1}\,T^{-2}$	Pa
p_v	Equilibrium vapour pressure	$M\,L^{-1}\,T^{-2}$	Pa
P	Power	$M\,L^2\,T^{-3}$	W
Po	Power number	—	
q	Heat input to control volume per unit mass	$L^2\,T^{-2}$	$J\,kg^{-1}$
Q	Quantity, variable	*n/a*	*n/a*
Q_v	Volumetric flow rate	$L^3\,T^{-1}$	$m^3\,s^{-1}$
r, r_0	Radial co-ordinate, pipe radius	$L\,T^{-1}$	$m\,s^{-1}$
Re	Reynolds number	—	—
u, U, u_{max}	Velocity in x-direction, velocity, maximum velocity	$L\,T^{-1}$	$m\,s^{-1}$
$\bar{u}, \tilde{u}$	Average velocity across a cross-section, mean at a point	$L\,T^{-1}$	$m\,s^{-1}$
u_∞	Velocity relative to infinity or to undisturbed flow	$L\,T^{-1}$	$m\,s^{-1}$
u^+	Dimensionless velocity	—	
w_x	Shaft work out from control volume per unit mass	$L^2\,T^{-2}$	$J\,kg^{-1}$
W_x	Rate shaft work out from control volume	$M\,L^2\,T^{-3}$	W
x	Distance in x-direction, distance along pipe	L	m
y	Distance in y-direction, distance from (pipe) wall	L	m
y^+	Dimensionless distance in y-direction	—	
Y	Length	L	m
z	Distance in z-direction, elevation	L	m
ε	Voidage	—	—
η	Efficiency	—	—
λ, ϕ	Friction factors	—	—
μ	Dynamic viscosity	$M\,L^{-1}\,T^{-1}$	$Pa\,s,\,N\,s\,m^{-2}$
v	Kinematic viscosity	$L^2\,T^{-1}$	$m^2\,s^{-1}$
Π	Dimensionless group	—	—
ρ	Fluid density	$M\,L^{-3}$	$kg\,m^{-3}$
τ, τ_w	Shear stress, wall shear stress	$M\,L^{-1}\,T^{-2}$	$Pa,\,N\,m^{-2}$

SUBSCRIPTS

1, 2, 3	at point 1, 2, 3
e	Equivalent hydraulic
f	Frictional
I	Impeller
p	Projected, particle
T	Total

SYMBOLS

Δ	" the change in "
[=]	" has the dimensions of "

REFERENCES

1. J.M. Coulson, J.F. Richardson, with J.R. Backhurst and J.H. Harker, *Coulson & Richardson's Chemical Engineering*, 6th edn, **Vol 1**, Butterworth-Heinemann, Oxford, 1999.
2. J.M. Kay and R.M. Neddermann, *Fluid Mechanics and Transfer Processes*, Cambridge University Press, Chapters 1–12, Cambridge, 1985.
3. B.S. Massey, revised by J. Ward-Smith, *Mechanics of Fluids*, Chapter 5, 8th edn, Taylor & Francis, Abingdon, Oxford, 2006.
4. R.H. Perry and D. Green, *Perry's Chemical Engineers' Handbook*, 7th edn, McGraw Hill, Section 6, New York, 1997.
5. R.H. Perry and D. Green, *Perry's Chemical Engineers' Handbook*, 7th edn, McGraw Hill, Section 10, New York, 1997.
6. C. Schaschke, *Fluid Mechanics Worked Examples for Engineers*, IChemE, Rugby, 1998.
7. R.B. Bird, W.E. Stewart and E.N. Lightfoot, *Transport Phenomena*, 2nd edn, John Wiley & Sons Inc., New York, 2001.
8. C.R. Duhne, Viscosity–temperature correlations for liquids, *Chem. Eng.*, July 16, 1979, **86**, 83–91.
9. R.P. Chhabra and J.F. Richardson, *Non-Newtonian Flow in the Process Industries*, Butterworth-Heinemann, Oxford, 1999.
10. J.E. Edwards, Flow measurement practice, *The Chemical Engineer*, May 1979, **344**, 325–333.

11. D. Ginesi, Choices abound in flow measurement, *Chem. Eng.*, April, 1991, **98**, 88–100.
12. D. Ginesi, A raft of flowmeters on tap, *Chem. Eng.*, May 1991, **98**, 146–155.
13. BS 1042, Measurement of fluid flow in closed conduits, BSI.
14. BS EN ISO 5167-1 . . . -4, Measurement of fluid flow by means of pressure differential devices inserted in circular cross-section conduits running full, BSI.
15. I.J. Karassik, J.P. Messina, P. Cooper and C.C. Heald, *Pump Handbook*, 3rd edn, McGraw-Hill, New York, 2001.
16. V.W. Uhl and J.B. Gray, *Mixing: Theory and Practice*, Vols I–III, Academic Press, New York, 1966, 1967 and 1986.
17. S. Nagata, *Mixing: Principles and Applications*, Kodansha (Wiley), New York, 1975.
18. N. Harnby, M.F. Edwards and A.W. Nienow, *Mixing in the Process Industries*, 2nd edn, Butterworth-Heinemann, Oxford, 1992.
19. G.B. Tatterson, *Fluid Mixing and Gas Dispersion in Agitated Tanks*, McGraw Hill, New York, 1991.
20. J.J. Ulbrecht and G.K. Patterson, *Mixing of Liquids by Mechanical Agitation*, Gordon and Breach, New York, 1985.
21. L.F. Moody, *Trans. Am. Soc. Mech. Eng.*, 1944, **66**, 671–684, and *Mech. Eng.*, 1947, **69**, 1005–1006.
22. E.E. Ludwig, *Applied Process Design for Chemical and Petrochemical Plants*, Vol 1, Gulf Publishing Co, Houston, Texas, Chapter 2, 1964.
23. R.H. Perry and D. Green, *Perry's Chemical Engineers' Handbook*, 7th edn, McGraw Hill, Figure 10–31, New York, 1997.

CHAPTER 4

An Introduction to Heat Transfer

TIM ELSON AND PAOLA LETTIERI

4.1 INTRODUCTION AND OBJECTIVES

Heat is transferred from regions of high temperature to regions of low temperature and may take place by three different mechanisms or modes.

4.1.1 Modes of Heat Transfer

4.1.1.1 Conduction. Conduction of heat results from the greater motion of molecules and atoms at higher energy (temperature) levels imparting energy to adjacent molecules and atoms at lower energy (temperature) levels. There is no overall motion of the material and it occurs in gases, liquids and solids. The physical mechanisms of heat conduction are complex, but in gases and liquids comprise of molecular collisions and lattice vibrations in solids and flow of free electrons in metals.

4.1.1.2 Convection. Heat is transferred by moving fluids and therefore only occurs in gases and liquids. A moving fluid takes thermal energy with it, and when it contacts another fluid or a surface at a different temperature, heat is transferred. There are two types of convection:

- *Forced Convection*, where the flow is induced by external forces, *e.g.* from a pump or a fan.
- *Natural Convection*, where the flow is induced by buoyancy forces, which result from the thermal expansion of hotter fluids that rise relative to cooler, denser fluids.

4.1.1.3 Radiation. All bodies emit and absorb electromagnetic radiation, such as infrared, ultraviolet and visible light. The predominant frequency and the energy of the emitted radiation increase as the temperature of the body increases. Heat may therefore be transferred from hotter bodies to cooler bodies by electromagnetic radiation. This will only occur through materials that are transparent or partially transparent to electromagnetic radiation.

4.1.2 Scope and Objectives

We will look at these three modes of heat transfer in more detail and consider the laws that govern the rates of heat transfer by these mechanisms. We will look at some of the more common types of heat exchanger and how we may use these laws to specify and design a heat exchanger.

Heat transfer is covered in many chemical engineering texts. Kern,[1] although first published in 1950, still remains a standard reference. Coulson and Richardson's Chemical Engineering series covers heat transfer in general in volume 1[2] and the detailed design of heat exchangers in volume 6.[3] Perry[7] also gives a good coverage of the subject.

4.2 MODES OF HEAT TRANSFER

4.2.1 Conduction

4.2.1.1 Fourier's Law of Conduction. Experiments have shown that at steady state, the heat flux q_y, which is the rate of heat transfer, Q, per unit area, A, through a material due to conduction, is proportional to the temperature gradient in the direction of heat flow, y in this case.

$$q_y = \frac{Q}{A} = -k\frac{\mathrm{d}T}{\mathrm{d}y} \tag{1}$$

The proportionality constant k is called the *thermal conductivity* of the material. Heat is transferred from regions of high temperatures to low temperatures – hence the negative sign in front of k. Equation (1) holds for liquids and gases, as well as for solids, as long as convection and radiation are prevented.

4.2.1.2 Thermal Conductivity. Figure 1 shows the temperature dependence of a range of materials, and it can be seen that the values of thermal conductivities of materials vary widely. Typical values are as follows:

- Gases: $0.01 \sim 0.1$ W m^{-1} K^{-1}, with higher values for H_2 and He
- Liquids: $0.1 \sim 1$ W m^{-1} K^{-1}, with much higher values for liquid metals

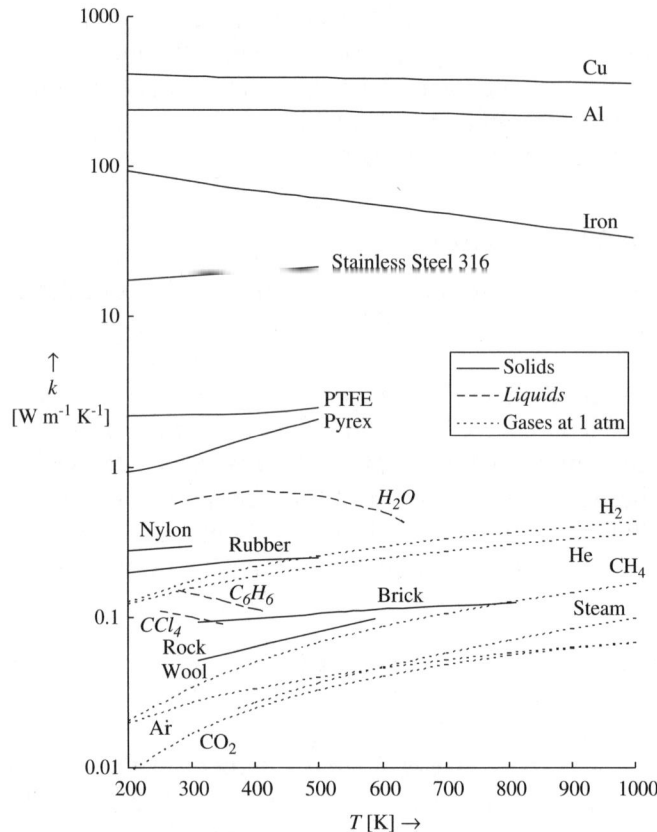

Figure 1   *Temperature variation of thermal conductivity for a number of materials*[7,8]

- Solids: $1 \sim 100$ W m^{-1} K^{-1}, with higher value for good conductors, *e.g.* Cu and Al, and much lower values for good insulators.

Some fibrous or lamellar materials have non-isotropic thermal conductivities. Examples are pine wood, with $k \approx 0.15$ W m^{-1} K^{-1} across the grain and ≈ 0.35 W m^{-1} K^{-1} parallel to the grain, and graphite at 300 K, $k \approx 10$ W m^{-1} K^{-1} parallel to the basal plane and ≈ 2000 W m^{-1} K^{-1} perpendicular to the basal plane.[7]

4.2.1.3   Rate of Heat Transfer.   Fourier's Law may be integrated and solved for a number of geometries to relate the rate of heat transfer by conduction to the temperature driving force. Equations are given below that allow the calculation of steady-state heat flux and temperature profiles for a number of geometries.

Large, Flat Plate. The heat flux through a flat plate, with thickness b in the y direction, is given by

$$q_y = \frac{Q}{A} = k\frac{(T_0 - T_b)}{b} \tag{2}$$

where the surfaces are being held at T_0 at $y = 0$ and at T_b at $y = b$. The temperature profile, $T(y)$, is found to be linear with distance through the plate, y:

$$T(y) = T_0 - \frac{y}{b}(T_0 - T_b) \tag{3}$$

Composite Plate. The heat flux through a flat plate comprising of, for example, three layers of material of different thickness b_1, b_2 and b_3, each with its own thermal conductivity, k_1, k_2 and k_3, respectively, and the outside surfaces being held at T_0 and T_3, is given by

$$q_y = \frac{T_0 - T_3}{(b_1/k_1) + (b_2/k_2) + (b_3/k_3)} = \frac{T_0 - T_3}{\Sigma(b/k)} = \frac{\Delta T}{\Sigma(b/k)} \tag{4}$$

The temperature profile, $T(y)$, in each layer is given by Equation (3) and illustrated in Figure 2.

Equation (4) is analogous to electrical current flowing through electrical resistances in series. Figure 3 illustrates such an electrical circuit, where an electrical current, i, flows through a series of three resistances, R_1, R_2 and R_3. The overall voltage difference of $V_0 - V_3$, is given by

$$i = \frac{V_0 - V_3}{R_1 + R_2 + R_3} = \frac{V_0 - V_3}{\Sigma R} = \frac{\Delta V}{\Sigma R} \tag{5}$$

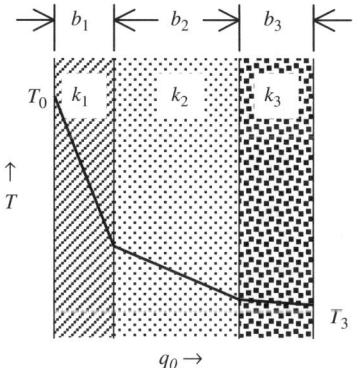

Figure 2 *Conduction through a composite plate (illustrated with $k_1 < k_2 < k_3$)*

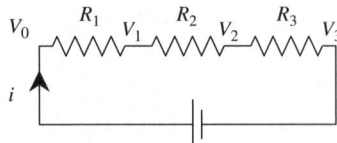

Figure 3 *Electrical resistances in series*

Comparing Equation (4) with Equation (5), $i \equiv q_y$, $\Delta V \equiv \Delta T$ and $R_n \equiv (b_n/k_n)$. Thus, we can think of the (b/k) values as thermal resistances, the resistance of the material layer to heat transfer by conduction. From Equation (4), it can be seen that the heat flux is equal to the overall temperature difference ΔT, divided by the sum of the thermal resistance of each layer $\Sigma(b/k)$. The concept of thermal resistance of a layer is very useful and we will return to it later.

Example Calculation of Heat Loss through a Composite Wall.   A furnace wall is constructed of a 3-cm thick, flat steel plate, with a firebrick insulation 30-cm thick on the inside, and rock wool insulation 6-cm thick on the outside. The inside surface temperature of the firebrick insulation is 700°C. If the temperature of the outer surface of the rock wool insulation is 50°C, what is the heat flux through the wall?

Thermal conductivities[7,8] are as follows: firebrick, 0.1 W m^{-1} K^{-1}; steel, 40 W m^{-1} K^{-1}; rock wool, 0.04 W m^{-1} K^{-1}.

From Equation (4) the heat flux is

$$q_y = \frac{T_0 - T_3}{(b_1/k_1) + (b_2/k_2) + (b_3/k_3)}$$

$$= \frac{700 - 50}{(0.3/0.1) + (0.03/40) + (0.06/0.04)}$$

$$= \frac{650}{3.000 + 0.001 + 1.5} = \frac{650}{4.501} = 144.4 \text{ W m}^{-2}$$

What is the temperature of the steel? From Equation (2)

$$T_1 = T_0 - q_y \frac{b_1}{k_1} = 700 - 144.4 \frac{0.3}{0.1} = 266.74 \,°\text{C}$$

This is the inner surface temperature of the steel. The steel outer surface temperature is then given by

$$T_2 = T_1 - q_y \frac{b_2}{k_2} = 266.74 - 144.43 \frac{0.03}{40} = 266.63 \,°\text{C}$$

We see that there is only a small temperature difference across the steel wall. This is due to the relatively small thermal resistance offered by this layer (b_2/k_2), approximately 0.02% of the total resistance. The wall is effectively at a uniform temperature of 267°C.

Hollow Cylinder. Radial heat flux, q_r, through the wall of a long cylinder of length L and inner radius r_i held at temperature T_i, and outer radius r_o held at temperature T_o, e.g. the wall of a pipe or tube in a heat exchanger, is given by

$$q_r = \frac{Q}{2\pi r L} = \frac{k(T_i - T_o)}{r \ln(r_o/r_i)} \tag{6}$$

The temperature profile, $T(r)$ is given by

$$T(r) = \frac{T_i \ln(r_o/r) + T_o \ln(r/r_i)}{\ln(r_o/r_i)} \tag{7}$$

Spherical Shell. Radial heat flux, q_r, at any radius r through a spherical shell of inner radius r_i held at temperature T_i, and outer radius r_o held at temperature T_o, e.g. heat loss through the wall of a storage sphere, is given by

$$q_r = \frac{Q}{4\pi r^2} = \frac{k(T_i - T_o)}{r^2(1/r_i - 1/r_o)} \tag{8}$$

The temperature profile, $T(r)$ is given by

$$T(r) = T_i - \frac{(T_i - T_o)(r - r_i)}{r r_i(1/r_i - 1/r_o)} \tag{9}$$

4.2.2 Convection

Convective heat transfer results from the motion of fluids of different temperature relative to each other and is therefore intimately related to fluid flow. In general, we would expect that the convective heat flux q to be proportional to the temperature difference driving force ΔT.

$$q = \frac{Q}{A} = h\Delta T \tag{10}$$

This proportionality constant is known as the convective heat transfer coefficient and is given by the symbol h. Equation (10) is the defining

Table 1   *Typical ranges of convective heat transfer coefficients*[3]

Typical values of h	$h \ (W \ m^{-2} \ K^{-1})$
Gases	10–200
Liquids	100–1000
Low viscosity liquids	1400–1800
Boiling liquids	1000–3000
Boiling water	1000–2000
Condensing vapours	1000–3000
Condensing steam	4000–10,000

equation for the convective heat transfer coefficient. Table 1 lists typical ranges of heat transfer coefficients.

Fluid flow is often turbulent, and so heat transfer by convection is often complex and normally we have to resort to correlations of experimental data. Dimensional analysis will give us insight into the pertinent dimensionless groups; see Chapter 6, *Scale-Up in Chemical Engineering*, Section 6.7.4.

We have noted earlier that fluid flow may be either natural or forced and different variables will come into play in each.

4.2.2.1   Forced Convection.   In forced convection, flow is generated by some external means, *e.g.* a pump, and heat transfer coefficients would be expected to be a function of

- fluid flow properties, density, ρ and viscosity, μ
- fluid thermal properties, specific heat capacity, c_p and thermal conductivity, k
- a characteristic length, d
- a characteristic velocity, u, generated by the external means

So, $h = f(\rho, \mu, c_p, k, d, u)$.

Dimensional analysis gives the following dimensionless groups:

$$Nu = \frac{hd}{k} = f\left(\frac{\rho u d}{\mu}, \frac{c_p \mu}{k}\right) = f(Re, Pr) \tag{11}$$

As we will see in Chapter 6, dimensionless groups may be considered as ratios. The Nusselt number *Nu* is a ratio of heat transfer resistances:

$$Nu = \frac{hd}{k} = \frac{d/k}{1/h} \equiv \frac{\text{conductive resistance}}{\text{convective resistance}}$$

The Reynolds number is a ratio of forces:

$$Re = \frac{\rho u d}{\mu} = \frac{\rho u^2 d^2}{\mu u d} \equiv \frac{\text{inertial forces}}{\text{viscous forces}}$$

and the Prandtl number is a ratio of diffusivities:

$$Pr = \frac{c_p \mu}{k} = \frac{\mu/\rho}{k/\rho c_p} \equiv \frac{\text{momentum diffusivity}}{\text{thermal diffusivity}}$$

4.2.2.2 Natural Convection. Natural convection arises from the differential thermal expansion of fluids at different temperatures. The resulting density differences give rise to buoyancy driven flows. Heat transfer coefficients would be expected to be a function of

- fluid flow properties, density, ρ and viscosity, μ
- fluid thermal properties, specific heat capacity, c_p and thermal conductivity, k
- a characteristic length, L
- buoyancy forces that may be characterised by
  - gravitational acceleration, g
  - temperature difference between surface and bulk fluid, ΔT
  - coefficient of thermal expansion, β, which is defined by $\rho = \rho_0(1 + \beta\Delta T)$ and for an ideal gas, $\beta = 1/T(K)$

Now, from the above, $h = f(\rho, \mu, c_p, k, L, g, \Delta T, \beta)$ and dimensional analysis gives

$$Nu = \frac{hL}{k} = f\left(\frac{\beta g \rho^2 L^3 \Delta T}{\mu^2}, \frac{c_p \mu}{k}\right) = f(Gr, Pr) \tag{12}$$

For natural convection, the Grashof number, Gr, is found instead of the Reynolds number for forced convection correlations. The Grashof number may be shown to be equivalent to a ratio of forces:

$$\frac{\beta g \rho^2 L^3 \Delta T}{\mu^2} \propto \frac{(\rho_0 - \rho)L^3 g \times \rho u^2 L^2}{(\mu u L)^2} \propto \frac{\text{buoyancy forces} \times \text{inertial forces}}{(\text{viscous forces})^2}$$

Table 2 lists heat transfer correlations for forced and natural convection for a number of geometries.

4.2.3 Radiation

All bodies emit and absorb electromagnetic radiation resulting in a transfer of energy. All materials are characterised by their ability to absorb, emit, transmit or reflect electromagnetic radiation.

Table 2 *Some heat transfer coefficient correlations[1,2,7]*

Flow and geometry	Conditions	Correlation
Forced convection		
Laminar inside a pipe	$(Re\,Pr\,d/L) > 10$	$Nu = 1.86\,(Re\,Pr\,d/L)^{0.33}\,(\mu_b/\mu_w)^{0.14}$
Turbulent inside a pipe	$Re > 10^4,\ 0.7 < Pr < 17{,}000,\ L/d > 60$	$Nu = 0.023\,Re^{0.8}\,Pr^{0.33}\,(\mu_b/\mu_w)^{0.14}$
Outside a horizontal cylinder	$1000 < Re < 1{,}00{,}000$	$Nu = 0.26\,Re^{0.6}\,Pr^{0.3}$
Across a bank of cylinders	$Re > 3000$ (characteristic length: d_e)	$Nu = 0.36\,Re^{0.55}\,Pr^{0.33}\,(\mu_b/\mu_w)^{0.14}$
Around a sphere	$1 < Re < 7 \times 10^4,\ 0.6 < Pr < 400$	$Nu = 2 + 0.6\,Re^{0.5}\,Pr^{0.33}$
Natural convection		
Outside a horizontal cylinder	$10^4 < Gr\,Pr < 10^9$	$Nu = 0.525\,(Gr\,Pr)^{0.25}$
	$Gr^{0.25}\,Pr^{0.33} < 2000$	$Nu = 2 + 0.6\,Gr^{0.25}\,Pr^{0.33}$
Around a sphere	$10^4 < Gr\,Pr < 10^9$	$Nu = 0.59\,(Gr\,Pr)^{0.25}$
Vertical flat plate	$10^9 < Gr\,Pr < 10^{13}$	$Nu = 0.13\,(Gr\,Pr)^{0.33}$
Horizontal flat plate	$10^5 < Gr\,Pr < 2 \times 10^7$	$Nu = 0.54\,(Gr\,Pr)^{0.25}$
Facing upwards	$2 \times 10^7 < Gr\,Pr < 3 \times 10^{10}$	$Nu = 0.14\,(Gr\,Pr)^{0.33}$
Facing downwards	$3 \times 10^5 < Gr\,Pr < 3 \times 10^{10}$	$Nu = 0.27\,(Gr\,Pr)^{0.25}$

The absorptivity, α, is the ratio of the absorbed energy to the incident energy from the source. A perfect absorber is called a *black body*. Non-perfect absorbers, $1 \geq \alpha \geq 0$, are known as *grey bodies*.

The transmittance, τ, is the ratio of the transmitted energy to the incident energy. The fraction of energy reflected will be $1 - \alpha - \tau$. For a black body, $\tau = 0$.

The emissivity, ε, is the ratio of the emitted energy to that of a perfect emitter. A perfect emitter is also a perfect absorber, a black body. For grey bodies, $\varepsilon = \alpha = $ constant is often assumed. Values of ε for many materials are given in the literature, *e.g.* ref. 7.

The energy emitted by a body is given by Stefan's law:

$$q = \frac{Q}{A} = \sigma \varepsilon T^4 \tag{13}$$

where $\sigma = 5.67 \times 10^{-8} \text{ W m}^{-2} \text{ K}^{-4}$ and is known as the Stefan–Boltzmann constant.

It can be seen that the energy emitted is a very strong function of temperature. All other things being equal, a hot surface will radiate more energy than a cold surface and therefore there is a net heat transfer from hot to cold surfaces. The surfaces do not need to be in physical contact, but must be in *thermal contact, i.e.* they must be able to "see" each other, and material inbetween the surfaces must be transparent to the radiation.

The net rate of heat transfer from surface 1 to surface 2 is

$$Q_{1 \to 2} = \sigma \varepsilon_1 A_1 \left(T_1^4 - T_2^4 \right) \tag{14}$$

This assumes that all energy emitted by surface 1 reaches surface 2. However, normally only a fraction of the energy emitted by surface 1 will reach surface 2. This fraction is called the *view factor*, F_{12}. Thus

$$Q_{1 \to 2} = \sigma \varepsilon_1 A_1 F_{12} \left(T_1^4 - T_2^4 \right) \tag{15}$$

Since for two black bodies at equal temperatures the absorbed energy must equal the emitted energy, it can be shown that $A_1 F_{12} = A_2 F_{21}$.

The detailed analysis of radiative heat transfer can easily become extremely complicated when transmission, reflection and complex geometries are taken into account.[4] An important conclusion which may be reached is that the heat flux corresponding to the surrounding black body is the maximum radiative heat flux that may be achieved, *i.e.* for $F_{12} - 1$ and $\varepsilon - 1$:

$$q_{\text{RadB}} = \frac{Q_{\text{RadB}}}{A_1} = \sigma \left(T_1^4 - T_2^4 \right) \tag{16}$$

As a rough approximation, one can define a radiative heat transfer coefficient and describe the heat flux using

$$q = \frac{Q}{A_1} = h_{\text{Rad}}(T_1 - T_2) \tag{17}$$

where

$$h_{\text{Rad}} = \frac{\sigma\left(T_1^4 - T_2^4\right)}{T_1 - T_2} \tag{18}$$

Figure 4 shows the predictions of Equation (18) for various values of T_1 and T_2.

4.3   THE OVERALL HEAT TRANSFER COEFFICIENT, U

When considering heat transfer by conduction, we have neglected the mechanism by which heat is transferred to and from the outer solid surfaces. In convective heat transfer, heat is transferred to and from flowing fluids.

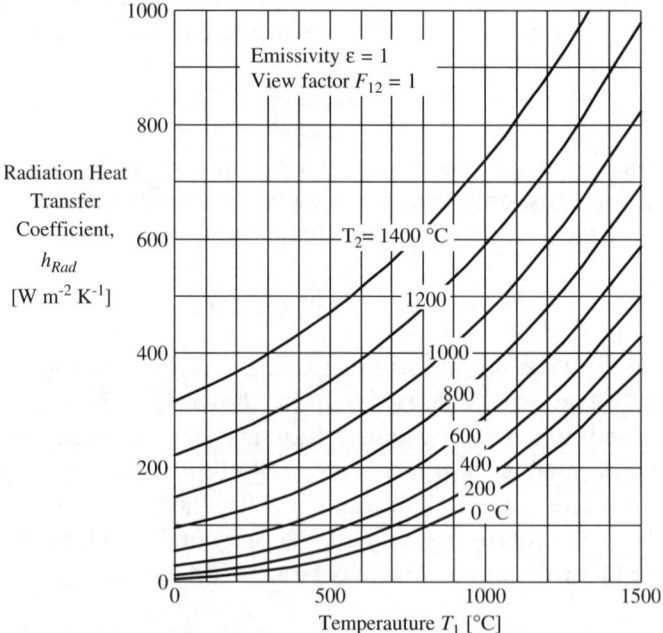

Figure 4   *Maximum obtainable values for h_{Rad}*

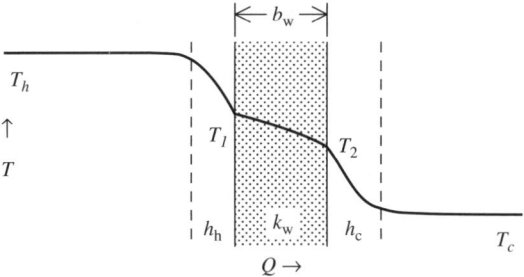

Figure 5 *Heat transfer from a hot fluid, though a wall, to a cold fluid*

When a fluid is present in contact with each solid wall, there will be an additional resistance to heat transfer in each fluid boundary layer or "film". The combined mechanism of heat transfer from a hot fluid through a dividing wall to a cold fluid has many similarities to conduction through a composite slab reviewed earlier.

Figure 5 shows hot and cold fluids separated by a flat, conducting, dividing wall. Overall, heat is transferred from the bulk of the hot fluid to the bulk of the cold fluid. In detail, hot fluid in the turbulent bulk will be at a uniform temperature T_h. Next to the wall will be a boundary layer through which heat is transferred to the wall. Heat is then transferred through the dividing wall by conduction to the cold fluid, and temperature drops from T_1 to T_2. Next to the wall on the cold side will be another boundary layer, through which heat is transferred into the cold side turbulent bulk, temperature T_c.

The boundary layers are often called "*films*", and heat transfer by conduction predominates in them giving rise to significant resistances to heat transfer. We assume that all the resistance to heat transfer on the hot side is confined to the hot film and all that on the cold side is confined to the cold film.

The associated heat transfer coefficients for these films are known as "*film heat transfer coefficients*", h_h and h_c for the hot and cold side, respectively. The heat flux through the hot film is given by $q = h_h(T_h - T_1)$, through the wall by $q = (k_w/b_w)(T_1 - T_2)$ and through the cold film by $q = h_c(T_2 - T_c)$. At steady state, the heat fluxes are constant and it is easy to show that

$$q = \frac{Q}{A} = \frac{(T_h - T_c)}{(1/h_h) + (b_w/k_w) + (1/h_c)} = U(T_h - T_c) \qquad (19)$$

U is known as the "*overall heat transfer coefficient*" and relates the heat flux to the overall temperature difference $T_h - T_c$.

The overall heat transfer coefficient is related to the individual film coefficients by

$$\frac{1}{U} = \frac{1}{h_h} + \frac{b_w}{k_w} + \frac{1}{h_c}$$ (20)

Equation (20) may be compared to Equation (4) for the heat flux through a composite plate. As we have seen, (b_w/k_w) is the thermal resistance of the wall to heat transfer by conduction. Similarly, $(1/h_h)$ and $(1/h_c)$ are the resistances to heat transfer offered by the hot and cold films, respectively.

The reciprocal of the overall heat transfer coefficient, $1/U$, is equivalent to the overall resistance to heat transfer, being equal to the sum of the resistances in the hot film, the wall and the cold film.

4.3.1 Example Calculation of U

Water flows through a mild steel pipe: $k = 45$ W m^{-1} K^{-1} and wall thickness 1 mm. Air flows around the outside of the pipe. If $h_{air} = 50$ W m^{-2} K^{-1} and $h_{water} = 1000$ W m^{-2} K^{-1}, estimate the overall heat transfer coefficient U.

$$U = \frac{1}{(1/h_{air}) + (b_w/k_w) + (1/h_{water})}$$

$$= \frac{1}{(1/50) + (0.001/45) + (1/1000)}$$

$$= \frac{1}{0.02 + 0.000022 + 0.001} = 47.6 \text{ W m}^{-2} \text{ K}^{-1}$$

It can be seen that in this example the largest resistance, the controlling film, is that of the air, $1/U \approx 1/h_{air}$, *i.e.* $U \approx h_{air}$, and so the air film controls the overall heat transfer process. If we wish to increase the overall rate of heat transfer by increasing U, we should look at the dominating resistance, the air film, and try to reduce it, *e.g.* by increasing the air-side velocity. Reducing the water film resistance has very little effect upon U.

4.3.2 Overall Heat Transfer Coefficients with Curvature, for Example, Transfer through a Pipe Wall

Here we will consider the problem of heat transfer through a cylindrical wall, for example, the wall of a pipe or tube in a shell-and-tube heat exchanger (Figure 6).

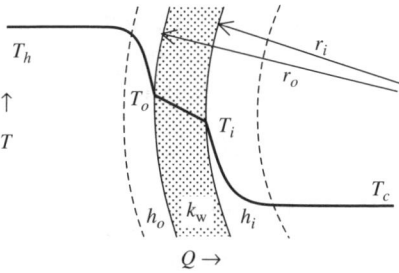

Figure 6 *Heat transfer through a cylindrical pipe wall*

The area for heat transfer varies with radius; therefore, we define the overall heat transfer coefficient based upon the outer area of the cylinder, $A_o = 2\pi r_o L$, where L is the length of the cylinder, giving

$$Q = U_o A_o (T_h - T_c) \tag{21}$$

$$\frac{1}{U_o} = \frac{1}{h_o} + \frac{r_o \ln(r_o/r_i)}{k_w} + \frac{r_o}{r_i h_i} \tag{22}$$

We again see that the reciprocal of the overall heat transfer coefficient is equivalent to the overall resistance to heat transfer, which is equal to the sum of resistances in the hot film ($1/h_o$), the wall $[r_o \ln(r_o/r_i)]/k_w$ and the cold film ($r_o/r_i h_i$), as shown in Equation (22). However in this case, the expressions for the resistances in the wall and the cold film are modified to take the curvature into account. No such modification is required for the outside film resistance, as the overall heat transfer coefficient is defined based on the outside surface area of the cylinder.

4.3.3 Overall Heat Transfer Coefficients with Convection and Radiation

For processes in parallel, the conductance is additive; so for combined convective and radiative heat transfer at a surface

$$h_{Tot} = h_{Conv} + h_{Rad} \tag{23}$$

For example, for heat transfer through a flat wall, where heat is transferred by convection and radiation from the wall, the overall heat transfer coefficient is related to the individual coefficients by

$$\frac{1}{U} = \frac{1}{h_h} + \frac{b_w}{k_w} + \frac{1}{h_{Conv} + h_{Rad}} \tag{24}$$

Since the two contributions are additive, we need to consider radiation only when the two coefficients are comparable. Therefore, it is often not necessary to carry out the detailed analysis of radiative heat transfer, since it is possible to obtain the value of h_{Conv} from correlations and compare it to the maximum value of h_{Rad}, as shown in Figure 4.

4.4 TRANSIENT HEAT TRANSFER

When manufacturing goods or sterilising lines, one often needs to know how long it will take for a certain part of a body to reach a specified temperature. In order to address this issue, it is necessary to consider the dynamic behaviour of heat transfer and write the balance equations introducing time. In the case of conduction in a solid with plate geometry, this becomes

$$\frac{\partial T}{\partial t} = \frac{k}{\rho c_p} \frac{\partial^2 T}{\partial x^2} \tag{25}$$

or for conduction in three dimensions

$$\frac{\partial T}{\partial t} = \frac{k}{\rho c_p} \left(\frac{\partial^2 T}{\partial x^2} + \frac{\partial^2 T}{\partial y^2} + \frac{\partial^2 T}{\partial z^2} \right) \tag{26}$$

These partial differential equations need to be solved with the associated initial and boundary conditions. A comprehensive list of solutions is available in Carslaw and Jaeger.[5]

Williamson and Adams[6] presented data, shown in Figure 7, for the dimensionless centre temperature $(T_c - T_{surf})/(T_0 - T_{surf})$ of variously shaped bodies as a function of dimensionless time Fo. Here T_c is the centre point temperature, T_0 the initial body temperature and T_{surf} the surface temperature. Fo is the Fourier number, which is given by $Fo = \alpha t/\delta^2$, where α is the thermal diffusivity, t is the time and δ is the characteristic length for conduction, the distance of the centre point or centreline of the body to the nearest part of the surface. The data given in Figure 7 neglects the thermal resistance at the surface.

When the heat transfer resistance is concentrated at the surface of a body of area A, the *thermally thin body* approximation can be applied, and the temperature throughout the solid, T, is assumed to be independent of position. In this case, the heat balance becomes

$$m_{solid} c_{p_{solid}} \frac{dT}{dt} = h_{fluid} A (T_{surf} - T) \tag{27}$$

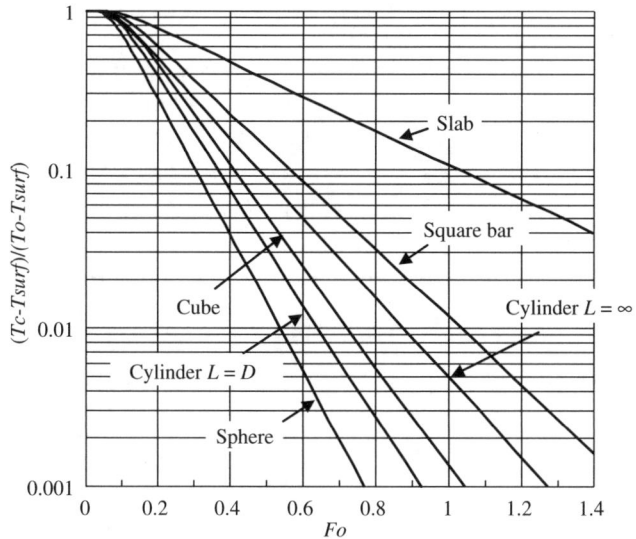

Figure 7 *Central dimensionless temperature variation with dimensionless temperature for a number of bodies, with negligible surface resistance*[6]

or

$$\frac{T - T_{\text{surf}}}{T_0 - T_{\text{surf}}} = \exp\left(-\frac{h_{\text{fluid}} A}{m_{\text{solid}} c_{p_{\text{solid}}}} t\right) \tag{28}$$

To determine whether the thin body approximation may be used, one should compare the surface heat transfer coefficient, h_{fluid} and the thermal conductance of the solid, k_{solid}/δ. Their ratio is the Biot number,

$$Bi = \frac{h_{\text{fluid}} \delta}{k_{\text{solid}}} = \frac{\delta/k_{\text{solid}}}{1/h_{\text{fluid}}} = \frac{\text{conductive resistance of solid}}{\text{convective resistance of fluid}} \tag{29}$$

When Bi is very small, the thin body approximation can be used, *e.g.* a small metal sphere cooled by air (k_{solid} large, δ and h_{fluid} small). When Bi is very large, the external resistance is negligible, *e.g.* large body with low thermal conductivity immersed in a flowing liquid (k_{solid} small, δ and h_{fluid} large).

4.4.1 Example Calculation, Small *Bi* Case: Cooling of a Copper Sphere in Air

A copper sphere 1 cm in diameter initially at 100°C is cooled in a stream of air at 20°C, flowing at 5 m s⁻¹. Estimate the time taken for the sphere to reach 25°C.

4.4.1.1   Physical Properties[7,8]:

	Copper	*Air*
Density (ρ, kg m^{-3})	8300	1.3
Specific heat (c_p, J kg^{-1} K^{-1})	419	812
Thermal conductivity (k, W m^{-1} K^{-1})	372	0.0234
Viscosity (μ, Pa s)	–	0.000018

4.4.1.2   Estimate h and, hence, Bi.

$$Re_{\text{air}} = \frac{\rho u d}{\mu} = \frac{1.3 \times 5 \times 0.01}{0.000018} = 3611$$

$$Pr_{\text{air}} = \frac{c_p \mu}{k} = \frac{812 \times 0.000018}{0.0234} = 0.62$$

From Table 2 for forced convection to a sphere under these conditions,

$$Nu_{\text{air}} = 2 + 0.6Re^{0.5}Pr^{0.33} = 2 + 0.6 \times 3611^{0.5} \times 0.62^{0.33} = 32.9$$

Heat transfer coefficient,

$$h_{\text{air}} = \frac{Nu_{\text{air}} \times k_{\text{air}}}{d} = \frac{32.9 \times 0.0234}{0.01} = 76.9 \text{ W m}^{-2} \text{ } K^{-1}$$

Biot number,

$$Bi = \frac{h_{\text{air}}\delta}{k_{\text{Cu}}} = \frac{76.9 \times 0.01/2}{372} = 0.001$$

Bi is small and so the thermal resistance of the air film dominates the overall heat transfer process.

4.4.1.3   Estimate Time.   Rearranging Equation (28) and solving for time *t* when $T = 25°C$

$$t = -\frac{m_{\text{Cu}}c_{p_{\text{Cu}}}}{h_{\text{air}}A} \ln\left(\frac{T - T_{\text{surf}}}{T_0 - T_{\text{surf}}}\right)$$

$$= -\frac{0.004346 \times 419}{76.9 \times 0.000314} \ln\left(\frac{25 - 20}{100 - 20}\right) = 209 \text{ s} = 3.5 \text{ min}$$

4.4.2 Example Calculation, Large *Bi* Case: Cooling of a Perspex Plate in Water

A vertical Perspex plate, 3-cm thick, 10-cm high and initially at 100°C, is cooled by immersion in a large water bath at 20°C. Estimate the time taken for the centre of the plate to reach 25°C.

4.4.2.1 Physical Properties[7,8].

	Perspex	Water
Density (ρ, kg m^{-3})	1180	1000
Specific heat (c_p, J kg^{-1} K^{-1})	1440	4182
Thermal conductivity (k, W m^{-1} K^{-1})	0.184	0.6
Viscosity (μ, Pa s)	–	0.001
Coefficient of thermal expansion (β, K^{-1})	–	0.000888

4.4.2.2 Estimate h and, hence, Bi. As ΔT changes with time from 80 to 5°C, an average ΔT of 25°C is taken to estimate *Gr*.

$$Gr_{\text{Water}} = \frac{\beta g \rho^2 L^3 \Delta T}{\mu^2}$$

$$= \frac{0.000888 \times 9.81 \times 1000^2 \times 0.1^3 \times 25}{0.001^2} = 2.2 \times 10^8$$

$$Pr_{\text{water}} = \frac{c_p \mu}{k} = \frac{4182 \times 0.001}{0.6} = 6.97$$

From Table 2, for natural convection from a vertical plate with $GrPr = 1.533 \times 10^9$,

$$Nu_{\text{water}} = 0.13(GrPr)^{0.33} = 0.13 \times \left(1.533 \times 10^9\right)^{0.33} = 139$$

Heat transfer coefficient,

$$h_{\text{water}} = \frac{Nu_{\text{water}} \times k_{\text{water}}}{L} = \frac{139 \times 0.6}{0.1} = 836 \text{ W m}^{-2} \text{ K}^{-1}$$

Biot number,

$$Bi = \frac{h_{\text{water}} \delta}{k_{\text{perspex}}} = \frac{836 \times 0.03/2}{0.184} = 68$$

Bi is large, and so the thermal resistance of the water film may be neglected.

4.4.2.3   Estimate Time.   Dimensionless final temperature,

$$\frac{T - T_{\text{surf}}}{T_0 - T_{\text{surf}}} = \frac{25 - 20}{100 - 20} = 0.0625$$

From Figure 7, for a slab the equivalent Fourier number Fo at this dimensionless temperature is 1.22.

Now, $Fo = \alpha t/\delta^2$. Rearranging and solving for time t, when $T = 25°C$,

$$t = \frac{Fo\delta^2}{\alpha_{\text{perspex}}} = \frac{Fo\delta^2}{\left(k_{\text{perspex}}/\rho_{\text{perspex}}c_{p\text{perspex}}\right)}$$

$$= \frac{1.22 \times (0.03/2)^2}{0.184/(1180 \times 1440)} = 2535 \text{ s} = 42.2 \text{ min}$$

4.5   HEAT EXCHANGERS

A heat exchanger is any device that effects transfer of thermal energy between two fluids that are at different temperatures. The two fluids do not come in direct contact but are separated by a solid surface or tube wall. Common heat exchangers include

- Shell-and-tube
- Flat plate
- Finned tubes

Applications of heat exchangers include air conditioning systems and car/truck radiators, but they are also widely employed in the process industry whenever there is a need for cooling or heating up a process fluid.

In practice, engineers often have to do the following:

- *Select* a heat exchanger that will achieve a specific temperature change in a fluid stream of known mass flow rate.
- *Design* a heat exchanger, *i.e.* determine the surface area to transfer heat at a given rate for given fluid temperatures and flow rates.
- *Predict* the outlet temperatures of hot and cold fluid streams in a specified heat exchanger.

4.5.1   Double Pipe Heat Exchanger

The simplest form of a heat exchanger consists of two concentric cylindrical tubes, the *double pipe heat exchanger*, as schematically represented in Figure 8:

In the schematic in Figure 8, a hot fluid "A" flows in the inner tube and exchanges heat through an inner wall with a cooler fluid "B",

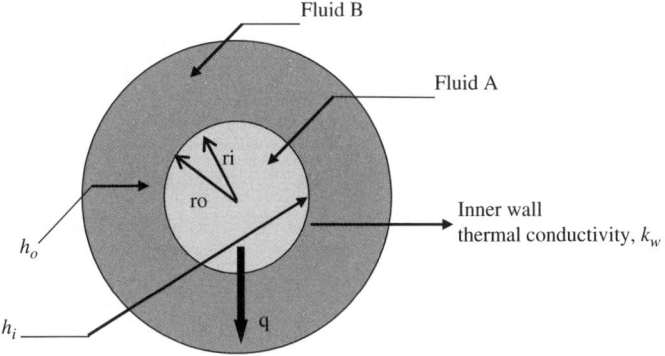

Direction of heat flow if fluid "A" is hotter than fluid "B"

Figure 8 *Schematic representation of a double pipe heat exchanger*

which flows in the outer tube. Heat transfer involves convection in each fluid (with h_o and h_i being the convective heat transfer coefficients of the fluid in the outer and inner tube, respectively) and conduction through the wall of thermal conductivity k_w separating the two fluids.

Two different types of heat exchangers can be identified. In the *co-current-flow* heat exchanger, both fluids flow in the same direction, as shown in Figure 9. In this case, the two fluids enter the heat exchanger from the same end.

In a *counter-current-flow* heat exchanger, the fluids flow in opposite directions, as shown in Figure 10. In this case, the hot and cold fluids enter the heat exchanger from opposite ends. In both the parallel and counter-flow configurations, the two fluids are forced to flow along the heat exchanger by either pumps or fans.

In the schematics depicted in Figures 9 and 10, the inlet and outlet temperatures of the hot fluid (T_{hi}, T_{ho}) and cold fluid (T_{ci}, T_{co}) are also shown. The temperature profiles for the two configurations are shown in Figures 11 and 12.

For the *co-current-flow* configuration, the temperature difference ΔT is large at the inlet but decreases exponentially towards the outlet. The temperature of the hot fluid decreases and the temperature of the cold fluid increases along the heat exchanger. The outlet temperature of the cold fluid can *never* exceed that of the hot fluid, no matter how long the heat exchanger is.

For the *co-current-flow* configuration, the outlet temperature of the cold fluid may exceed the outlet temperature of the hot fluid, the so-called "temperature cross". The outlet temperature of the cold fluid can never exceed the inlet temperature of the hot fluid.

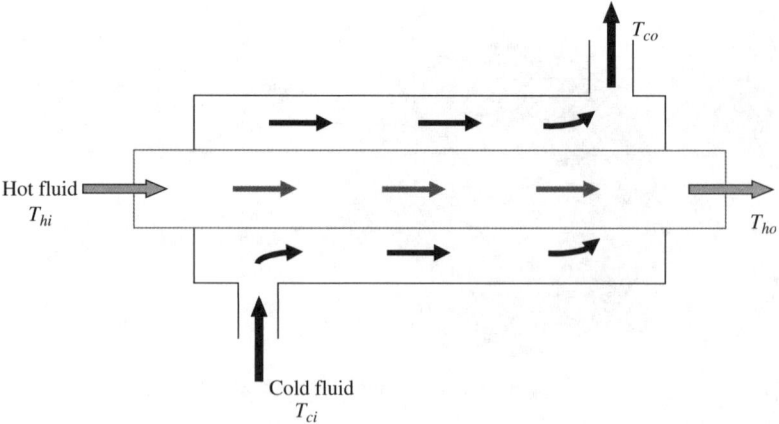

Figure 9   *Schematic representation of a co-current-flow heat exchanger*

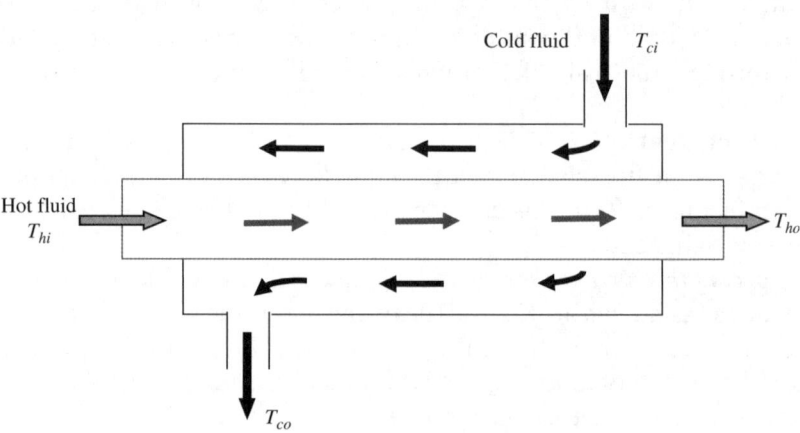

Figure 10   *Schematic representation of a counter-current--flow heat exchanger*

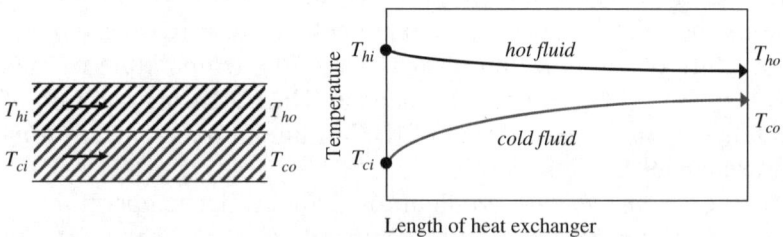

Figure 11   *Temperature profile – co-current-flow heat exchanger*

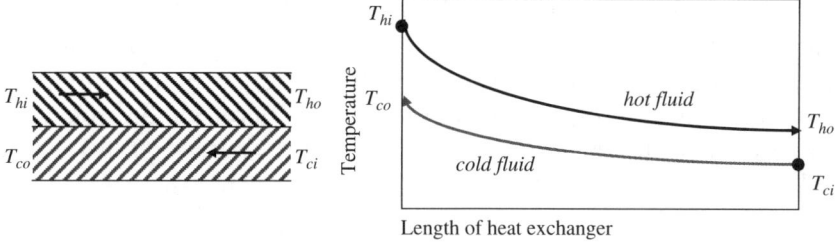

Figure 12 *Temperature profile – counter-current-flow heat exchanger*

4.5.2 Heat Transfer in Heat Exchangers

Heat exchangers operate for long periods with no change in the operating conditions; thus, they can be modelled as *steady-flow* devices, for which the following assumptions are valid:

- The overall heat transfer coefficient, U, is constant with time and is independent of the position along the exchanger.
- The mass flow rate of each fluid remains constant.
- The specific heats of the fluids are constant.
- The temperature of the two fluids is constant over a specific cross-section.
- The outer surface is perfectly insulated, so that any heat transfer occurs between the two fluids only, and there are no heat losses to the surroundings.

Under these assumptions, it follows that the *rate of heat transfer*, q, from the hot fluid is equal to the rate of heat transfer to the cold one. The basic design equations for heat exchangers are therefore the energy balances for each fluid:

Energy given up by hot fluid:

$$q = -G_h c_{p_h} (T_{ho} - T_{hi}) \tag{30}$$

Energy gained by cold fluid:

$$q = G_c c_{p_c} (T_{co} - T_{ci}) \tag{31}$$

where G (kg s^{-1}) is the mass flow rate and c_p (kJ kg^{-1} K^{-1}) is the specific heat, and the defining equation for U (Equation (19)) which is reiterated here:

$$q = UA\Delta T \tag{19}$$

4.5.2.1 Log Mean Temperature Difference, ΔT_{lm}. The temperature difference between the hot and cold fluids, ΔT, varies along the length of the heat exchanger. Therefore, we have to express ΔT in terms of a suitable mean temperature difference, the log mean temperature difference, ΔT_{lm}. This can be determined by simply applying the design equations (30), (31) and (19) to a cross-section of the heat exchanger, as represented in Figure 13 for the case of *co-current-flow*.

Based on the assumptions set out, the heat transfer rate for a differential length is:

$$\delta q = U(T_h - T_c)\delta A \tag{32}$$

Having assumed that the outer surface of the exchanger is perfectly insulated, the energy gained by the cold fluid is equal to that given up by the hot fluid. Thus, an energy balance on each fluid in the differential section is as follows:

Energy given up by hot fluid:

$$\delta q = -G_h c_{p_h} \delta T_h \tag{33}$$

Energy gained by cold fluid:

$$\delta q = G_c c_{p_c} \delta T_c \tag{34}$$

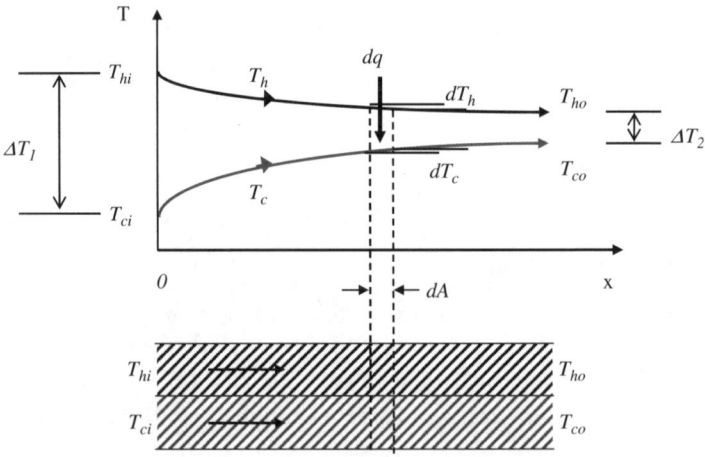

Figure 13 *Temperature profile along a co-current-flow heat exchanger*

Note that the temperature change of the hot fluid, δT_h, is a negative quantity to make the heat transfer a positive quantity. Solving the above equations for δT_h and δT_c gives

$$\delta T_h = -\frac{\delta q}{G_h c_{p_h}} \quad \text{and} \quad \delta T_c = \frac{\delta q}{G_c c_{p_c}}$$

Taking their difference, we get

$$\delta T_h - \delta T_c = \delta(T_h - T_c) = -\delta q \left(\frac{1}{G_h c_{p_h}} + \frac{1}{G_c c_{p_c}} \right) \tag{35}$$

The rate of heat transfer in the differential section of the exchanger is also given by Equation (32), reported below for convenience:

$$\delta q = U(T_h - T_c)\delta A \tag{32}$$

Substituting this into Equation (35) gives

$$\delta(T_h - T_c) = -U\delta A(T_h - T_c) \left(\frac{1}{G_h c_{p_h}} + \frac{1}{G_c c_{p_c}} \right) \tag{36}$$

Re-arranging Equation (36), we obtain

$$\frac{\delta(T_h - T_c)}{T_h - T_c} = -U\delta A \left(\frac{1}{G_h c_{p_h}} + \frac{1}{G_c c_{p_c}} \right) \tag{37}$$

Letting δA tend to zero and integrating from one end of the heat exchanger to the other we obtain

$$\ln \frac{(T_{ho} - T_{co})}{T_{hi} - T_{ci}} = -UA \left(\frac{1}{G_h c_{p_h}} + \frac{1}{G_c c_{p_c}} \right) \tag{38}$$

Finally, solving the energy balance on each fluid for c_p gives

$$G_h c_{p_h} = -\frac{q}{T_{ho} - T_{hi}} \quad \text{and} \quad G_c c_{p_c} = \frac{q}{T_{co} - T_{ci}} \tag{39}$$

Substituting Equation (39) into Equation (38), we obtain

$$\ln \frac{(T_{ho} - T_{co})}{(T_{hi} - T_{ci})} = -UA \frac{(T_{hi} - T_{ho}) + (T_{co} - T_{ci})}{q} \tag{40}$$

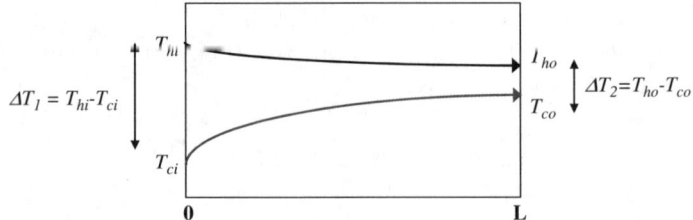

Figure 14   *Temperature profile showing ΔT_1 and ΔT_2 for a co-current-flow heat exchanger*

Or, in terms of the differences in end temperatures (see Figure 14)

$$\ln\frac{\Delta T_2}{\Delta T_1} = UA\frac{\Delta T_2 - \Delta T_1}{q} \tag{41}$$

Thus, the heat transfer rate is equal to

$$q = UA\ \frac{\Delta T_2 - \Delta T_1}{\ln(\Delta T_2/\Delta T_1)} = UA\ \Delta T_{lm} \tag{42}$$

where the *log mean temperature difference*, ΔT_{lm}, is the mean temperature along the heat exchanger:

$$\Delta T_{lm} = \frac{\Delta T_2 - \Delta T_1}{\ln(\Delta T_2/\Delta T_1)} \tag{43}$$

The ΔT_1 and ΔT_2 expressions in *co-current-flow* are given by the difference between the inner and outer temperatures of the two fluids, as shown in Figure 14.

The procedure to obtain ΔT_{lm} is the same for a *counter-current-flow* heat exchanger. It is again given by Equation (43); however in this case, ΔT_1 is the temperature difference between the inlet and outlet temperature of the hot fluid and the cold fluid respectively at one end of the exchanger, and ΔT_2 is the temperature difference between the hot fluid and the cold fluid at the other end, as shown in Figure 15.

Therefore, whether considering *co-current-flow* or *counter-current-flow*, the procedure to obtain ΔT_{lm} is the same.

4.5.2.2 Comparison between Co-Current-Flow and Counter-Current-Flow Heat Exchangers. The worked examples reported below will demonstrate the different efficiencies of the two heat exchanger configurations in terms of the heat transfer area required in both cases for the same U. Let us consider two fluids between which heat is being

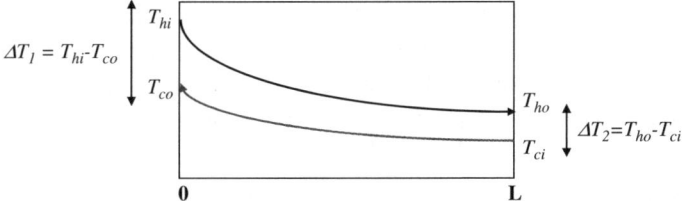

Figure 15 *Temperature profile showing ΔT_1 and ΔT_2 for a counter-current-flow heat exchanger*

exchanged. For a given amount of heat exchanged, q, the log mean temperature difference, ΔT_{lm}, will be calculated for both *co-current-flow* and *counter-current-flow* and the heat transfer area required in both cases compared, assuming that U is unchanged. Both inlet and outlet temperatures of each fluid are given:

Hot fluid	Cold fluid
$T_{hi} = 200°C$	$T_{ci} = 80°C$
$T_{ho} = 150°C$	$T_{co} = 120°C$

Using the given data, ΔT_{lm} for *co-current-flow* is given by

$$\Delta T_{lm} = \frac{\Delta T_2 - \Delta T_1}{\ln(\Delta T_2 / \Delta T_1)} = \frac{(150 - 120) - (200 - 80)}{\ln[(150 - 120)/(200 - 80)]} = 64.9\,°C$$

For *counter-current-flow*

$$\Delta T_{lm} = \frac{\Delta T_2 - \Delta T_1}{\ln(\Delta T_2 / \Delta T_1)} = \frac{(150 - 80) - (200 - 120)}{\ln[(150 - 80)/(200 - 120)]} = 74.9\,°C$$

For a given duty, q, and assuming that U is the same for parallel and counter-flow exchangers; the heat transfer area required is given by Equation (42):

$$A = \frac{q}{U \Delta T_{lm}}$$

From which

$$\frac{A_{parallel}}{A_{counter}} = \frac{\Delta T_{lm\ counter}}{\Delta T_{lm\ parallel}} = \frac{74.9}{64.9} = 1.15$$

The ratio between the heat transfer areas required for parallel and counter flow shows that the heat transfer area required with the parallel flow configuration is larger by 15% than with counter flow. This demonstrates the higher efficiency of the counter-flow heat exchanger compared to the parallel flow configuration, which would require a larger and, thus, more expensive heat exchanger to perform the same duty.

4.5.3 Performance of Heat Exchangers – Fouling

The performance of a heat exchanger depends upon the transfer surfaces being clean and uncorroded. The performance deteriorates with time due to accumulation of deposits on the heat transfer surfaces. The layer of deposits represents additional resistance to heat transfer. In the design of heat exchangers, this added resistance is accounted for by a *fouling factor* or *dirt factor*, R_d.

The most common type of fouling is the precipitation of solid deposits in a fluid on the heat transfer surfaces. A layer of calcium-based deposits forms after prolonged use on the surfaces at which boiling occurs, similar to what can be observed on the inner surface of a kettle. To avoid this potential problem, water in process plants is treated to remove its solid content.

Another form of fouling is corrosion or chemical fouling due to accumulation of the products of chemical reactions. This can be avoided by coating the metal surfaces with glass or by using plastic pipes. Heat exchangers may also be fouled by the growth of algae in warm fluids, biological fouling that can be prevented by chemical treatment.

Fouling has to be considered in the design and selection of heat exchangers. Larger and more expensive heat exchangers may be necessary to ensure the heat transfer requirements even after fouling occurs. Periodic cleaning of exchangers and the resulting down time are additional penalties associated with fouling.

Thermal resistance due to fouling has to be included in the calculation of the overall heat transfer coefficient. By definition, the overall heat transfer coefficient is the reciprocal of the overall thermal resistance. Thus, including the thermal resistance due to fouling, it follows

$$\frac{1}{U} = \frac{1}{(1/h_o) + R_{do} + (b_w/k_w) + R_{di} + (1/h_i)} \tag{44}$$

Typical values for R_d are given in the literature.[1,3,7]

4.5.4 Types of Heat Exchangers

4.5.4.1 Shell-and-Tube. The most widely used types of heat exchangers in the process industry are shell-and-tube heat exchangers. This can

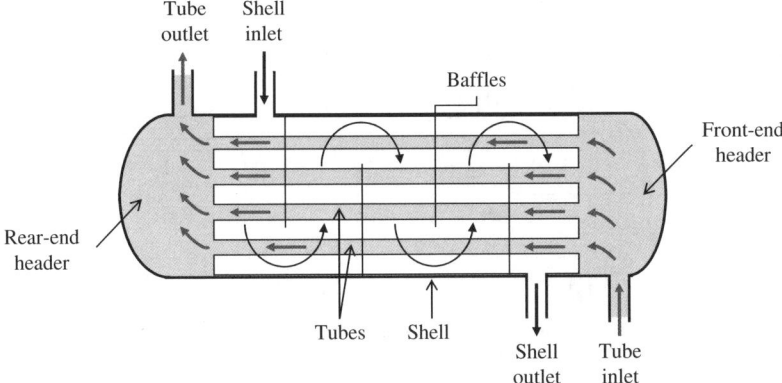

Figure 16 *Schematic of a shell-and-tube heat exchanger (one-shell pass and one-tube pass)*

be one-shell pass and one-tube pass (see Figure 16) or multi-pass shell-and-tube heat exchangers (as shown in Figure 17 for the case of one-shell pass and two-tube passes). Baffles are installed in the shell in order to promote mixing and cross-flow, thereby giving higher heat transfer coefficients.

The flow arrangement in shell-and-tube heat exchangers can involve both *co-current-flow* and *counter-current-flow*, as shown in the schematic in Figure 17.

Determination of the average temperature difference, ΔT_{lm}, is very complex for these types of heat exchangers. In the top half of the exchanger illustrated in Figure 17, there is parallel flow and in the bottom half, counter flow. For this reason, it is common practice to introduce a *correction factor*, F_t, into Equation (42). The heat transfer rate is therefore given by

$$q = UAF_t \Delta T_{lm} \qquad (45)$$

where ΔT_{lm} is that for the counter-flow double-pipe heat exchanger with the same fluid inlet and outlet temperatures as in the more complex design shown in Figure 17. F_t values for one-shell pass and two or more tube passes are given in Figure 18, where

$$R = \frac{T_{s1} - T_{s2}}{T_{t2} - T_{t1}} \quad \text{and} \quad S = \frac{T_{t2} - T_{t1}}{T_{s1} - T_{t1}} \qquad (46)$$

Derivations of F_t for various geometries are given by Kern,[1] and charts for F_t are given in the literature.[1,3,7]

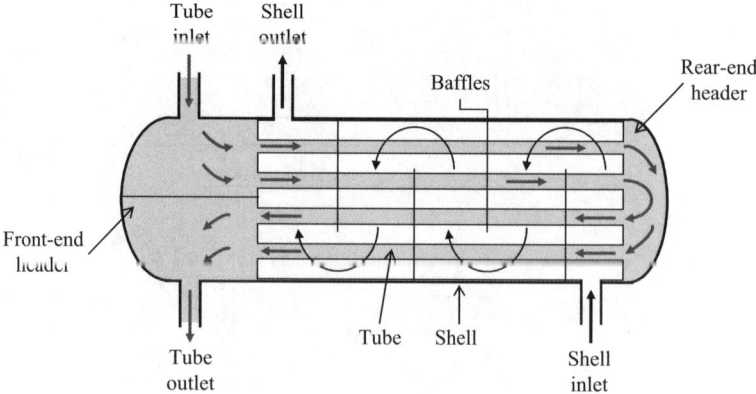

Figure 17   *Schematic of a shell-and-tube heat exchanger (one-shell pass and two-tube pass)*

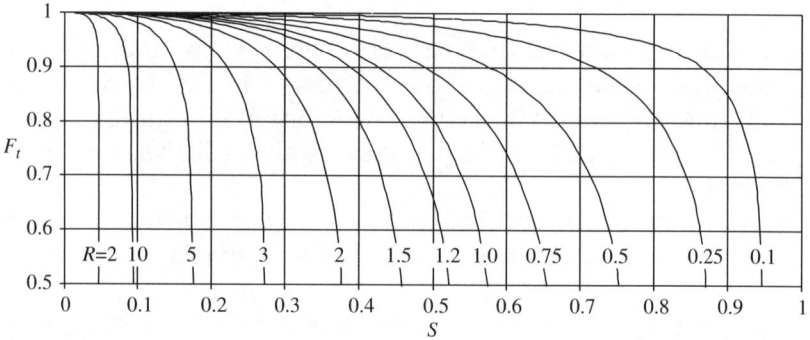

Figure 18   *LMTD correction factor F_t, for 1 shell pass and 2 or more tube passes[1]*

4.5.4.2 Plate Heat Exchangers.

Plate heat exchangers are built up from individual plates separated by gaskets. These plates are assembled into a pack and clamped in a frame, as shown in Figure 19. They are applied in the energy recovery section of many processes because of their *high efficiency* and *low costs*. In this type of heat exchanger, fluids are in *co-current-flow* and *counter-current-flow* in alternate channels, as shown in Figure 20.

Plates are made of thin sheet material, resulting in economic units, particularly when expensive material is involved. Plates are especially corrugated to promote turbulence even at low *Re*, resulting in

- very high heat transfer coefficients
- reduced fouling
- easy chemical cleaning

Figure 19 *Plate heat exchanger (Picture courtesy of WCR Inc of Dayton, Ohio)*

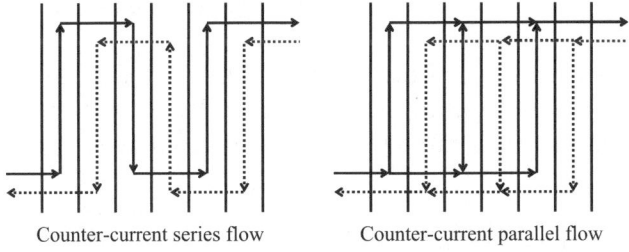

Counter-current series flow   Counter-current parallel flow

Figure 20 *Flow channels in a plate heat exchanger*

They are applied to various processes, such as the energy recovery sections of crystallisation and desalination plants.

4.5.4.3 Finned Heat Exchangers. Finned tube heat exchangers are used when heat is to be transferred between gases and liquids. The finned tubes provide a large heat transfer area, since gases, which are good thermal insulators, transfer smaller amounts of heat than liquids.

There are various types of finned tubes depending on the application. They are applied, for example, in large air conditioning systems, for radiators of vehicles or where cooling water is in short supply.

4.5.5 Numerical Examples

4.5.5.1 Cooling Hot Process Fluid in a Counter-Flow Heat Exchanger. A counter-flow double-pipe heat exchanger is used to cool a hot process fluid using water. The process fluid flows at 18 kg s^{-1} and is cooled from 105°C to 45°C. The water flows counter-currently to the process fluid, entering at 25°C and leaving at 50°C. Assuming no heat losses, calculate the required flow rate for the cooling water. Neglecting the tube wall curvature, calculate the required area for heat exchange.

The specific heat for water is 4.2 kJ kg^{-1} K^{-1} and that of the process fluid is 3.4 kJ kg^{-1} K^{-1}. The process fluid side film heat transfer coefficient is 2500 W m^{-2} K^{-1} and the cooling water side heat transfer coefficient is 1200 W m^{-2} K^{-1}. The tube wall thickness is 3 mm and the thermal conductivity is 220 W m^{-1} K^{-1}.

Problem Solution. Calculate the required flow rate for the cooling water: The heat given up by hot process fluid is equal to

$$Q_f = G_f c_{p_f}(T_{hi} - T_{ho}) = 18 \times 3.4 \times 10^3 \times (105 - 45)$$
$$= 3,672,000 \text{ W} = 3672 \text{ kW}$$

From the energy balance Equations (30) and (32), the heat given up by hot process fluid is equal to the heat gained by the cooling water; thus, we can obtain the flow rate required for the cooling water:

$$G_w = \frac{Q_w}{c_{p_w}(T_{co} - T_{ci})} = \frac{3672 \times 10^3}{4.2 \times 10^3 \times (50 - 25)} = 34.97 \text{ kg s}^{-1}$$

Calculate the required area for heat exchange: The heat exchanged is given by Equation (42), where the log mean temperature difference is given by (see Figure 21):

$$\Delta T_{lm} = \frac{\Delta T_2 - \Delta T_1}{\ln(\Delta T_2/\Delta T_1)} = \frac{(T_{ho} - T_{ci}) - (T_{hi} - T_{co})}{\ln[(T_{ho} - T_{ci})/(T_{hi} - T_{co})]}$$

$$= \frac{(45 - 25) - (105 - 50)}{\ln[(45 - 25)/(105 - 50)]} = \frac{-35}{-1.01} = 34.6 \,°C$$

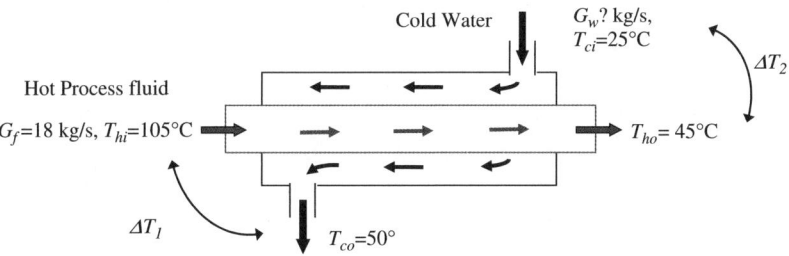

Figure 21 *Problem double-pipe heat exchanger*

Neglecting the effect of fouling, the overall heat transfer coefficient is

$$U = \frac{1}{R_{tot}} = \frac{1}{(1/h_f) + (x/k) + (1/h_w)}$$

$$= \frac{1}{(1/2500) + (0.003/220) + (1/1200)}$$

$$= 802 \ W \ m^{-2} \ K^{-1}$$

Thus, the heat transfer area is therefore equal to

$$A = \frac{q}{U\Delta T_{lm}} = \frac{3672 \times 1000}{802 \times 34.6} = 132.3 \ m^2$$

4.5.5.2 Shell-and-Tube Heat Exchanger Rating. A 20,000 kg h^{-1} of light hydrocarbon liquid stream is to be cooled from 200°C to 100°C by 65,000 kg h^{-1} of a heavy hydrocarbon liquid stream with a feed temperature of 35°C.

A shell-and-tube heat exchanger is available with one-shell pass and two-tube passes (illustrated schematically in Figure 17) and with dimensions listed below. Is it suitable for the duty?

Heat Exchanger Geometry:

Shell:

- Inside diameter, D_{iS}, 0.54 m
- Number of baffles, N_B, 38
- Number of shell passes, N_{PS}, 1

Tubes:

- Outside diameter, D_{oT}, 0.025 m
- Inside diameter, D_{iT}, 0.020 m
- Length, L_T, 4.4 m
- Number of tubes, N_T, 158
- Number of tube passes, N_{PT}, 4
- Pitch, P_T, 0.032 m
- Layout pattern, square

Heat Balance. The first task is to calculate the required duty of the exchanger and the outlet temperature of the heavy hydrocarbon stream. Heat lost by light hydrocarbon stream:

$$Q_{duty} = G \times c_p \times \Delta T = G \times c_p \times (T_{in} - T_{out})$$

$$= \frac{20,000}{3600} \times 2470 \times (200 - 100) = 1372 \text{ kW}$$

Outlet temperature of heavy hydrocarbon stream:

$$T_{out} = T_{in} + \Delta T = T_{in} + \frac{Q_{duty}}{G \times c_p} = 35 + \frac{1372 \times 10^3}{(65,000/3600) \times 2050} = 72 \,^\circ\text{C}$$

It is proposed to put the heavy hydrocarbon stream on the tube-side as it has the higher fouling factor. The tube-side is easier to clean than the shell-side.

Fluid Physical Properties at Mean Bulk Temperatures[1,3]

	Light h/c Shell-side	Heavy h/c Tube-side
Mean bulk temperature (T_b, °C)	150	53.5
Density (ρ, kg m^{-3})	730	820
Viscosity (μ, Pa s)	0.0004	0.0025
Specific heat (c_p, J kg^{-1} K^{-1})	2470	2050
Thermal conductivity (k, W m^{-1} K^{-1})	0.1329	0.1333
Fouling factor (R_d, m^2 K W^{-1})	0.0003	0.0005

The calculation procedure shown is based on Kern's method,[1] with modifications to include the heat transfer correlations presented earlier in this chapter. The individual film heat transfer coefficients will be estimated first and then combined to give the overall heat transfer coefficient and the overall rate of heat transfer.[1,3]

Tube-Side Calculations. The tube-side velocity is given by

$$v_T = \frac{G}{\rho} \times \frac{1}{\text{flow area}} = \frac{G}{\rho} \times \frac{4}{\pi D_{iT}^2} \times \frac{N_{PT}}{N_T}$$

$$= \frac{65,000}{3600 \times 820} \times \frac{4}{\pi(0.020)^2} \times \frac{4}{158} = 1.77 \text{ m s}^{-1}$$

It is seen that the flow area through the tubes is the cross-sectional area for flow of one tube multiplied by the total number of tubes and divided by the number of tube passes.

Reynolds Number Equation (11):

$$Re_T = \frac{\rho v_T D_{iT}}{\mu} = \frac{820 \times 1.77 \times 0.02}{0.0025} = 11,640$$

Prandtl Number Equation (11):

$$Pr_T = \frac{c_p \mu}{k} = \frac{2050 \times 0.0025}{0.1333} = 38.5$$

Nusselt Number:

$$Nu_T = 0.023 Re^{0.8} Pr^{0.33} (\mu_b/\mu_w)^{0.14}$$

See Table 2 (note all conditions for this correlation are met: $Re_T > 10^4$; $0.7 < Pr_T < 17,000$ and $L_T/D_{iT} = 244 > 60$). $Nu_T = 0.023 \times 11640^{0.8} \times 8.5^{0.33} = 137$, neglecting the viscosity correction $(\mu_b/\mu_w)^{0.14}$ as the wall temperature is not yet known.

The uncorrected tube-side heat transfer coefficient in Equation (11) is

$$h_T = \frac{Nu_T k}{D_{iT}} = \frac{139 \times 0.1333}{0.02} = 915 \text{ W m}^2 \text{ K}^{-1}$$

Shell-Side Calculations. In the shell, kerosene flows between the baffles over the tubes. The area for cross-flow between the tubes is given by the length of tube between baffles, $L_T/(N_B + 1)$, multiplied by the shell internal diameter, D_{iS}, adjusted by the fractional available area for flow between the tubes, $(P_T - D_{oT})/P_T$. Thus, the shell-side cross-flow area is given by

$$A_S = \frac{L_T}{N_B + 1} \times D_{iS} \times \frac{P_T - D_{oT}}{P_T}$$

$$= \frac{4.4}{38 + 1} \times 0.54 \times \frac{0.032 - 0.025}{0.032} = 0.0133 \text{ m}^2$$

Hence, the shell-side velocity becomes

$$v_S = \frac{G}{\rho A_S} = \frac{20,000}{3600 \times 730 \times 0.0133} = 0.57 \text{ m s}^{-1}$$

To estimate a Reynolds number for the flow on the shell-side, we use the equivalent hydraulic diameter

$$D_{eS} = \frac{4 \times \text{flow area}}{\text{wetted perimeter}}$$

(see Chapter 3). For a square tube layout, consider a square section of the pipe bundle, whose four corners are the centre-lines of four adjacent tubes, as illustrated in Figure 22.

The flow area, shaded in Figure 22, is given by the area of a square of side equal to the tube pitch, minus four quarter pipe cross-sectional areas. The wetted perimeter is four-quarter pipe perimeters.

$$D_{eS} = \frac{4 \times \text{flow area}}{\text{wetted perimeter}} = \frac{4\left[P_T^2 - \left(\pi D_{oT}^2/4\right)\right]}{\pi D_{oT}}$$

$$= \frac{4\left[0.0320^2 - \left(\pi \times 0.025^2/4\right)\right]}{\pi \times 0.025} = 0.0272 \text{ m}$$

Reynolds Number Equation (11):

$$Re_S = \frac{\rho v_S D_{eS}}{\mu} = \frac{730 \times 0.57 \times 0.0272}{0.0004} = 28,297$$

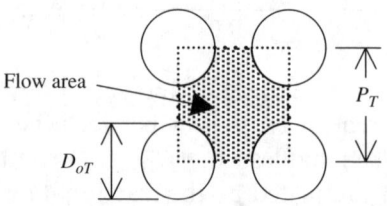

Flow area

P_T

D_{oT}

Figure 22   *Flow area for a square tube layout*

Prandtl Number Equation (11):

$$Pr_S = \frac{c_p\mu}{k} = \frac{2470 \times 0.0004}{0.1329} = 7.44$$

Nusselt Number:

$$Nu_S = 0.36 Re^{0.55} Pr^{0.33} (\mu_b/\mu_w)^{0.14}$$

See Table 2 (note flow is turbulent, $Re \gg 3000$)
$Nu_S = 0.36 \times 28,297^{0.55} \times 7.44^{0.33} = 196$, again neglecting the viscosity correction.

Shell-side heat transfer coefficient:

$$h_S = \frac{Nu_S k}{D_{eS}} = \frac{196 \times 0.1329}{0.0272} = 959 \text{ W m}^2\text{K}^{-1}$$

Overall Heat Transfer Coefficients. Overall resistance to heat transfer based upon the outside area for the clean exchanger:

$$\frac{1}{U_o} = \frac{1}{h_o} + \frac{r_o \ln(r_o/r_i)}{k_w} + \frac{r_o}{r_i h_i} \tag{22}$$

$$\frac{1}{U_o} = \frac{1}{959} + \frac{(0.025/2) \times \ln(0.025/0.020)}{42} + \frac{(0.025/2)}{(0.020/2) \times 926}$$

$$= 0.00104 + 0.00007 + 0.00135 = 0.00246 \text{ m}^2 \text{ K W}^{-1}$$

Therefore, the overall heat transfer coefficient for the clean exchanger is $U_{oc} = 407 \text{ W m}^{-2} \text{ K}^{-1}$.

The overall resistance to heat transfer based upon the outside area, including the fouling coefficients for the dirty exchanger, is

$$\frac{1}{U_o} = \frac{1}{h_o} + R_{do} + \frac{r_o \ln(r_o/r_i)}{k_w} + \frac{r_o R_{di}}{r_i} + \frac{r_o}{r_i h_i} \tag{44}$$

$$\frac{1}{U_o} = \frac{1}{959} + 0.0003 + \frac{(0.025/2) \times \ln(0.025/0.020)}{42}$$

$$+ \frac{(0.025/2) \times 0.0005}{(0.020/2)} + \frac{(0.025/2)}{(0.020/2) \times 926}$$

$$= 0.00104 + 0.00030 + 0.00007 + 0.00063 + 0.00135$$

$$= 0.00338 \text{ m}^2 \text{ K W}^{-1}$$

Therefore, the overall heat transfer coefficient for the dirty exchanger is $U_{oc} = 293$ W m^{-2} K^{-1}.

Viscosity Correction for the Dirty Exchanger. These overall coefficients are based upon the uncorrected film heat transfer coefficients, *i.e.* those without the viscosity correction parameter $(\mu_b/\mu_w)^{0.14}$. As the temperature difference across a layer is proportional to the layer's thermal resistance (see Equations (2) and (10)), the relative resistances, above, allow estimation of the mean tube wall temperature, and hence evaluation of the viscosity corrections $(\mu_b/\mu_w)^{0.14}$ to the film coefficients. Mean bulk temperature difference:

$$\Delta T_b = T_{bS} - T_{bT} = 150.0 - 53.5 = 96.5°C$$

Mean tube-side wall temperature:

$$T_{wT} = T_{bT} + \frac{\text{tube-side film resistance}}{\text{total resistance}} \times \Delta T_b$$

$$= 53.5 + \frac{0.00135}{0.00338} \times 96.5 = 92.0\,°C$$

Mean shell-side wall temperature:

$$T_{wS} = T_{bS} - \frac{\text{shell-side film resistance}}{\text{total resistance}} \times \Delta T_b$$

$$= 150.0 - \frac{0.00104}{0.00338} \times 96.5 = 120.3\,°C$$

Note, in the above analysis for the dirty exchanger the fouling has been included as part of the wall. Neglecting fouling, *i.e.* for a clean exchanger, these temperatures become 106.5°C and 109.2°C, respectively.

	Shell-side	Tube-side
Mean wall temperature (T_w, °C)	120.3	92.0
Viscosity[1] (μ_w, Pa s)	0.0006	0.001
Viscosity correction factor $(\mu_b/\mu_w)^{0.14}$	0.95	1.14
Corrected film coefficients (h, W m^{-2} K^{-1})	913	1040

Corrected overall heat transfer coefficient for the dirty exchanger: $U_{od} = 304$ W m^{-2} K^{-1}

For the clean exchanger, the overall heat transfer coefficient: $U_{oc} = 423$ W m^{-2} K^{-1}.

Calculation of Log Mean Temperature Difference. Log mean temperature difference:

$$\Delta T_{lm} = \frac{(200 - 72) - (100 - 35)}{\ln[(200 - 72)/(100 - 35)]} = 92.9 \text{ K}$$

ΔT_{lm} correction factor F_t, using $R = (200 - 100)/(72 - 35) = 2.70$ and $S = (72 - 35)/(200 - 35) = 0.225$, gives $F_t = 0.922$, from Figure 18.

Calculation of Rate of Heat Transfer. Heat transfer area:

$$A_o = N_T \times \pi \times D_{oT} \times L = 158 \times \pi \times 0.025 \times 4.4 = 54.6 \text{ m}^2$$

Clean exchanger:

$$Q_c = U_{oc} \times A_o \times F_t \times \Delta T_{lm} = 423 \times 54.6 \times 0.922 \times 92.9 = 1979 \text{ kW}$$

Dirty exchanger:

$$Q_d = U_{od} \times A_o \times F_t \times \Delta T_{lm} = 304 \times 54.6 \times 0.922 \times 92.9 = 1422 \text{ kW}$$

Thus, the clean exchanger comfortably exceeds the required duty. The dirty exchanger also exceeds the required duty and so this exchanger seems suitable.

Pressure Drop. All that remains is to check the pressure drops during flow through the exchanger. Kern[1] states that it is customary to allow a pressure drop of 30 to 70 kPa for an exchanger. If the pressure drop is too low, there will be maldistribution of the flow through the exchanger. If it is too high, the pumping/compression costs will become significant.

The flow on the shell-side is complex and the prediction of shell-side pressure drop is beyond the scope of this chapter.[1,3] However, we will estimate the tube-side pressure losses, considering only the losses during flow through the tubes and the tube entrance and exit losses.

It will be assumed that the tubes have an absolute roughness, e, of 0.0015 mm (drawn tubing; Chapter 3, Table 4). At $Re_T = 11,640$ and $e/D_{iT} = 0.000075$, from Chapter 3 Equation (32), or from Chapter 3 Figure 9, $c_f = 0.0075$.

From Chapter 3, Equation (25), the frictional pressure loss in the tubes is

$$\Delta p_f = 2c_f \frac{N_{PT} \times L_T}{D_{iT}} \rho v_T^2$$

$$= 2 \times 0.0075 \times \frac{4 \times 4.4}{0.020} \times 730 \times 1.77^2 = 34.0 \text{ kPa}$$

For each tube pass there will be one entrance loss and one exit loss, 0.5 and 1.0 velocity heads, respectively, (Chapter 3, Table 5). From Chapter 3, Equation (33), these losses are

$$\Delta p_{loss} = N_{PT} \times (0.5 + 1.0) \times \frac{\rho v_T^2}{2} = 4 \times 1.5 \times \frac{730 \times 1.77^2}{2} = 7.7 \text{ kPa}$$

Total tube-side pressure loss $= \Delta p_f + \Delta p_{loss} = 34.0 + 7.7 = 41.7$ kPa

This is a reasonable pressure loss, within the range recommended by Kern[1] and noted above.

The above, simplified analysis used example correlations given earlier in this chapter. Coefficients, assumed constant above, can change significantly with temperature, and especially if there is a change of phase, with boiling and/or condensing on one or both sides of the exchanger. Hence, coefficients can change with position within an exchanger. Often exchangers are divided into sections and heat transfer coefficients are evaluated in each section.

The design of heat exchangers, rather than the rating of an existing one as shown, is an iterative process:

- Guess a value for U and estimate the required heat transfer area.
- Select an exchanger geometry that gives this area.
- Calculate film coefficients, U, and estimate pressure drops.
- Compare calculated U with the assumed value.
- Iterate upon the assumed U and geometry until converged and pressure drops are reasonable.

More detailed analyses and detailed design procedures for various types of heat exchangers are given in the literature.[1,3,7,8]

4.6 CONCLUSIONS

In this chapter, we have looked at the three modes of heat transfer, *viz.*, conduction, convection and radiation, and the laws relating heat fluxes to temperature driving forces for each mode. From these laws we have shown how to estimate heat transfer rates in a range of geometries and situations.

We have introduced the concept of thermal resistances and shown how these may be combined to give overall resistances to heat transfer and overall heat transfer coefficients.

Finally, we looked at a number of common types of heat exchanger along with some example design calculations.

NOMENCLATURE

Variables

Symbol	Description	Dimensions	SI unit
A	Area, surface area for heat transfer	L^2	m^2
A_o	Outside surface area for heat transfer	L^2	m^2
b	Layer/pipe wall thickness	L	m
Bi	Biot number	$-$	$-$
c_f	Fanning friction factor	$-$	$-$
c_P	Heat capacity at constant pressure	$H\,M^{-1}\,\Theta^{-1}$	$J\,kg^{-1}\,K^{-1}$
D	Diameter	L	m
e	Absolute roughness	L	mm
f	Function	n/a	n/a
F_{12}	View factor from 1 to 2	$-$	$-$
Fo	Fourier number	$-$	$-$
F_t	LMTD correction factor	$-$	$-$
g	Gravitational acceleration	$L\,T^{-2}$	$m\,s^{-2}$
G	Mass flow rate	$M\,T^{-1}$	$kg\,s^{-1}$
Gr	Grashof number	$-$	$-$
h	Heat transfer coefficient	$H\,L^{-2}\,T^{-1}\,\Theta^{-1}$	$W\,m^{-2}\,K^{-1}$
i	Electrical current	A	A
k	Thermal conductivity	$H\,L^{-1}\,T^{-1}\,\Theta^{-1}$	$W\,m^{-1}\,K^{-1}$
L	Characteristic length scale, pipe/tube length	L	m
L_T	Pipe/tube length	L	m
N_B	Number of baffles	$-$	$-$
N_{PT}	Number of tube passes	$-$	$-$
N_T	Number of tubes	$-$	$-$
Nu	Nusselt number	$-$	$-$
p	Pressure	$M\,L^{-1}\,T^{-2}$	Pa
Pr	Prandtl number	$-$	$-$
P_T	Tube pitch	$-$	$-$

(*continued*)

q	Heat flux	$H\,L^{-2}\,T^{-1}$	$W\,m^{-2}$
Q	Rate of heat flow	$H\,T^{-1}$	W
R	Electrical resistance	$H\,A^{-2}$	Ω
r	Radius, radial co-ordinate	$L\,T^{-1}$	$m\,s^{-1}$
R	Thermal resistance	$H^{-1}\,L^2\,T\,\Theta$	$m^2\,K\,W^{-1}$
	Parameter in calculation of F_t		
R_d	Thermal resistance of dirt, fouling factor	$H^{-1}\,L^2\,T\,\Theta$	$m^2\,K\,W^{-1}$
Re	Reynolds number	–	–
S	Parameter in calculation of F_t		
T	Temperature	Θ	K
t	Time	T	s
U	Overall heat transfer coefficient	$H\,L^{-2}\,T^{-1}\,\Theta^{-1}$	$W\,m^{-2}\,K^{-1}$
U_o	Overall heat transfer coefficient based upon outside area	$H\,L^{-2}\,T^{-1}\,\Theta^{-1}$	$W\,m^{-2}\,K^{-1}$
V	Electrical potential	$H\,A^{-1}$	V
v	Velocity	$L\,T^{-1}$	$m\,s^{-1}$
V	Voltage	$L\,T^{-1}$	volts
y	Distance in y-direction, distance from (pipe) wall	L	m
α	Absorptivity	–	–
	Thermal diffusivity	$L^2\,T^{-2}$	$m^2\,s^{-1}$
β	Coefficient of thermal expansion	Θ^{-1}	K^{-1}
δ	Distance,	L	m
	Small difference	*n/a*	*n/a*
ε	Emissivity	–	–
μ	Dynamic viscosity	$M\,L^{-1}\,T^{-1}$	$Pa\,s,\,N\,s\,m^{-2}$
ρ	Fluid density	$M\,L^{-3}$	$kg\,m^{-3}$
σ	Stefan–Boltzmann constant	$H\,L^{-2}\,T^{-1}\,\Theta^{-4}$	$W\,m^{-2}\,K^{-4}$
τ	Transmittance	–	–

Subscripts

0	At time $= 0$
1, 2, *etc.*	At points 1, 2, *etc.*

(*continued*)

b	At bulk conditions (temperature)
c	Cold-side, clean, centre
Conv	Convective
d	Dirt, dirty
e	Equivalent hydraulic
f	Frictional
h	Hot-side
i	Inside, inlet
lm	Log-mean
o	Outside, outlet
Rad	Radiatative
surf	Surface
S	Shell, shell-side
T	Tube, tube-side
Tot	Total
w	Of the wall, at wall conditions (temperature)
x	in the x-direction
y	in the y-direction
z	in the z-direction

Symbols

Δ	"The change in"
δ	"Small difference in"

Fundamental Dimensions

M	Mass
L	Length
T	Time
Θ	Temperature
H	Heat
A	Electrical current

REFERENCES

1. D.Q. Kern, *Process Heat Transfer*, McGraw-Hill, New York, 1965.
2. J.M. Coulson, J.F. Richardson, J.R. Backhurst and J.R. Harker, *Coulson & Richardson's Chemical Engineering*, 6th edn, Vol 1, Butterworth-Heinemann, Oxford, 1999.
3. R.K. Sinnott, *Coulson & Richardson's Chemical Engineering*, 3rd edn, Vol 6, Butterworth-Heinemann, Oxford, 1999.

4. J.P. Holman, *Heat Transfer*, McGraw-Hill, New York, 1989.
5. H.S. Carslaw and J.C. Jaeger, *Conduction of Heat in Solids*, 2nd edn, Oxford University Press, Oxford, 1959.
6. E.D. Williamson and H.A. Adams, *Phys. Rev.*, 1919, **14**, 99–114.
7. R.H. Perry and D.W. Green, *Perry's Chemical Engineers' Handbook*, 7th edn, McGraw-Hill, New York, 1997.
8. A. Bejan and A.D. Kraus, *Heat Transfer Handbook*, Wiley, Hoboken, New Jersey, 2003.

CHAPTER 5

An Introduction to Mass-Transfer Operations

EVA SORENSEN

5.1 INTRODUCTION

A significant number of chemical engineering unit operations are concerned with the problem of changing the compositions of solutions and mixtures through methods not necessarily involving chemical reactions. Usually these operations are directed towards separating a substance into its component parts. For mixtures, such separations may be entirely mechanical, *e.g.* the filtration of a solid from a suspension in a liquid, or the separation of particles of a ground solid according to their density. On the other hand, if the operations involve changes in composition of *solutions*, they are knows as *mass-transfer operations* and it is these which concern us in this chapter.

The importance of mass-transfer operations in chemical engineering is profound. There is scarcely any chemical process which does not require either a preliminary purification of raw materials or a final separation of products from by-products, and for these, mass-transfer operations are commonly used. Frequently, the separations constitute the major part of the costs of a process.

The chemical engineer faced with the problem of separating the components of a solution must ordinarily choose from several possible methods. While the choice is usually limited by the specific physical characteristics of the materials to be handled, the necessity for making a decision nevertheless almost always exists. The principle basis for choice in any situation is cost; that method which costs the least is usually the one to be used. Occasionally, other factors also influence the decision, however. The simplest operation, while it may not be the least costly, is sometimes desired because it will be more trouble-free. Sometimes a method will be discarded because of imperfect knowledge of design methods or unavailability of data for design, as results cannot be guaranteed. Favourable previous experience with one method may be given strong consideration.

The choice of the most suitable separation method for a given separation depends on the mixture at hand and the required purity specifications. Although this is the most important decision in the design and operation of a separation process, it will therefore not be considered in this chapter.

In the following, the principles of mass-transfer separation processes will be outlined first. Details of mass-transfer calculations will be introduced next and examples will be given of both equilibrium-stage processes and diffusional rate processes. The chapter will then conclude with a detailed discussion of the two single most applied mass-transfer processes in the chemical industries, namely distillation and absorption.

5.2 MECHANISMS OF SEPARATION

The separation of chemical mixtures into their constituents has been practised as an art for millennia. Separations, including enrichment, concentration, purification, refining and isolation, are important to both chemists and chemical engineers. Chemists use analytical separation methods, such as chromatography, to determine compositions of complex mixtures *quantitatively*. Chemists also use small-scale preparative separation techniques, often similar to analytical methods, to recover and purify chemicals. Chemical engineers, however, are more concerned with the *manufacture* of chemicals using economical, large-scale separation methods, which may differ considerably from laboratory techniques. For example, in a laboratory, chemists separate and analyse light-hydrocarbon mixtures by chromatography, while in a large manufacturing plant, a chemical engineer uses distillation to separate the same mixtures.

The mixing of chemicals to form a mixture is a spontaneous, natural process that is accompanied by an increase in entropy and randomness. The inverse process, the separation of that mixture into its constituent chemical species, is not a spontaneous process; it requires an expenditure of energy. A mixture to be separated usually originates as a single, homogeneous phase (solid, liquid or gas). If it exists as two or more immiscible phases, it is often best to first use some mechanical means, based on gravity, centrifugal force, pressure reduction or an electrical and/or magnetic field, to separate the phases. Then, appropriate separation techniques are applied to each phase.

A schematic diagram of a general separation process is shown in Figure 1. The feed mixture can be vapour, liquid or solid, while the two

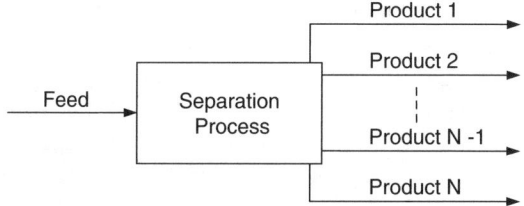

Figure 1 *General separation process*

products may differ in composition and phase state from each other, as well as from the feed.

The separation is accomplished by forcing the different chemical species (components) in the feed into different spatial locations by any of five general separation techniques, or combinations thereof, as shown in Figure 2.[1]

The most common industrial technique involves the creation of a second phase (vapour, liquid or solid) that is immiscible with the feed phase (Figure 2a). The creation is accomplished by energy transfer (heat and/or shaft work) to or from the process or by pressure reduction.

In the second technique, separation by phase addition, the second phase (MSA) is introduced into the system in the form of a solvent that selectively dissolves some of the species in the feed mixture (Figure 2b).

Less common, but of growing importance, is the use of a barrier such as a membrane, which restricts and/or enhances the movement of certain chemical species with respect to other species (Figure 2c).

Also of growing importance are techniques that involve the addition of a solid agent, which acts directly or as an inert carrier for other substances so as to cause separation (Figure 2d).

Finally, external fields of various types are sometimes applied to specialised separations, although this is rare (Figure 2e).

For all of the general techniques of Figure 2, the separations are achieved by enhancing the rate of mass transfer by diffusion of *certain* species relative to mass transfer of *all* species by bulk movement within a particular phase. The driving force and direction of mass transfer by diffusion is governed by thermodynamics, with the usual limitations of equilibrium. Thus, both transport and thermodynamic considerations are crucial in separation operations. The *rate* of separation is governed by mass transfer, while the *extent* of separation is limited by thermodynamic equilibrium. Fluid mechanics also plays an important role, and applicable principles are included in other chapters.

In laboratory analysis, the *extent* of separation is of major importance, *i.e.* is it thermodynamically possible to separate the components,

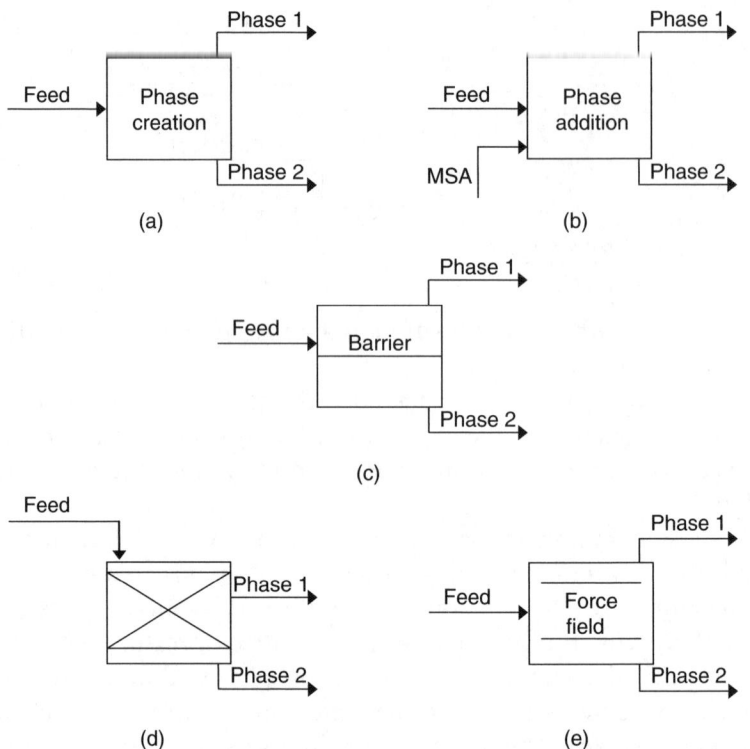

Figure 2   *General separation techniques: (a) separation by phase creation; (b) separation*
by phase addition (MSA: mass separating agent); (c) separation by barrier;
(d) separation by solid agent; (e) separation by force field or gradient

however long this may take? In manufacturing, the extent of separation
will determine if the separation is possible, but the *rate* of mass transfer
will determine if the separation is fast enough to be economically viable.
Chemical engineers strive towards designing separation processes in
such a way as to maximise the rate of mass transfer, as it is generally not
possible to change the thermodynamics, *i.e.* the extent of the separation.

The *extent of separation* achieved between or amongst the product
phases for each of the chemical species present in the feed depends on
the exploitation of differences in molecular, thermodynamic and transport properties of the species in the different phases present. Some
properties of importance are as below.

(i) Molecular Properties: Molecular weight, van der Waals volume,
polarisability, electrical charge.
(ii) Thermodynamic and Transport Properties: Vapour pressure,
solubility, adsorptivity, diffusivity.

Values of these properties for many substances are available in handbooks, specialised reference books and journals. Some of these properties can also be estimated using commercial programmes. When the values are not available, the properties must be estimated or determined experimentally if a successful application of the appropriate separation operation(s) is to be designed.

5.3 GENERAL SEPARATION TECHNIQUES

In the following, the first four separation techniques illustrated in Figure 2 are outlined with examples of separation methods which fall within each category. Most, if not all, of the methods mentioned are described in detail in standard chemical engineering textbooks.[1-6]

5.3.1 Separation by Phase Addition or Phase Creation

The first two separation techniques are very similar and will be considered together. If the feed mixture is a homogeneous, single-phase solution (gas, liquid or solid), a second immiscible phase must often be developed or added before separation of chemical species can be achieved (see Figure 2a and b). This second phase is created by an *energy-separating agent* and/or added as a *mass-separating agent*.

Application of an energy-separating agent involves heat transfer and/or shaft work to or from the mixture to be separated. Alternatively, vapour may be created from a liquid phase by reducing the pressure.

A mass-separating agent may be partially immiscible with one or more of the species in the mixture. In this case, the mass-separating agent frequently remains the constituent with the highest concentration in the added phase. Alternatively, the mass-separating agent may be completely miscible with a liquid mixture to be separated, but may selectively alter the *partitioning* of species between vapour and liquid phases. This facilitates a more complete separation when used in conjunction with an energy-separating agent, such as in extractive distillation.

Table 1 is a compilation of the more common industrial separation operations based on inter-phase mass transfer between two phases, either created by an energy-separating agent or added as a mass-separating agent. A more comprehensive table is given by Seader and Henley.[1] In the following, the operations listed in Table 1 will be outlined briefly. The first two methods, distillation and absorption, will be discussed in more detail later.

Distillation is the most widely utilised industrial separation method. Distillation involves multiple contacts between counter-currently

Table 1 *Common separation operations based on phase creation or addition (MSA: mass-separating agent; ESA: energy-separating agent)*

Separation operation	Initial or feed phase	Created or added phase	Separating agent	Industrial example
Distillation	Vapour and/or liquid	Vapour and liquid	ESA	Purification of styrene
Absorption	Vapour	Liquid	MSA	Separation of CO_2 from combustion products
Liquid–liquid extraction	Liquid	Liquid	MSA	Recovery of aromatics
Drying	Liquid and often solid	Vapour	Gas (MSA) and/or heat transfer (ESA)	Removal of water from polyvinyl-chloride
Crystallisation	Liquid	Solid (and vapour)	ESA	Crystallisation of p-xylene from a mixture with m-xylene
Desublimation	Vapour	Solid	ESA	Recovery of phthalic anhydride from non-condensible gas
Leaching	Solid	Liquid	Liquid solvent	Extraction of sucrose from sugar beets with hot water

flowing liquid and vapour phases. Each contact consists of mixing the two phases to promote rapid partitioning of species by mass transfer, followed by phase separation. The contacts are often made on horizontal trays (referred to as stages) arranged in a vertical column. Vapour, while flowing up the column, is increasingly enriched with respect to the more volatile species. Correspondingly, liquid flowing down the column is increasingly enriched with respect to the less volatile species. Feed to the distillation column enters on a tray somewhere between the top and bottom trays, most often near the middle of the column. The portion of the column above the feed entry is called *the enriching or rectifying* section; and that below is the *stripping* section. Feed vapour will move up the column; feed liquid will move down. Liquid is required for making contacts with vapour above the feed tray, and vapour is required for making contacts with liquid below the feed tray. Normally, vapour from the top of the column is condensed in a condenser by cooling water or a refrigerant to provide contacting liquid, called *reflux*. Similarly, liquid at the bottom of the column passes through a reboiler, where it is heated by condensing steam or some other heating medium to provide contacting vapour, called *boilup*.

In *absorption*, a soluble vapour is absorbed by means of an added liquid mass-separating agent. The constituents of the vapour feed dissolve in the absorbent liquid to varying extents, depending on their solubilities. Vapourisation of a small fraction of the absorbent also generally occurs. The solute is subsequently recovered from the liquid absorbent by distillation, and the absorbing liquid can either be discarded or reused. The operation may not require an energy-separating agent and is frequently conducted at ambient temperature and high pressure.

Liquid–liquid extraction, using one or two solvents respectively, is widely used when distillation is impractical, especially when the mixture to be separated is temperature-sensitive and/or more than 100 distillation stages would be required. When one solvent is used, it selectively dissolves only one or a fraction of the components in the feed mixture. In a two-solvent extraction system, each solvent has its own specific selectivity for dissolving the components of the feed mixture. Additional separation operations are generally required to recover, for recycling, solvent from streams leaving the extraction operation.

Since many chemicals are produced wet but sold in dry, solid forms, one of the more common manufacturing steps is a *drying* operation, which involves removal of a liquid from a solid by vapourisation of the liquid. Although the only basic requirement in drying is that the vapour pressure of the liquid to be evaporated is higher than its partial pressure in the gas stream, the design and operation of dryers represents a complex problem in heat transfer, fluid mechanics and mass transfer.

Crystallisation is carried out in many organic, and almost all inorganic, chemical manufacturing plants, where the desired product is a finely divided solid. Since crystallisation is essentially a purification step, the conditions in the crystalliser must be such that impurities do not precipitate with the desired product.

Sublimation is the transfer of a substance from the solid to the gaseous state without formation of an intermediate liquid phase, usually at a relatively high vacuum. The reverse process, *desublimation*, is also practised, for example, in the recovery of phthalic anhydride from gaseous reactor effluent.

Liquid–solid extraction, often referred to as *leaching*, is widely used in the metallurgical, natural product, and food industries under batch, semi-continuous or continuous operating conditions. The major problem in leaching is to promote diffusion of the solute out of the solid and into the liquid solvent. The most effective way of doing this is to reduce the dimensions of the solid to the smallest feasible particle size. For large-scale applications, large open tanks are used in counter-current operation. The major difference between solid–liquid and liquid–liquid systems centers around the difficulty in transporting the solid, or the solid slurry, from stage to stage. For this reason, the solid may be left in the same tank, with only the liquid transferred from tank to tank. In the pharmaceutical, food and natural-product industries, countercurrent solid transport is often provided by fairly complicated mechanical devices.

5.3.2 Separation by Barrier

The third main class of separation methods, the use of micro-porous and non-porous membranes as semi-permeable barriers (see Figure 2c) is rapidly gaining popularity in industrial separation processes for application to difficult and highly selective separations. Membranes are usually fabricated from natural fibres, synthetic polymers, ceramics or metals, but they may also consist of liquid films. Solid membranes are fabricated into flat sheets, tubes, hollow fibres or spiral-wound sheets. For the micro-porous membranes, separation is effected by differing rates of diffusion through the pores, while for non-porous membranes, separation occurs because of differences in both the solubility in the membrane and the rate of diffusion through the membrane. Table 2 is a compilation of the more common industrial separation operations based on the use of a barrier. A more comprehensive table is given by Seader and Henley.[1]

Table 2   *Common separation operations based on a barrier*

Separation operation	Initial or feed phase	Separating agent	Industrial example
Reverse osmosis	Liquid	Non-porous membrane with pressure gradient	Desalination of sea water
Microfiltration	Liquid	Porous membrane with pressure gradient	Removal of bacteria from drinking water
Ultrafiltration	Liquid	Microporous membrane with pressure gradient	Separation of whey from cheese
Pervaporation	Liquid	Non-porous membrane with pressure gradient	Separation of azeotropic mixtures
Gas permeation	Vapour	Non-porous membrane with pressure gradient	Hydrogen enrichment

Osmosis involves the transfer, by a concentration gradient, of a solvent through a membrane into a mixture of solute and solvent. The membrane is almost non-permeable to the solute. In *reverse osmosis*, transport of solvent in the opposite direction is effected by imposing a pressure, higher than the osmotic pressure, on the feed side. Using a non-porous membrane, reverse osmosis successfully desalts water.

Microporous membranes can be used in a manner similar to reverse osmosis to selectively allow small solute molecules and/or solvents to pass through the membrane and to prevent large dissolved molecules and suspended solids from passing through. *Microfiltration* refers to the retention of molecules typically in the size range from 0.05 to 10 μm. *Ultrafiltration* refers to the range from 1 to 100 nm. To retain even smaller molecules, reverse osmosis, sometimes called hyperfiltration, can be used down to less than 2 nm.

Although reverse osmosis can be used to separate organic and aqueous–organic liquid mixtures, very high pressures are required. Alternatively, *pervaporation* can be used in which the species being absorbed by, and transported through, the non-porous membrane are evaporated. This method, which uses much lower pressures than reverse osmosis, but where the heat of vapourisation must be supplied, is used to separate azeotropic mixtures.

The separation of gas mixtures by selective *gas permeation* through membranes using pressure as the driving force is a relatively simple process. Non-porous polymer membranes are being used commercially to enrich gas mixtures containing hydrogen, to recover hydrocarbons from gas streams and to produce nitrogen-enriched and oxygen-enriched air.

Table 3   *Common separation operations based on a solid agent*

Separation operation	Initial or feed phase	Separating agent	Industrial example
Adsorption	Vapour or liquid	Solid adsorbent	Purification of p-xylene
Chromatography	Vapour or liquid	Solid adsorbent or liquid adsorbent on a solid support	Separation of xylene isomers and ethylbenzene
Ion exchange	Liquid	Resin with ion-active sites	Demineralisation of water

5.3.3   Separation by Solid Agent

Typical separation operations that use solid mass-separating agents are listed in Table 3 (see Figure 2d). The solid, usually in the form of a granular material or packing, acts as an inert support for a thin layer of absorbent, or alternatively, enters directly into the separation operation by selective adsorption of, or chemical reaction with, certain species in the feed mixture. Adsorption is confined to the surface of the solid adsorbent, unlike absorption, which occurs throughout the bulk of the absorbent. In all cases, the active separating agent eventually becomes saturated with solute and must be regenerated or replaced periodically. Such separations are often conducted batch-wise or semi-continuously. However, equipment is available to simulate continuous operation, *e.g.* Simulated Moving Bed (SMB) chromatography.

Adsorption is used to remove components present in low concentrations in non-adsorbing solvents or gases and to separate the components in gas and liquid mixtures by selective adsorption on solids, followed by desorption to regenerate the adsorbent. Typical adsorbents are activated carbon, aluminium oxide, silica gel and synthetic sodium or calcium aluminiosilicate zeolite adsorbents (molecular sieves). Regeneration is accomplished by one of four methods: (1) vapourise the adsorbate with a hot purge gas (thermal swing adsorption), (2) reducing the pressure to vapourise the adsorbate (pressure-swing adsorption), (3) inert purge stripping without change in temperature or pressure and (4) displacement adsorption by a fluid containing a more strongly adsorbed species.

Chromatography is a method for separating the components of a feed gas or liquid mixture by passing the feed through a bed of packing. The feed may be volatilised into a carrier gas, and the bed may be a solid adsorbent (gas or liquid–solid chromatography) or a solid inert support that is coated with a very viscous liquid that acts as an absorbent (gas or liquid–liquid chromatography). Because of selective adsorption on the solid adsorbent surface or absorption into the liquid absorbent, followed by desorption, the different components of the feed mixture move through the bed at different rates, thus effecting the separation.

Ion exchange resembles adsorption in that solid particles are used and regeneration is necessary. However, a chemical reaction is involved. In water softening, a typical ion-exchange application, an organic or inorganic polymer in its sodium form removes calcium ions by exchanging calcium for sodium.

5.4 MASS TRANSFER CALCULATIONS

Mass-transfer calculations, such as the analysis or design of separation units, can be solved by two distinctly different methods, based on either the concept of: (1) Equilibrium stage processes or (2) Diffusional rate processes.

The choice of method depends on the kind of equipment in which the operation is carried out. Distillation, leaching and sometimes liquid extraction, are performed in equipment such as mixed-settler trains, diffusional batteries or plate towers, which contain a series of discrete processing units, and problems in these areas are commonly solved by equilibrium stage calculations. Gas absorption and other operations which are carried out in packed towers or similar equipment, are usually handled using the concept of a diffusional process.

Strictly speaking, all separation processes are diffusional, as one or more of the components in the feed are diffusing from one phase or region to another due to a concentration difference. However, chemical engineers often assume that certain regions of the mass-transfer equipment are at equilibrium. It is important to note that the two calculation methods are just two different ways of *describing* the mass-transfer taking place inside the separation units; the distinction is purely in terms of the way the calculations are carried out when analysing or designing these units. In the following, descriptions will be given of both types of calculations, although examples will only be provided for equilibrium-stage calculations.

5.4.1 Equilibrium-Stage Processes

A large class of mass-transfer devices consists of assemblies of individual units, or stages, that are interconnected so that the materials being processesed pass through each stage in turn (see Figure 3). The two streams move through the assembly either counter-currently, co-currently or cross-currently. In each stage, they are brought into contact, mixed and then separated. Such multistage systems are called *cascades*. The

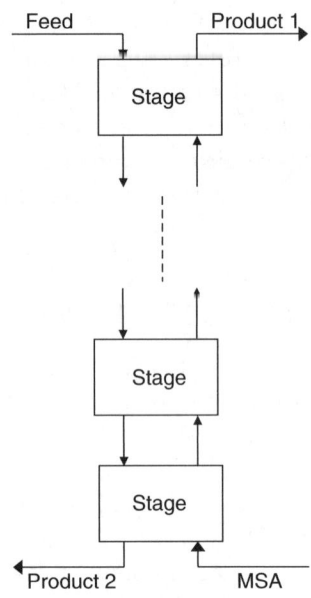

Figure 3   *Cascade of contacting stages (MSA)*

counter-current arrangement is normally preferred, because it generally results in a higher degree of separation than the other two (Figure 3).

For mass transfer to take place, the streams entering each stage must not be in equilibrium with each other, for it is the *departure*, or *deviation*, from equilibrium conditions that provides the driving force for transfer. The leaving streams are usually not in equilibrium either but are much closer to being so than the entering streams are. The closeness of the approach to equilibrium depends on the effectiveness of mixing and mass transfer between the phases. To simplify the calculations for a cascade, the streams leaving each stage are often assumed to be in equilibrium, which, by definition, makes each stage ideal. A correction factor, or *efficiency* factor, is applied later to account for any actual departures from equilibrium.

The cascade shown in Figure 3 consists of a series of single sections or stages with streams entering and leaving only from the ends. Such cascades are used to recover components from a feed stream but are not generally useful for making a sharp separation between two selected feed components. To do this, it is best to provide a cascade consisting of two sections and the counter-current cascade of Figure 4 is often used. It consists of one section above the feed and one below. We will see later how this arrangement is used in distillation.

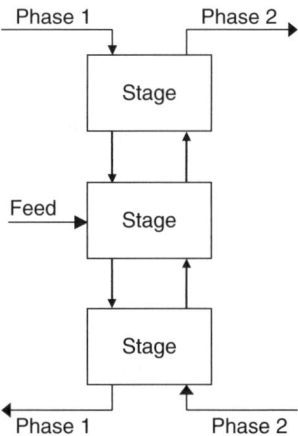

Figure 4 *Two section counter-current cascade*

5.4.2 Stage Calculations

The solution to a multi-component, multi-phase, multi-stage separation problem is found in the simultaneous or iterative solution of the material balances, the energy balance and the phase equilibrium equations (see Chapter 1). This implies that a sufficient number of design variables are specified so that the number of remaining unknown variables exactly equals the number of independent equations. When this is done, a separation process is said to be *specified*.

When considering multistage systems as a whole, attention is focused on the streams passing between the individual stages. An individual unit in a cascade normally receives two streams from the two units adjacent to it (*e.g.* one from the stage above and the other from the stage below), brings them into close contact, and delivers two streams respectively (*e.g.* one to the stage above and one to the stage below). The fact that the contact units may be arranged either one above the other, as in a distillation column, or side by side, as in a stage leaching plant, is important mechanically and may affect some of the details of the operation of individual units. However, the same balance equations are used for either arrangement.

The individual contact units in a cascade are numbered sequentially, starting from one end. In this chapter, the stages are numbered from the bottom and up. A general stage in the system is the nth stage, which is number n counting from the bottom. The stage immediately ahead of stage n in the sequence is stage $n-1$ and that immediately following it is stage $n+1$. To designate the streams pertaining to any one stage, all streams originating in that stage carry the number of the unit as a

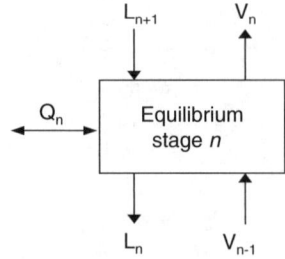

Figure 5   *Equilibrium stage with heat addition*

subscript. Thus, for a two-component system, y_{n-1} is the mole fraction of the lightest component (*i.e.* the one with the lowest boiling point) in the *vapour* phase leaving stage $n-1$, and L_n is the molar flow rate of the L phase leaving the nth stage. For a single equilibrium stage with two entering streams and two exit streams, as shown in Figure 5, the variables are those associated with the four streams plus the heat transfer rate to or from the stage. (Note that some textbooks number the stages from the top and down but the principle of numbering remains the same.)

The total input of material to stage n is $L_{n+1} + V_{n-1}$ units (mass or moles) per time and the total output is $L_n + V_n$ units per time. Since, under steady flow, there is neither accumulation nor depletion, the input and the output are equal and the material balance can be written as:

$$L_{n+1} + V_{n-1} = L_n + V_n$$

This equation is the *total material balance* for the stage. Another balance can be written by equating input to output for component i giving the *component i balance*:

$$L_{n+1}\, x_{i,n+1} + V_{n-1}\, y_{i,n-1} = L_n x_{i,n} = V_n y_{i,n}$$

Obviously, if all the component balances for a system were added up, the sum would be equal to the total material balance.

In most equilibrium stage processes, the general *energy balance* can be simplified by neglecting potential energy and kinetic energy. If, in addition, the process is workless and adiabatic, a simple enthalpy balance applies:

$$L_{n+1}\, H_{L,n+1} + V_{n-1}\, H_{V,n-1} + Q_n = L_n H_{L,n} + V_n H_{V,n}$$

where H_L and H_V are the enthalpies per unit (mass or moles) of the L phase and V phase, respectively, and Q_n is the heat added (negative value if heat is removed).

Examples of stage calculations for the most common chemical engineering separation methods are given later in this chapter.

5.4.3 Graphical Methods

For systems containing only two components it is possible to solve certain simple mass-transfer problems graphically. The methods are based on material balances and equilibrium relationships; some more complex methods require enthalpy balances as well. A typical example is a McCabe–Thiele diagram used for staged distillation calculations as shown in Figure 6. This diagram can be used to find the number of ideal stages necessary for a distillation column to achieve a given separation, as well as the stage number where the feed is to enter, given the binary feed composition and column operating pressure. The method was published in 1925 and, although computer-aided methods are more accurate and easier to apply, the graphical construction of the McCabe–Thiele method greatly facilitates the visualisation of many of the important aspects of multistage distillation, and therefore the effort required to learn the method is well justified. A description of the method is given in all major chemical engineering textbooks.[1,2,4-6]

5.4.4 Diffusional Rate Processes

The second main class of mass-transfer operations take place in packed towers or similar equipment and are usually handled using the concept of a diffusional process. When considering this class of processes, the assumption of staging can no longer be made and the material and

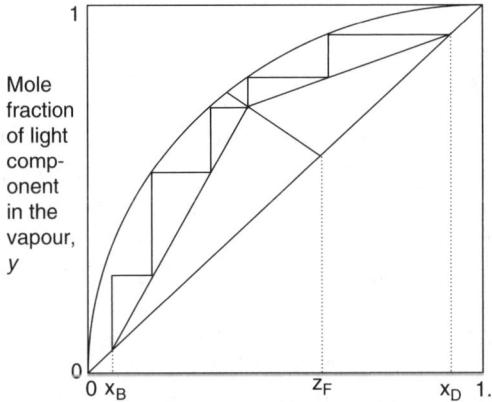

Figure 6 *McCabe–Thiele diagram*

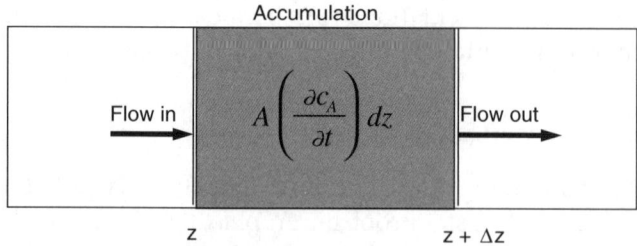

Figure 7   *Diffusion through a differential volume Adz*

energy balances are instead derived for a differential element of the equipment as shown in Figure 7. The solution to the resulting equation set, which consists of several/many partial differential equations, is generally much more complex than those based on the concept of equilibrium stages, which is why chemical engineers normally simplify their calculations by assuming equilibrium stages.

Diffusion is the movement, under the influence of a physical stimulus, of an individual component through a mixture. The most common cause of diffusion is a concentration gradient of the diffusing component. A concentration gradient tends to move the component in such a direction as to equalise concentrations and destroy the gradient. When the gradient is maintained by constantly supplying the diffusing component to the high-concentration end of the gradient and removing the low-concentration end, the flow of the diffusing component is continuous. This movement is exploited in mass-transfer operations. Mass-transfer occurs by two mechanisms: (1) *molecular diffusion* by random and spontaneous movement of individual molecules in a gas, liquid or solid as a result of thermal motion; and (2) *eddy (turbulent) diffusion* by random macroscopic fluid motion.

5.4.5   Diffusion Calculations

In most calculations involving diffusion, attention is focused on diffusion in a direction *perpendicular* to the interface between the two phases and at a definite location in the equipment. Steady state is often assumed, and the concentrations at any point do not change with time. Five inter-related concepts are used in diffusion theory:

- Velocity u, defined as usual by length/time;
- Flux across a plane N, moles/area-time;
- Flux relative to a plane of zero velocity J, moles/area-time;
- Concentration c and molar density ρ_M, moles/volume (mole fraction may also be used);

– Concentration gradient dc/dz, where z is the length of the path perpendicular to the area across which diffusion is occurring.

Fick's law of molecular diffusion states that, for a binary mixture of components A and B, the molar flux of *component* A by ordinary molecular diffusion relative to the molar average velocity of *the mixture* in the positive z direction, is proportional to the concentration gradient dc_A/dz, which is negative in the direction of ordinary molecular diffusion:

$$J_{Az} = -D_{AB} \frac{dc_A}{dz}$$

$$J_{Bz} = -D_{BA} \frac{dc_B}{dz}$$

Here, $D_{AB} = D_{BA}$ are the mutual diffusion coefficients of component A in component B and B in A, respectively.

In all mass-transfer operations, diffusion occurs in at least one phase and often in both phases. The material nature of diffusion and the resulting flow lead to three types of situations:

- *Unimolar Counter-Diffusion.* Only one component A of the mixture is transferred to or from the interface between the two phases, and the total flow is the same as the flow of A, thus:

$$N_B = 0 \text{ and } N = N_A$$

  Absorption of a single component from a gas into a liquid is an example of this type.

- *Equimolar Counter-Diffusion.* The diffusion of component A in a mixture to and from the interface is balanced by an equal and opposite molar flow of component B, so that there is no net molar flow, thus:

$$N = N_A + N_B = 0 \text{ and } J_A = -J_B$$

  This is generally the case in distillation, and it means there is no net volume flow in the gas phase. However, there is generally a net volume or mass flow in the liquid phase because of the difference in molar densities.

- *Uneven Counter-Diffusion.* Diffusion of A and B takes place in opposite directions, but the molar fluxes are unequal. This situation often occurs in diffusion of chemically reacting species to and from a catalyst surface.

In most common separation processes, the main mass transfer is across an interface between a gas and a liquid or between two liquid phases. At fluid–fluid interfaces, turbulence may persist to the interface. A simple theoretical model for turbulent mass transfer to or from a fluid-phase boundary was suggested in 1904 by Nernst, who postulated that the entire resistance to mass transfer in a given turbulent phase lies in a thin, stagnant region of that phase at the interface, called a film, hence the name *film theory*.[2,4,5] Other, more detailed, theories for describing the mass transfer through a fluid–fluid interface exist, such as the *penetration theory*.[1,4]

As mentioned earlier, calculations of diffusional rate processes are difficult as they involve the solution of partial differential equations. Even for processes which are clearly diffusional controlled, such as absorption, chemical engineers normally simplify the calculations by assuming equilibrium stages and may instead correct for possible deviations by using efficiency factors afterwards. Most commercial process design software, such as HYSYS, AspenPlus and ChemCAD, make the assumption of staged equilibrium processes.

5.5 SEPARATION BY DISTILLATION

Distillation is the most mature separation method in industry and its design and operation procedures are well established. Only when vapour–liquid equilibrium or other data are uncertain, are laboratory and/or pilot plant studies necessary prior to the design of a commercial unit.

In distillation, a feed mixture of two or more components is separated into two or more products, including, and often limited to, an overhead distillate and a bottom product, whose compositions differ from that of the feed (see Figure 8). Most often, the feed is a liquid or a vapour–liquid mixture. The bottom product is always a liquid, but the distillate may be a liquid or a vapour or both. The separation requires that: (1) a second phase be formed so that both liquid and vapour phases are present and can contact each other; (2) the components have different volatilities so that they will partition between the two phases to different extents and (3) the two phases can be separated by gravity or other mechanical means.

Distillation differs from absorption in that the second fluid phase is created by thermal means (ESA), vapourisation and condensation, rather than by the introduction of a second phase that usually contains an additional component or components not present in the feed mixture (MSA).

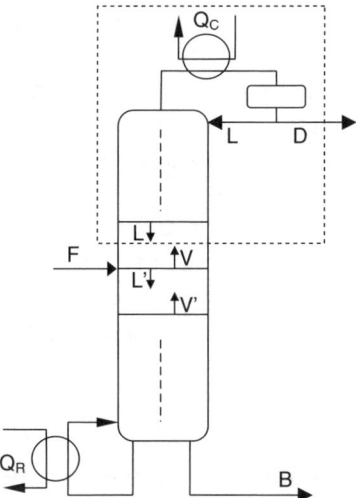

Figure 8   *General distillation column*

Throughout the twentieth century, multistage distillation has been by far the most widely used method for separating liquid mixtures of chemical components. Unfortunately, distillation is a very energy intensive technique, especially when the relative volatility of the components being separated is low (<1.5). The energy consumption for distillation alone in the US for 1976 totalled nearly 3% of the entire national energy consumption.[1] Approximately two-thirds of the distillation energy was consumed by petroleum refining, where distillation is widely used to separate crude oil into petroleum fractions, light hydrocarbons and aromatic chemicals.

5.5.1   Trays and Packing

Distillation has traditionally been carried out in trayed columns. However, more and more frequently, additional distillation capacity is being achieved with existing trayed columns by replacing all or some of the trays by structured packing. The choice between a packed column and a tray-type column is based mainly on economics when factors of contacting efficiency, loadability, and pressure drop must be considered.

Cross flow plates are the most common type of plate contactors used in tray distillation column. In a cross-flow plate, the liquid flows across and down from the plate and the vapour flows up through the plate. The flowing liquid is transferred from plate to plate through vertical channels called downcomers. A pool of liquid is retained on the plate by an outlet weir. Other types of plates, such as bubble-cap and valve trays, are also used.[7–10]

For a packed column, the principle requirements for the packing is that it should (1) provide a large surface area between the gas and liquid; (2) promote uniform liquid distribution on the packing surface; (3) promote uniform vapour gas flow across the column cross-section and (4) have an open structure to give a low resistance to gas flow. Many diverse types and shapes of packing have been developed to satisfy these requirements and details of the most recent ones can be found on the web sites of column internals vendors, *e.g.* Refs. 9 or 10.

5.5.2 Design Range

When designing a distillation column, the main degrees of freedom are:

(i) The height of the column section
(ii) The diameter of the column section

The height of the column section is determined mainly by the number of stages or plates required which in turn depends on how easy or difficult the mixture is to separate, *i.e.* the extent of separation. If the boiling points of the components to be separated are close, then a large number of stages are required leading to a very tall column. The column height is limited, however, by factors such a wind conditions, *etc.*

The diameter of the column section is determined mainly by the amount of material flowing in the column, which in turn depends on the reflux ratio, *i.e.* how much liquid is returned from the top of the column, and the boilup, *i.e.* how high the heat input is at the bottom.

5.5.3 Operating Range

When operating a distillation column, the main *degrees of freedom* are:

(i) The flow rate, composition, temperature, pressure and phase condition of the feed.
(ii) The column pressure.
(iii) The degree of subcooling, if any, in the condenser.
(iv) The reflux ratio, *i.e.* the ratio between the amount of distillate returned to the column as liquid and that taken off as product ($R = L/D$).
(v) The vapour rate given by the heat input to the reboiler.
(vi) The amount of product taken off at the bottom or at the top.

Of these, the feed mixture may or may not vary, but is generally taken as given. The column pressure and the degree of subcooling are normally fairly constant. The main operational variables are the reflux ratio R and the heat input to the reboiler Q_R and once these are set, the amount of product withdrawal at the bottom or at the top will also be given by the product specifications. An optimum exists for the reflux ratio in terms of operating costs, and normally a number of ratios are tested, and the economics of each scenario is investigated, before a decision is reached.

Note that the reflux ratio and the boilup rate cannot be chosen independently as there is a strong interaction between them. The upper limit to the vapour flow is set by the condition of *flooding*. At flooding there is a sharp drop in plate efficiency and increase in pressure drop. Flooding is caused either by the excessive carryover of liquid to the next plate by entrainment, or by liquid backing up in the downcomer. The lower limit of the vapour flow is set by the condition of *weeping*. Weeping occurs when the vapour flow is insufficient to maintain a level of liquid on the plate. *Coning* occurs at low liquid rates, and is the term given to the condition where the vapour pushes the liquid back from the holes and jets upwards with poor liquid contact.[2,3]

5.5.4 Design Calculations

The basis for any design calculations for distillation is the thermodynamic data, *i.e.* the vapour–liquid equilibrium characteristics of the mixture which is to be separated. As mentioned earlier, it is the *extent of separation* which determines whether or not a separation is feasible and how easy or difficult it will be to separate the mixture. Vapour–liquid equilibrium (VLE) data can be found either tabulated in textbooks[7], given by equations or correlations such as Antoine's equation[1,2,4,5] or calculated using commercial programmes. For hand-calculations, VLE data is needed to generate the McCabe–Thiele diagram. If designing using commercial software, most programmes will have VLE databases built in. It is important to note that accurate thermodynamic data is vital for design. If the data is wrong, the design will be wrong, regardless of how accurate the calculations are. Spending time getting reliable thermodynamic data is, therefore, a good investment and is highly recommended.

The equilibrium stage calculations discussed earlier can easily be applied to distillation in order to find the required operating and design parameters. Consider the following problem specification.

5.5.4.1 Problem. A mixture of methanol and water containing 40 mol per cent of methanol is to be separated to give a product of at least 90 mol per cent of methanol at the top, and a bottom product with no more than 10 mol per cent of methanol. The feed flow rate is 100 kmol h^{-1} and the feed is heated so that it enters the column at its boiling point. The vapour leaving the column is condensed, but not sub-cooled, and provides reflux and product. Since all the vapour from the column is condensed, the composition of the vapour from the top plate must equal that of the top product as well as that returned as reflux.

It is proposed to operate the unit with a reflux ratio of 3 kmol kmol^{-1} product (reflux ratio $R = L/D$, where L is the flow returned to the column and D is the distillate product). It is required to find the top and bottom product flow rates, as well as the liquid and vapour flow rates above and below the feed point.

5.5.4.2 Solution. A total material balance over the whole column, where F, D and B are the feed, top product and bottom product flow rates, respectively, yields:

$$F = D + B$$

The component balance for the most volatile component (methanol), where x_F, x_D and x_B are the feed, top product and bottom product mole fractions (of methanol), respectively, gives:

$$F\, x_F = D\, x_D + B\, x_B$$

Combining the equations:

$$B = F(x_F - x_D)/(x_B - x_D) = 100\,(0.4 - 0.9)/(0.1 - 0.9) = 62.5 \text{ kmol h}^{-1}$$

and

$$D = F - B = 100 - 62.5 = 37.5 \text{ kmol h}^{-1}$$

A total mass balance over the top of the column as given by the dotted area in Figure 8 gives:

$$V_n = L_{n+1} + D$$

The reflux ratio is given by $R = L_{n+1}/D = 3$.

Combining the top total material balance with the equation for the reflux ratio, gives an expression for the internal liquid flow L above the feed stage (assuming $L_n = L_{n+1} = L$):

$$V = L + D = RD + D = (R+1)D$$

Hence the internal vapour and liquid flow rates *above* the feed plate are:

$$V = 150.0 \text{ kmol h}^{-1} \text{ and } L = 112.5 \text{ kmol h}^{-1}$$

Since the feed is liquid at its boiling point, it will run down the column together with the liquid from the top part of the column:

$$L' = L + F$$

Hence the internal liquid flow rate *below* the feed plate is:

$$L' = 212.5 \text{ kmol h}^{-1}$$

The vapour flow is unchanged (as the feed is all liquid):

$$V' = 150.0 \text{ kmol h}^{-1} \ (= V)$$

Given the heat of vapourisation of the mixture, the required heat input Q_R (in kW) can be found from the vapour flow rate V. In the calculations above, the reflux ratio was given. As mentioned earlier, different values are typically considered, and the calculations repeated, to determine the reflux ratio which gives the most economical column operation and design.

5.5.5 Distillation Column Height

The height of a distillation column depends on the feed conditions, the product purity specifications and the extent of separation through the vapour–liquid equilibrium relationship, but also on the type of tray or packing used in the column as this affects the *rate* of separation. Column vendors will normally provide information on tray or packing efficiencies.[9,10]

Using design software or McCabe–Thiele's method, the number of stages required for the separation outlined above is five plates. As equilibrium has been assumed when performing the calculations, an efficiency factor needs to be applied, which can range from 0.3 to 1 depending on the type of trays or packing chosen. With an efficiency of 0.5, 10 actual stages would be required in the column section.

The diameter of the column is determined mainly by the vapour flow rate and correlations for this can be found in textbooks. If using commercial programmes, suitable correlations for different trays/packings will normally be incorporated.

5.6   SEPARATION BY ABSORPTION

The removal of one of more selected components from a mixture of gases by absorption into a suitable solvent (Mass Separating Agent, MSA) is the second major operation of chemical engineering after distillation. Absorption is based on interface mass transfer controlled largely by rates of diffusion. It is worth noting that absorption followed by a chemical reaction in the liquid phase is often used to get more removal of a solute from a gas mixture.

In considering the design of equipment to achieve gas absorption, the main requirement is that the gas is brought into intimate contact with the liquid, and the effectiveness of the equipment will largely be determined by the success with which it promotes contact between the two phases. The general form of equipment is similar to that described earlier for distillation and both tray and packed columns are generally used for large installations.[9,10] The method of operation, however, is not the same. In absorption, the feed is a gas and is introduced at the bottom of the column, and the solvent (MSA) is fed to the top as a liquid. The solvent with the absorbed gas leaves at the bottom as liquid and the unabsorbed components leave as gas from the top (see Figure 9).

The essential difference between distillation and absorption is that in the former, the vapour has to be introduced in each stage by partial vapourisation of the liquid, which is therefore at its boiling point, whereas in absorption the liquid is well below its boiling point. In distillation, there is diffusion of molecules in both directions, so that for an ideal system, equimolar counter-diffusion exists. In absorption, gas

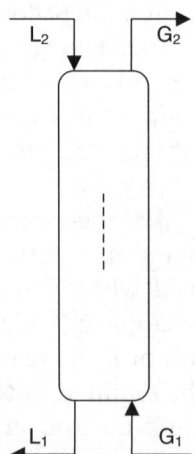

Figure 9   *General absorption column*

molecules are diffusing into the liquid and the movement in the reverse direction is negligible. In general, the ratio of the liquid flow rate to the gas flow rate is considerably greater in absorption than in distillation, with the result that the layout of the columns is different in the two cases. Furthermore, with the higher liquid rates in absorption, packed columns are much more commonly used.

5.6.1 Design Range

When designing an absorption column, the main degrees of freedom are:

(1) The choice of solvent.
(2) The height of the column section.
(3) The diameter of the column section.

One of the main degrees of freedom when designing absorption processes is the choice of solvent. The main criteria are that: (1) the component which is to be removed absorbs well, and fast, into the solvent and (2) the absorption of the other components present in the vapour is negligible. The choice, therefore, depends on the thermodynamic properties of the components in the feed and will not be considered here.

The same issues apply to the height and diameter of an absorption column as for a distillation column; the more difficult the separation, the taller the column and the more liquid and vapour flowing inside the column, the wider the column diameter.

5.6.2 Operating Range

When operating an absorption column, the main *degrees of freedom* are:

(i) The flow rate, composition, temperature and pressure of the gas feed.
(ii) The column pressure.
(iii) The amount of solvent introduced at the top.

The conditions of the gas feed may or may not vary, but is normally taken as given. The column pressure is normally fairly constant. The main operational variable is the amount of MSA required – the more solvent that is used, the more gas will be absorbed. However, the more liquid is present in the column, the wider the column needs to be, which will make the column more expensive. As for distillation, a trade-off therefore exists between operation and design.

5.6.3   Design Calculations

Most absorption operations are carried out in counter-current flow contactors in which the gas phase is introduced at the bottom of the column and the liquid solvent is introduced at the top of the column.

The material balances for an absorption column is normally set up per unit cross-sectional area, *i.e.*

$$G_1 + L_2 = G_2 + L_1$$

where G_1 and G_2 are the vapour flow rate per unit cross-sectional area into the column at the bottom and out of the column at the top, respectively; and L_1 and L_2 are the liquid flow rate per unit cross-sectional area out of the column at the bottom and into the column at the top, respectively. (In other words, index 1 denotes the bottom of the column and index 2 the top of the column. This may vary from textbook to textbook, so always check which is which before using an equation.)

Note that the material balances for an absorption column are normally written on a *solute-free* basis (the solute is the component being absorbed). In other words, we give the flow rates in terms of the components which are *not* being absorbed. This makes the calculations easier as the solute-free flow rates of both gas and liquid in or out of the column are constant.

The solute-free liquid and gas flow rates are given by:

$$L = L_i(1 - x_{A,i}) \text{ where } i = 1 \text{ or } 2, \text{ } i.e. \text{ bottom or top of column;}$$

$$G = G_i(1 - y_{A,i}) \text{ where } i = 1 \text{ or } 2, \text{ } i.e. \text{ bottom or top of column;}$$

where L and G are the solute-free liquid and gas flow rate in or out of the column, respectively; and x_{Ai} and y_{Ai} are the liquid and gas mole fraction of component A (the solute which is absorbed) in or out of the column, respectively.

When setting up the component material balance, we express the amounts of each component in terms of mole *ratios* X_A and Y_A instead of mole *fractions* x_A and y_A:

$$X_A = \frac{x_A}{1 - x_A} = \frac{\text{mole fraction of component A in the liquid}}{\text{mole fraction of non-A components in the liquid}}$$

$$Y_A = \frac{y_A}{1 - y_A} = \frac{\text{mole fraction of component A in the vapour}}{\text{mole fraction of non-A components in the vapour}}$$

The component material balance then becomes:

$$G \ Y_{A1} + L \ X_{A2} = G \ Y_{A2} + L \ X_{A1}$$

5.6.3.1 Problem. A hydrocarbon gas stream is to be purified by continuous counter-current contact with a liquid organic solvent in an absorption column. The inlet gas contains 1.5% by volume of toxic DMSO of which 95% is to be removed. The gas flows at a rate of $G = 0.1 \ \text{kmol} \ \text{s}^{-1}$ on a DMSO-free basis. Calculate

    (i) the mol fractions of vapour y_{Ai} and the corresponding mol ratios
        of gas Y_{Ai}
    (ii) the total gas flow rates (DMSO *and* hydrocarbon), G_i
    (iii) the component A vapour flow rates (DMSO only), G_{Ai}

at the bottom ($i = 1$) and at the top ($i = 2$) of the column.

5.6.3.2 Solution. The conditions of the inlet vapour feed are given. The mole *fraction* of A at the inlet is therefore:

$$y_{Ai} = 1.5 \ \text{vol}\% = 0.015$$

The mole *ratio* of A is:

$$Y_{A1} = \frac{y_{A1}}{1 - y_{A1}} = \frac{0.015}{1 - 0.015} = 0.0152$$

The total vapour flow rate (hydrocarbon *and* DMSO) into the column at the base is:

$$G_1 = G/(1 - y_{A1}) = 0.1 \ \text{kmol} \ \text{s}^{-1}/(1 - 0.015) = 0.10155 \ \text{kmol} \ \text{s}^{-1}$$

The component A gas flow rate (DMSO only) into the column at the base is:

$$G_{A1} = G_1 \ y_{A1} = 0.10155 \ \text{kmol} \ \text{s}^{-1} \ 0.015 = 0.00152 \ \text{kmol} \ \text{s}^{-1}$$

For the top of the column, it is specified that 95% of the DMSO must be removed, *i.e.* only 0.5% remains:

$$G_{A2} = G_{A1} \ 0.05 = 0.00152 \ \text{kmol} \ \text{s}^{-1} \ 0.05 = 7.6 \times 10^{-5} \ \text{kmol} \ \text{s}^{-1}$$

The total gas flow rate (hydrocarbon *and* DMSO) out of the column is then:

$$G_2 = G + G_{A2} = 0.1 \ \text{kmol} \ \text{s}^{-1} + 7.6 \times 10^{-5} \ \text{kmol} \ \text{s}^{-1} = 0.100076 \ \text{kmol} \ \text{s}^{-1}$$

The mole fraction of component A at the top of the column is then:

$$y_{A2} = G_{A2}/G_2 = 7.6 \times 10^{-5} \text{ kmol s}^{-1}/0.100076 \text{ kmol s}^{-1} = 7.6 \times 10^{-4}$$

and the mole ratio of A is:

$$Y_{A2} = \frac{y_{A2}}{1 - y_{A2}} = \frac{7.6 \times 10^{-4}}{1 - 7.6 \times 10^{-4}} = 7.6 \times 10^{-4}$$

5.6.4 Amount of Solvent

As previously mentioned, one of the main degrees of freedom in absorption operations is the amount of solvent which is required to achieve the required absorption. This amount depends on how well the undesired gas component (the solute) is absorbed into the solvent. The vapour–liquid equilibrium relationship for absorption of a gas component into a liquid solvent is expressed as:

$$Y^* = f(X)$$

where Y^* and X are the mole ratio of the absorbing component in the gas and liquid phases at equilibrium, respectively. For some separations, the equilibrium relationship is linear, which greatly simplifies the calculations. For non-linear relationships, graphical calculation methods exists.[2,4,5]

As for distillation, the thermodynamic data form the basis for the design for absorption and the absorption equilibrium data can be found either tabulated, from correlations or from commercial software. Again, good data is vital to ensure good design.

We know that the column will not operate at equilibrium, but to avoid having to use complex calculation methods based on diffusion, we find the solvent flow rate which *would have been* required to achieve the separation had the column operated at equilibrium, and then assume that the actual flow rate will be a fraction larger than this, typically 1.4–1.5 times larger.

5.6.4.1 Problem. Consider the same separation as previously, where a hydrocarbon gas stream is to be purified by continuous counter-current contact with a liquid organic solvent in an absorption column. The inlet gas contains 1.5% by volume of toxic DMSO of which 95% is to be removed. The gas flows at a rate of $G = 0.1$ kmol s^{-1} on a DMSO-free basis.

The organic liquid solvent initially contains 0.001 mole fraction DMSO. The required solvent flow rate is to be 1.5 times the minimum. The gas–liquid equilibrium relationship is given by:

$$Y^* = 0.2X$$

where Y^* is the mole fraction of DMSO in the gas phase and X is the mole ratio of DMSO in the liquid phase.

Calculate

(i) the required solvent flow rate L
(ii) the actual mole fraction of the solvent at the bottom of the column x_{A1}.

5.6.4.2 Solution. The mole ratio of the solvent entering the column at the top is:

$$X_{A2} = \frac{x_{A2}}{1 - x_{A2}} = \frac{0.001}{1 - 0.001} = 0.001001$$

Had the column operated at equilibrium, then the maximum amount of solute would have been removed and we would have used the minimum amount of liquid. Given the gas–liquid equilibrium relationship:

$$Y^* = 0.2X$$

the maximum mole ratio of solute in the solvent, $X_{A1,max}$, is given by:

$$Y_{A1} = Y_{A1}* = 0.2X_{A1,max}$$

therefore (with Y_{A1} found earlier):

$$X_{A1,max} = Y_{A1}/0.2 = 1.52 \times 10^{-2}/0.2 = 7.6 \times 10^{-2}$$

From an overall mass balance between the top and bottom of the column:

$$G\,Y_{A1} + L_{min}\,X_{A2} = G\,Y_{A2} + L_{min}\,X_{A1,max}$$

$$G(Y_{A1} - Y_{A2}) = L_{min}(X_{A1,max} - X_{A2})$$

$$L_{min} = 0.1(1.52 \times 10^{-2} - 7.6 \times 10^{-4})/(7.6 \times 10^{-2} - 1.001 \times 10^{-3})$$

$$L_{min} = 1.9 \times 10^{-2}\ \text{kmol s}^{-1}$$

The actual liquid solvent flow rate which is required is then given by:

$$L = 1.5 \, L_{min} = 2.9 \times 10^{-2} \, \text{kmol s}^{-1}$$

The *actual* liquid mole ratio X_{A1} can then be found from an overall mass balance between the top and bottom of the column:

$$G \, Y_{A1} + L \, X_{A2} = G \, Y_{A2} + L \, X_{A1}$$

$$G(Y_{A1} - Y_{A2}) = L(X_{A1} - X_{A2})$$

$$X_{A1} = X_{A2} + G/L(Y_{A1} - Y_{A2})$$

$$X_{A1} = 1.001 \times 10^{-3} + 0.1/(2.9 \times 10^{-2})(1.52 \times 10^{-2} - 7.6 \times 10^{-4})$$

$$X_{A1} = 0.051$$

5.6.5 Absorption Column Height

The height of an absorption column depends on the feed conditions, the product purity specifications, the solvent used and the extent of separation through the absorption equilibrium relationship, but also on the *rate* of separation. If the rate of mass transfer of the gaseous component from the gas phase into the liquid phase is slow, then the column needs to be longer to ensure that the required amount is removed. The rate of mass transfer depends on the *mass-transfer coefficient*, normally denoted k_G or k_L. The value of the mass-transfer coefficient depends on the components in the gas feed and on the solvent used and is often determined experimentally. The type of packing used in the column will also have an impact on the column height as for distillation.

The diameter of the column is determined mainly by the vapour flow rate and correlations for this can be found in textbooks. If using commercial programmes, suitable correlations for different trays/packings will normally be incorporated.

5.7 FURTHER READING

This chapter has mainly been based on textbooks generally used in undergraduate courses on mass transfer and separation processes.[1–6] Most of these textbooks contain sections on both basic calculations as well as some design considerations. Coulson *et al.*[3] is devoted entirely to design. Perry,[7] in addition to basic theory, also includes thermodynamic data for many common mixtures which may be used for preliminary calculations. Most of the more recent textbooks, such as Seader & Henley,[1] also consider less common separation methods, such as membrane separations, ion exchange and chromatography, in some detail.

Stichlmair and Fair[11] and Rose[8] are textbooks devoted entirely to design and operation of distillation columns, Ruthven[12] considers absorption and Astarita *et al.*[13] absorption with chemical reaction. Ho[14] is a handbook for membrane separations and Guiochon[15] considers chromatographic separations.

5.8 RELEVANT MATERIAL NOT CONSIDERED IN THIS CHAPTER

5.8.1 Combined Reaction and Separation

The possibility of combining a reaction unit and a separation unit into one reactive separation unit has not been considered in this chapter. However, some of the textbooks, Coulson *et al.*[2] and Seader and Henley,[1] give a brief discussion of this type of equipment for distillation and absorption, respectively. A lot of research has also been devoted to this type of operation in recent years and it is generally believed that it will become more widely used in the future.

5.8.2 Combined Separation Units

The possibility of combining two different separation units into one, hybrid, process has not been considered in this chapter. Hybrid processes are quite novel and have only very recently been considered by industry and have, therefore, so far not made it into the standard textbooks. A hybrid process has the combined benefits of both of the component units and the benefits should theoretically outweigh the disadvantages. An example is a hybrid of a distillation column and a pervaporation unit for azeotropic separation, where the distillation unit alone is limited by the azeotropic point. Again, a lot of research is currently devoted to this type of operation and it is generally believed that it will become more widely used in the future.

5.8.3 Batch Operation

In this chapter, continuous operation has been implicitly assumed. However, batch operation is very often used in both the fine chemical and pharmaceutical industries. Analysing and designing batch equipment is generally different from the equivalent continuous units, in that the usual assumption of steady-state operation does not apply. However, the main characteristics of the processes are still the same, such as suitability of plates or packing, trade-offs between reflux ratio and boilup rate in a distillation column, *etc.* Unfortunately, most textbooks, including Perry and Green,[7] contain very little information on batch equipment or batch design.

REFERENCES

1. J.D. Seader and E.J. Henley, *Separation Process Principles*, Wiley, New York, 1998.
2. J.M. Coulson, J.F. Richardson, J.R. Backhurst and J.H. Harker, *Coulson & Richardson's Chemical Engineering, Vol. 2 (Particle Technology and Separation Processes)*, 4th edn (revised), Pergamon Press, Oxford, 1993.
3. J.M. Coulson, J.F. Richardson and R.K. Sinnott, *Coulson & Richardson's Chemical Engineering, Vol. 6 (Design)*, 1st edn (revised), Pergamon Press, Oxford, 1986.
4. W.L. McCabe, J.C. Smith and P. Harriott, *Unit Operations of Chemical Engineering*, 5th edn, McGraw-Hill, Singapore, 1993.
5. R.E. Treybal, *Mass-Transfer Operations*, 3rd edn, McGraw-Hill, New York, 1981.
6. J.R. Welty, C.E. Wicks and R.E. Wilson, *Fundamentals of Momentum*, Heat and Mass Transfer, 3rd edn, Wiley,1984.
7. R.H. Perry and D.W. Green, *Perry's Chemical Engineers' Handbook*, 7th edn, McGraw-Hill, New York, 1997.
8. L.M. Rose, *Distillation Design in Practice*, Elsevier, New York, 1985.
9. Sulzer Chemstech at www.sulzer.com.
10. Koch-Glitsch at www.koch-glitsch.com.
11. J.G. Stichlmair and J.R. Fair, *Distillation – Principles and Practice*, Wiley, New York, 1998.
12. D.M. Ruthven, *Principles of Adsorption and Adsorption Processes*, Wiley, New York, 1984.
13. G. Astarita, D.W. Savage and A. Bisio, *Gas Treating With Chemical Solvents*, Wiley, New York, 1983.
14. W. Ho and K. Sirkar (eds), *Membrane Handbook*, 1st edn, van Nostrand Reinhold, New York, 1992.
15. G. Guiochon, *Fundamentals of Preparative and Nonlinear Chromatography*, Academic Press, Boston, 1994.

Scale-Up in Chemical Engineering

TIM ELSON

6.1 INTRODUCTION AND OBJECTIVES

In only a few cases can theoretical analysis alone provide full description of chemical engineering process units. Normally it is necessary to turn to experimental work, often upon small- or pilot-scale models, to complete such a study. Even if a complete quantitative theory is available, experimental results are still necessary to verify it, since theories are invariably based on assumptions that may not be completely satisfied in the real systems.

Although it is desirable to conduct the experimental work in the system for which the results are required, this is not always easy. The system of interest, or *prototype*, may be hazardous or expensive to build and run, while the fluids involved may be corrosive or toxic. In this case scale *models* are used, which overcome the above problems and allow extensive experimentation. In the majority of cases the model is smaller in size than the prototype. In a number of cases, however, the model and the prototype may be of the same size but the fluids involved are different.

Examples of the use of models for the design of large-scale systems include the measurement of pressure drop and heat transfer in model heat exchangers, the mixing and rate of reaction in a bench-top batch reactor and the prediction of pressure drops in pipelines.

The results from the model will have to be applied to the prototype. Since the model is usually smaller than the prototype, this procedure is called *scale-up*. Scale-up is only possible when both the model and the prototype are *physically similar*.[1-3] Similar comments apply to scale-down, where the model is larger than the prototype.

Engineering data are often presented in dimensionless form. Applying this data to a particular problem in hand is a form of scale-up, the literature data or correlation being the model here. The dimensionless form is obtained by *dimensional analysis*.

The objectives of this chapter are

- to show how dimensionless variables (groups) are obtained from primary variables;
- to show some of the advantages and limitations of dimensional analysis; and
- to show how to obtain and use engineering data in dimensionless form.

6.2   UNITS AND FUNDAMENTAL DIMENSIONS

Chemical engineers and scientists will know that the data they use are expressed in a great variety of different units. The length of a rod may be variously described as 12 in, 1 ft, 0.3048 m, 304.8 mm, *etc*. These lengths are all equivalent. Inch, foot, metre and millimetre define the size of the *unit* and 12, 1, 0.3048 and 304.8 define the number of the units in each system. There are many useful references in the literature that discuss conversion between units and tabulate conversion factors.[1,4,5]

Although the units for each of the measurements above are different, they share the same *fundamental dimension*, length **L**. Each of these measurements has the dimensions of length.

Velocity is the distance travelled per unit time and so has the dimensions of length, **L**, over time, **T**, or $\mathbf{LT}^{-1}$. The quantity of material, given by its mass, has the dimensions of mass, **M**.

Mass, **M**, length, **L** and time, **T**, are called *fundamental dimensions*. With heat transfer problems, temperature, **Θ**, and heat, **H**, are introduced as fundamental dimensions (see Section 6.7.4).

Astarita[6] presents a brief history of the concept of dimensions and dimensional analysis from Euclid to the present day.

6.3   PHYSICAL SIMILARITY

Two systems are said to be *physically similar* in respect to certain specified physical quantities when the ratio of corresponding magnitudes of these quantities between the two systems is the same everywhere.

Table 1 *Types of physical similarity*

Type	Physical quantity	Example system
Geometric	Lengths	Stirred tank
Kinematic	Lengths + time intervals, or velocities	Planetarium, tidal models
Dynamic	Forces	Flow models, wind tunnel
Thermal	Temperature differences	Pilot-plant heat exchanger
Chemical	Concentration differences	Bench scale reactor

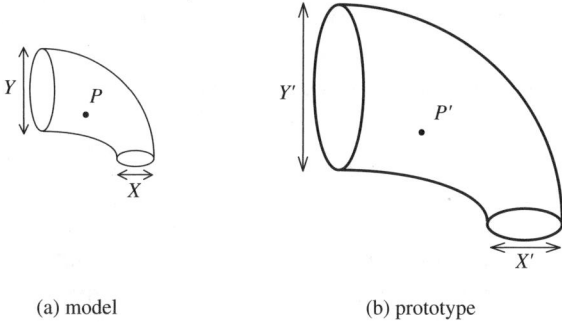

(a) model   (b) prototype

Figure 1 *Geometric similarity*

Table 1 lists the types of physical similarity, and the associated similar quantities, commonly used in chemical engineering.

We will discuss two of these types of physical similarities in the following sections.

6.3.1 Geometric Similarity

Geometric similarity is the most obvious requirement in a model system designed to correspond to a given prototype system. It is the similarity of shape. *Two systems are geometrically similar when the ratio of any length in one system to the corresponding length in another system is everywhere the same.*

Figure 1 illustrates two geometrically similar objects. Here P and P' are known as *corresponding points*; X and X', Y and Y' are known as *corresponding lengths*. The ratio of corresponding lengths is known as the *scale factor*:

$$\text{Scale factor} = \frac{X}{X'} = \frac{Y}{Y'} \tag{1}$$

Perfect geometric similarity is not always easy to achieve and difficulties can arise due to

- scaling of surface roughness or finish,
- scaling of surface tension (*e.g.* scale-down of a river will result in a thin water layer where surface tension effects will be more important than in the river).

6.3.2 Dynamic Similarity

Dynamic similarity is similarity of forces. *A model and prototype are dynamically similar when all forces acting at corresponding points, on fluid elements or corresponding boundaries, form a constant ratio between model and prototype.*

Forces of the same kind, *e.g.* gravitation, centripetal, viscous, *etc.*, acting at *corresponding points* at *corresponding times* are *corresponding forces*.

Dynamic similarity is especially important in fluid flow systems. A fluid's flow pattern is determined by the forces acting on the fluid elements. The net force acting on a fluid element gives the acceleration of that element over the next time interval and, hence, determines its motion. If the net forces on corresponding fluid elements at corresponding times are similar, then their motion will be similar.

Many forces may be involved, *e.g.*

- viscous forces – due to the fluid's viscosity;
- inertial forces – due to its density and velocity;
- pressure forces – due to pressure difference acting over an area of fluid;
- body forces, *e.g.* gravitational – due to earth's gravitational field;
- surface tension forces – due to the presence of interfaces;
- boundary forces, *e.g.* forces imparted by a moving boundary, a rotating impeller.

6.3.3 Dimensionless Groups

Considering all the different types of physical quantities involved in a system, it is obvious that complete physical similarity is very difficult to attain and to ensure. For this reason, *dimensionless groups* are used for scale-up purposes.

If the model and prototype illustrated in Figure 1 are geometrically similar, then the ratios of the lengths, X/Y and X'/Y', are equal. Now, X/Y [=] (length/length), *i.e.* a ratio of lengths and is therefore dimensionless, a dimensionless group. [=] stands for "... *has the dimensions of*...". Thus, "*X* has the dimensions of length" may be abbreviated to "*X* [=] length" or "*X* [=] **L**".

Dimensionless groups allow us to perform scale-up, and we will see a number of examples of this in later sections of this chapter.

6.4 DIMENSIONAL HOMOGENEITY

Take, for illustrative purposes, the equation relating the distance *s*, covered by an object in time *t*, travelling at an initial velocity *u*, when subject to a constant acceleration *a*:

$$s = ut + \frac{1}{2}at^2 \qquad (2)$$

Each term in Equation (2), *s*, *ut* and $\frac{1}{2}at^2$, has the same dimensions,

s [=] length, **L**;
ut [=] velocity × time [=] length/time × time [=] $\mathbf{LT^{-1}}$ × **T** [=] **L**;
and $\frac{1}{2}at^2$ [=] acceleration × time^2 [=] length/time^2 × time^2 [=] $\mathbf{LT^{-2}}$ × $\mathbf{T^2}$ [=] **L**.

Thus Equation (2) is said to be dimensionally homogeneous or dimensionally consistent.

Any equation relating physical quantities must be dimensionally homogeneous or dimensionally consistent. It is not permissible to add a length to a velocity because they are quantities of different types. Even Einstein's $E = mc^2$ is dimensionally homogeneous.

Note that in some equations, the values of constants are a function of the system of units used. These constants have dimensions and are known as *dimensional constants*. For example, many flowmeters measure volumetric flow rates, Q_v, by measuring heads of fluids, Δh. Typically, $Q_v \propto \sqrt{\Delta h}$ or $Q_v = c\sqrt{\Delta h}$. Here, as $Q_v [\neq] \sqrt{\Delta h}$, *c* is a dimensional constant. Dimensionally, $\mathbf{L^3 T^{-1}} [=] c\sqrt{\mathbf{L}}$ or *c* [=] $\mathbf{L^{2.5} T^{-1}}$, and the value of *c* will depend upon the units of both Q_v and Δh.

6.5 DIMENSIONAL ANALYSIS AND DIMENSIONLESS GROUPS

Chemical process systems can be described by a number of physical parameters. In the majority of cases, however, these systems are very complex and the number of parameters needed to describe them very large. Dimensional analysis is used to reduce the number of parameters needed to describe a system without destroying the generality of the relationship.[1,2]

6.5.1 Example: Dimensional Analysis

The distance that an accelerating object covers in a certain time is given by Equation (2), $s = ut + \frac{1}{2}at^2$. So, four variables, s, u, a and t, are needed to describe this system. If we are interested in the distance travelled, we can write

$$s = f(u, t, a) \tag{3}$$

Figure 2 represents Equation (3) graphically, showing curves of s versus t for various values of a at one particular value of u.

To represent the effects of all the possible values of a, we require an infinite number of curves on this figure as a can take an infinite number of values. Now, this is just for one value of u, and u may itself take an infinite number of values. Therefore to represent s for all the possible values of t, a and u, we need an infinite number of figures (all possible u values), each with an infinite number of curves (all possible a values). This, clearly, is not very convenient.

We saw earlier that Equation (2), $s = ut + \frac{1}{2}at^2$, was dimensionally homogeneous; each term has the same dimensions. Ratios of the terms in this equation will therefore be dimensionless. So we can make the original expression dimensionless by, for example, dividing each term

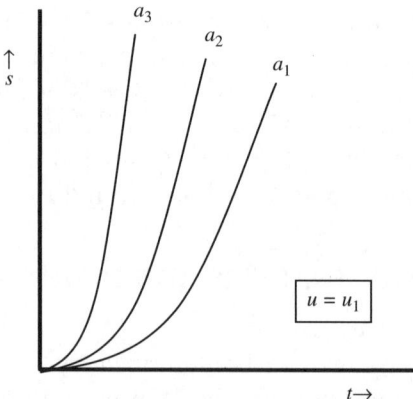

Figure 2   *Distance–time plots*

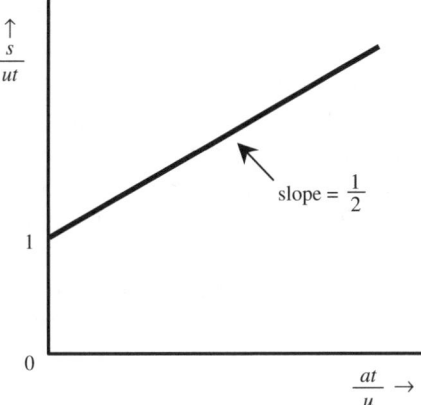

Figure 3 *Dimensionless distance–time plots*

by ut, giving

$$\frac{s}{ut} = 1 + \frac{1}{2}\frac{at}{u} \quad \text{or} \quad \frac{s}{ut} = f\left(\frac{at}{u}\right) \tag{4}$$

where s/ut and at/u are *dimensionless groups*. Plotting s/ut versus at/u, we get Figure 3.

Here we have reduced the number of variables required to describe this system from four to two and the number of curves needed to one (Figure 3). The new variables are the two dimensionless groups in Equation (4). This form is useful when we want to see how distance travelled, s, varies with acceleration, a. Each variable of interest only occurs once and in separate groups.

6.5.1.1 Alternate Form. If instead we were interested in distance travelled s, as a function of initial velocity u, we would wish s and u to appear only once each and in separate groups. So, taking $s = ut + \frac{1}{2}at^2$ and dividing each term by at^2 gives

$$\frac{s}{at^2} = \frac{u}{at} + \frac{1}{2} \quad \text{or} \quad \frac{s}{at^2} = f\left(\frac{u}{at}\right) \tag{5}$$

where s/at^2 and u/at are new dimensionless groups.

6.5.1.2 Points to Note.

- With dimensional analysis an equation can take several equally valid forms, *e.g.* $s/ut = f(at/u)$ or $s/at^2 = f(u/at)$.

- Although four dimensionless groups have been found here, there are only two independent dimensionless groups. The extra, alternate, groups are just products of the powers of the original groups, *i.e.* $s/at^2 = (s/ut) \times (at/u)^{-1}$ and $u/at = (at/u)^{-1}$.
- Dimensional analysis gives fewer independent dimensionless groups (two in this case) than the number of variables (four in this case).
- Dimensional analysis does not give the functional relationship. It does not say what form the function f takes. The functional relationship between the groups must be determined by other theory, experiment or combination of the two.

6.6 BUCKINGHAM Pi THEOREM AND METHOD

In the previous example, the physical relation that described the system was known and dimensional analysis helped to represent experimental data more effectively and concisely. Dimensional analysis is more useful when there is no governing equation for the system. In this case, the dimensionless groups are formed by following a more general procedure known as the *Buckingham method*. The objective of this method is to assemble all the parameters describing the problem into a smaller number of dimensionless groups (Πs). The relation connecting the different groups can be obtained experimentally or, in some cases, from mathematical analysis.

6.6.1 Buckingham Pi Theorem

The Buckingham method is based on the *Buckingham Pi Theorem*, which states

"If a system is characterised by m variables, which are completely described by n fundamental dimensions (**M**, **L**, **T**, *etc.*), then these m variables can be grouped into $m - n$ independent dimensionless groups Πs which describe the same situation."

Thus, if a system is described by m variables Q_1, Q_2, ..., Q_m, then $Q_1 = f(Q_2, Q_3, ..., Q_m)$ or

$$f(Q_1, Q_2, Q_3, ..., Q_m) = 0 \tag{6}$$

and these m variables possess n fundamental dimensions, mass M, length L and time T. With dimensional analysis, the system can be written as

$$f = (\Pi_1, \Pi_2, ..., \Pi_{m-n}) = 0 \tag{7}$$

where, Π_1, Π_2, ..., Π_{m-n} are dimensionless groups.

The number of variables needed to describe the system has been reduced from m to $m-n$.

6.6.2 Buckingham's Method – Procedure

(1) List all m variables describing the problem and their fundamental dimensions.

(2) Find the number of fundamental dimensions, n, that appear and the number, $m - n$, of the dimensionless groups Πs that can be formed.

(3) Select n variables, the *recurring set*. How?
   - Recurring sets must, among them, include all n fundamental dimensions.
   - Selected variables should not have the same dimensions or the same dimensions but to a different power.
   - Experience.
   - Exclude the dependent variable and those variables whose effect one tries to isolate.
   - If a different recurring set is chosen, the resulting dimensionless groups will be different, although these new groups will be products of the powers of the original groups.

(4) The dimensionless groups, Πs, are found one at a time. Take the product of one of the remaining variables to a known power (usually 1) and recurring variables to unknown powers. Determine the exponents of the recurring variables, using the principle of dimensional homogeneity, so that the group becomes dimensionless.

(5) Repeat the procedure in the previous step (4) $m - n$ times to define all the $m - n$ dimensionless groups.

6.6.3 Example: Drag Force on a Sphere

6.6.3.1 Problem Statement. A sphere, illustrated in Figure 4, has a diameter of d and is held inside a duct where fluid flows with a relative velocity u_∞. The fluid has density ρ and viscosity μ. Owing to the relative motion between the fluid and the sphere, a force F, called drag force, is exerted on the sphere. Which are the dimensionless groups that describe this system?

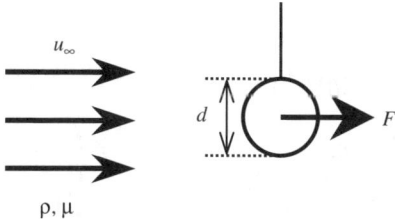

Figure 4 *Drag on a sphere*

6.6.3.2   Buckingham Method.

(1) List all variables m, describing the problem and their fundamental dimensions:

$$\begin{array}{ccccccc}
\textit{Variable} & F & = f & (\ u_\infty & d & \mu & \rho \ ) \\
\textit{Dimensions} & \mathbf{MLT^{-2}} & & \mathbf{LT^{-1}} & \mathbf{L} & \mathbf{ML^{-1}T^{-1}} & \mathbf{ML^{-3}}
\end{array}$$

(2) Find the number of fundamental dimensions, n, that appear and the number, $m - n$, of the dimensionless groups, Πs, that can be formed:

$m = 5$ (F, u_∞, d, μ and ρ), $n = 3$ ($\mathbf{M}$, $\mathbf{L}$ and $\mathbf{T}$); therefore, Πs $= m-n = 5-3 = 2$

(3) Select n variables, the *recurring set*:

Select ρ, d and u_∞, dimensions $[\mathbf{ML^{-3}}]$, $[\mathbf{L}]$ and $[\mathbf{LT^{-1}}]$, respectively

- Recurring set includes all n fundamental dimensions, $\mathbf{M}$, $\mathbf{L}$ and $\mathbf{T}$.
- Set excludes F (the dependent variable) and μ.
- Selected variables do not have the same dimensions.

(4) Find the first dimensionless group:

Take the product of one of the remaining variables (F the dependent variable) to a known power (1) and recurring variables to unknown powers:

$$\Pi_1 = F^1 \times \rho^{a_1} \times d^{b_1} \times u_\infty^{c_1}$$
$$[=](\mathbf{MLT^{-2}}) \times (\mathbf{ML^{-3}})^{a_1} \times (\mathbf{L})^{b_1} \times (\mathbf{LT^{-1}})^{c_1}$$

Equating the powers of each dimension in turn and solving simultaneously for the unknown powers:

$$\left.\begin{array}{lll}
\mathbf{M}: & 0 = 1 + a_1 & a_1 = -1 \\
\mathbf{L}: & 0 = 1 - 3a_1 + b_1 + c_1 & b_1 = -2 \\
\mathbf{T}: & 0 = -2 - c_1 & c_1 = -2
\end{array}\right\} \text{ and so } \Pi_1 = \frac{F}{\rho d^2 u_\infty^2}$$

(5) Repeat to find the second group by taking the one remaining variable (μ):

$$\Pi_2 = \mu^1 \times \rho^{a_2} \times d^{b_2} \times u_\infty^{c_2}$$
$$[=](\mathbf{ML^{-1}T^{-1}}) \times (\mathbf{ML^{-3}})^{a_2} \times (\mathbf{L})^{b_2} \times (\mathbf{LT^{-1}})^{c_2}$$

$$\left.\begin{array}{lll}
\mathbf{M}: & 0 = 1 + a_2 & a_2 = -1 \\
\mathbf{L}: & 0 = -1 - 3a_2 + b_2 + c_2 & b_2 = -1 \\
\mathbf{T}: & 0 = -2 - c_2 & c_2 = -2
\end{array}\right\} \text{ and so } \Pi_2 = \frac{F}{\rho d u_\infty}$$

Now,

$$f(\Pi_1, \Pi_2) = 0 \text{ or } f\left(\frac{F}{\rho d^2 u_\infty^2}, \frac{\mu}{\rho d u_\infty}\right) = 0$$

Thus,

$$\frac{F}{\rho d^2 u_\infty^2} = f\left(\frac{\mu}{\rho d u_\infty}\right) \tag{8}$$

Or, in conventional form,

$$C_D = \frac{8F}{\pi d^2 \rho u_\infty^2} = f\left(\frac{\rho u_\infty d}{\mu}\right) = f(Re) \tag{9}$$

C_D is known at the drag coefficient. Drag coefficients are normally expressed in the form $C_D = (F/A)/(\frac{1}{2}\rho u_\infty^2)$, where A is the projected or cross-sectional area of the particle in the direction of flow. Hence the particular form shown in Equation (9). Re is the particle Reynolds number, $Re = (\rho u_\infty d/\mu)$, and is the reciprocal of the second dimensionless group found above.

Most dimensionless groups can be thought of as ratios:

- Drag coefficient:

$$C_D = \frac{8F}{\mu d^2 \rho u_\infty^2} \equiv \frac{\text{drag forces on particle}}{\text{fluid inertial forces}}$$

- Particle Reynolds number:

$$Re = \frac{\rho u_\infty d}{\mu} \equiv \frac{\text{fluid inertial forces}}{\text{fluid viscous forces}}$$

The Reynolds number characterises the type of flow (see Chapter 3, "*Concepts of Fluid Flow*", for more information).

6.6.3.3 Experimental Results. Figure 5 shows the variation of C_D with Re for a sphere. The curve shown is valid for spheres of any diameter. The form of this chart and the related flow regimes around the sphere are discussed in Chapter 7: "*An Introduction to Particle Systems*" and in many texts.[1,3]

6.7 USE OF DIMENSIONLESS GROUPS IN SCALE-UP

Since the prototype (P) and the model (M) should be physically similar systems, they should be governed by the same equations. These

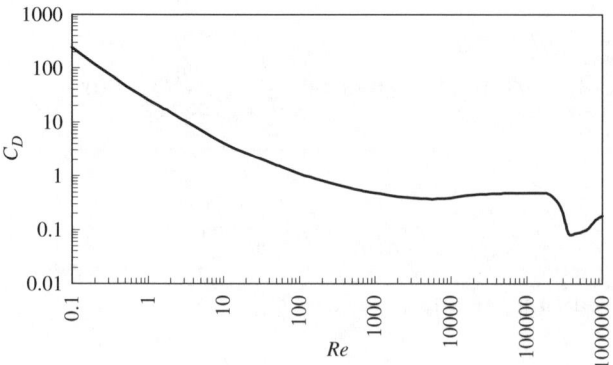

Figure 5   *Drag coefficients for a sphere*

equations can be expressed in terms of dimensionless groups in Equation (7):

$$f(\Pi_1, \Pi_2, \dots, \Pi_{m-n}) = 0 \qquad (7)$$

For the sake of argument, let $m - n = 3$ and so $f(\Pi_1, \Pi_2, \Pi_3) = 0$ or $\Pi_1 = f(\Pi_2, \Pi_3)$. So, if $\Pi_{2_M} = \Pi_{2_P}$ and $\Pi_{3_M} = \Pi_{3_P}$, then Π_{1_M} must equal Π_{1_P}, even though we do not know the form of the function f.

Thus, if we arrange for the values of all but one of the dimensionless groups (Π_2 and Π_3, above) to be equal between model and prototype, then the remaining dimensionless group (Π_1 above) must be equal between model and prototype.

The following sections show examples of scale-up using dimensionless groups.

6.7.1   Drag Force on a Particle

Dimensional analysis (see drag force on a sphere example) yields $C_D = f(Re)$. If we ensure that the value of Re is the same from prototype to model, then we will have dynamic similarity, and therefore C_D must have the same value from model to prototype.

6.7.1.1   Problem.   Calculate the drag force exerted on a spherical particle, diameter 25 mm, as it moves through water at 2 m s^{-1}.

6.7.1.2   Solution.   It is often convenient to tabulate the calculations, showing clearly the values of the variables for prototype and model in separate columns. The solution to this problem is given in Table 2.

Table 2 *Solution, drag force on a particle problem*

Variable	Prototype	Model – graphical data
U	$u_P = 2$ m s^{-1}	
D	$d_P = 0.025$ m	
ρ	$\rho_P = 1000$ kg m^{-3}	
μ	$\mu_P = 0.001$ Pa s	
$Re = \dfrac{\rho \bar{u} d}{\mu}$	$Re_P = \dfrac{1000 \times 2 \times 0.025}{0.001} = 50,000$	$Re_M = 50,000$

If we have dynamic similarity, *i.e.* Re = constant, then C_D must be constant between model (M) and prototype (P)

$$C_D = \frac{8F}{\pi \rho d^2 u^2}$$

Literature data: from Figure 5, at $Re_M = 50,000$, $C_{DM} = 0.48$

$$C_{Dp} = \frac{8F_P}{\pi \rho_P d_P^2 u_P^2}$$

$$0.48 = \frac{8F_P}{\pi \times 1000 \times 0.025^2 \times 2^2}$$

| F | Answer: $F_P = 0.47$ N | |

6.7.1.3 Achievements. Calculation of drag force on prototype based upon graphical experimental model data.

6.7.2 Pressure Drop in a Pipe

The frictional pressure drop per unit length of smooth pipe, $\Delta p_f/l$, is expected to be a function of the pipe diameter d, mean flow velocity $\bar{u}$ across the cross-section and fluid density ρ and viscosity μ (neglects end effects, *i.e.* $l \gg d$). Thus,

$$\frac{\Delta p_f}{l} = f(d, \bar{u}, \rho, \mu) \tag{10}$$

These five variables (considering $\Delta p_f/l$ as a single variable), $m = 5$, may be expressed in terms of $n = 3$ fundamental dimensions, *i.e.* mass, length and time. Using Buckingham's Π theorem, these five variables may be rearranged into $m - n = 5 - 3 = 2$ new dimensionless variables:

$$c_f = \frac{\Delta p_f}{l} \frac{d}{2\rho u^2} = f\left(\frac{\rho \bar{u} d}{\mu}\right) = f(Re) \tag{11}$$

where c_f is the "Fanning" friction factor and *Re* here is the pipe flow Reynolds number.

- Fanning friction factor:

$$c_f = \frac{\Delta p_f}{l}\frac{d}{2\rho u^2} \equiv \frac{\text{frictional forces}}{\text{inertial forces}}$$

- Pipe flow Reynolds number:

$$Re = \frac{\rho \bar{u} d}{\mu} \equiv \frac{\text{inertial forces}}{\text{viscous forces}}$$

Note for different flow situations, the Reynolds number may take different forms. See Chapter 3, "*Concepts of Fluid Flow*" for a discussion of the role of Reynolds number.

6.7.2.1 Problem. Methanol at 50°C flows at a mean velocity of 2 m s^{-1} through a smooth pipe of 25 mm internal diameter and 12 m length. Calculate the frictional pressure drop.

6.7.2.2 Solution. The solution to this problem is given in Table 3.

6.7.2.3 Achievements. Calculation of pipe frictional pressure drop based upon correlated experimental model data.

6.7.3 Modelling Flow Around a Body Immersed in a Fluid

6.7.3.1 Problem. In this example, we consider the flow around a body. Air, at atmospheric pressure, flows at 20 m s^{-1} across a bank of heat exchanger tubes. A 1/10th-scale model is built. At what velocity must air flow over the model bank of tubes to achieve dynamic similarity?

6.7.3.2 Solution. From dynamic similarity,

$$Re_M = Re_P \text{ or } \left(\frac{\rho u l}{\mu}\right)_M = \left(\frac{\rho u l}{\mu}\right)_P \tag{12}$$

where *l* is the characteristic length scale of the body, giving

$$u_M = u_P\left(\frac{l_P}{l_M}\right)\left(\frac{\rho_P}{\rho_M}\right)\left(\frac{\mu_M}{\mu_P}\right) \tag{13}$$

Table 3 *Solution, pressure drop in a pipe problem*

Variable	Prototype	Model–literature correlation
$\bar{u}$	$\bar{u}_P = 2$ m s^{-1}	
D	$d_P = 0.025$ m	
ρ	$\rho_P = 790$ kg m^{-3} (from charts[1])	
μ	$\mu_P = 0.00055$ Pa s (from charts[1])	
$Re = \dfrac{\rho \bar{u} d}{\mu}$	$Re_P = \dfrac{790 \times 2 \times 0.025}{0.00055} = 71,800$	$Re_M = 71,800$

If we have dynamic similarity, *i.e.* Re = constant, then c_f must be constant between model and prototype

$$c_f = \frac{\Delta p_f d}{2\rho u^2 l}$$

Literature data: for smooth pipes at moderate Reynolds numbers (see Chapter 3, *"Concepts of Fluid Flow"*): $c_f = 0.079\ Re^{-0.25}$ giving $c_{f_M} = 4.8 \times 10^{-3}$

$$c_{f_P} = \frac{\Delta p_{f_P} d_P}{2\rho_P u_P^2 l_P}$$

$$4.8 \times 10^{-3} = \frac{\Delta p_{f_P} \times 0.025}{2 \times 790 \times 2^2 \times 12}$$

Δp_f	Answer: $\Delta p_{f_P} = 14,600$ Pa $= 14.6$ kPa

Using the same fluid, air gives $u_M = u_P(l_P/l_M) = 20(1/0.1) = 200$ ms^{-1}.

A u_M value of 200 m s^{-1} is too high. It is approximately 60% of the speed of sound and compressibility effects will be significant, and so we cannot use air.

Let us keep a value of 20 m s^{-1} for u_M and look for a different fluid with suitable properties. From Equation (12),

$$\left(\frac{\mu_M}{\rho_M}\right) = \left(\frac{u_M}{u_P}\right)\left(\frac{l_M}{l_P}\right)\left(\frac{\mu_P}{\rho_P}\right) = \left(\frac{20}{20}\right)\left(\frac{0.1}{1}\right)\left(\frac{\mu_P}{\rho_P}\right) = 0.1\left(\frac{\mu_P}{\rho_P}\right)$$

Thus, we require a fluid with a kinematic viscosity (μ/ρ) one tenth that of air at atmospheric pressure for our experiments. Water at moderate temperatures should be suitable. See Chapter 3, *"Concepts of Fluid Flow"* for information on kinematic viscosity.

6.7.3.3 *Achievements.*

- Found that use of the same fluid for model and prototype was not feasible.
- Determined the fluid physical properties required for the model fluid to achieve dynamic similarity between prototype and model.

6.7.4 Heat Transfer

A novel shell-and-tube heat exchanger is being designed. It is decided to build a small-scale model to determine the tube-side process fluid heat transfer characteristics of this type of heat exchanger.

The rate of heat transfer per unit area of heat exchanger (heat flux), q, will be a function of the temperature driving force ΔT, tube diameter d, the mean fluid flow velocity $\bar{u}$, fluid flow properties – density ρ and viscosity μ – and fluid thermal properties – specific heat capacity c_p and thermal conductivity k.

$$q = f(\Delta T, d, \bar{u}, \rho, \mu, c_p, k) \tag{14}$$

$m = 8$ and $n = 5$ (**M, L, T, H, Θ**), therefore $m - n = 3$.

Note that temperature may be expressed in terms of kinetic energy of molecules. Heat, a form of energy, may be expressed in terms of **M, L** and **T**, $[\mathbf{M\ L^2\ T^{-2}}]$. However, in most heat transfer problems, heat is conserved and is not transformed into other forms of energy. Here we consider heat and temperature as new fundamental dimensions, **H** and **Θ**. This has the advantage of increasing the number of fundamental dimensions, thus reducing the number of dimensionless groups required to describe the problem.

Dimensional analysis yields

$$Nu = f(Re, Pr) \tag{15}$$

- Nusselt number:

$$Nu = \frac{q}{\Delta T}\frac{d}{k} = \frac{hd}{k} \equiv \frac{\text{convective heat transfer rate}}{\text{conductive heat transfer rate}}$$

- Tube flow Reynolds number:

$$Re = \frac{\rho \bar{u} d}{\mu} \equiv \frac{\text{inertial forces}}{\text{viscous forces}}$$

- Prandtl number:

$$Pr = \frac{c_p \mu}{k} \equiv \frac{\text{momentum diffusivity}}{\text{thermal diffusivity}}$$

In the Nusselt number, the term $(q/\Delta T)$, the rate of heat transfer per unit area of heat exchanger per unit temperature driving force, is known as the heat transfer coefficient and is given the symbol h. The heat transfer coefficient is used to characterise heat transfer rates. Heat transfer processes are described in more detail in Chapter 4, "*An Introduction to Heat Transfer*".

6.7.4.1 Problem. (a) A geometrically similar 1/10th-scale model of a prototype heat exchanger has been built and tests are to be performed upon it using the same process fluid, operating at similar temperatures as in the prototype. The prototype is designed to operate at a process fluid flow rate of 300 m³ h⁻¹. What flow rate of fluid through the model will give similar flow conditions to those in the prototype?

6.7.4.2 Solution. Table 4 tabulates the solution to part (a).

6.7.4.3 Alternative Representation. Dynamic similarity, $Re_M = Re_P$, or $(\rho \bar{u} d/\mu)_M = (\rho \bar{u} d/\mu)_P$. Since the physical properties are the same in model and prototype, the above equation becomes $\bar{u}_M d_M = \bar{u}_P d_P$, and as $d_M = 0.1 d_P$, so $\bar{u}_M = (\bar{u}_P d_P / d_M) = (\bar{u}_P d_P / 0.1 d_P) = 10 \bar{u}_P$:

$$Q_M = \bar{u}_M \frac{\pi d_M^2}{4} = 10 \bar{u}_P \frac{\pi (0.1 d_P)^2}{4} = \frac{1}{10} \bar{u}_P \frac{\pi d_P^2}{4}$$

$$= \frac{1}{10} Q_P = \frac{300}{10} = 30 \, \text{m}^3 \, \text{h}^{-1}$$

6.7.4.4 Problem Continued. (b) At the model flow rate obtained above, the tube-side heat transfer coefficient for the model was found to be 250 W m⁻² K⁻¹. What will be the heat transfer coefficient for the prototype?

6.7.4.5 Solution. Table 5 tabulates the solution to part (b).

6.7.4.6 Alternative Representation. As $Re_M = Re_P$ (dynamic similarity achieved with $Q_M = 30$ m³ h⁻¹) and as $Pr_M = Pr_P$ (same fluid), then $Nu_M = Nu_P$, or $(hd/k)_M = (hd/k)_P$. Thus, as we are using the same fluid

Table 4   *Solution to heat transfer problem, part (a)*

Variable	Prototype	Model
Q	$Q_P = 300 \ m^3 \ h^{-1}$	Q_M = unknown
d	d_P	$d_M = 0.1d_P$ 1/10th scale
$\bar{u} = \dfrac{Q}{csa} \propto \dfrac{Q}{d^2}$	$\bar{u}_P \propto \dfrac{Q_P}{d_P^2}$	$\bar{u}_M \propto \dfrac{Q_M}{d_M^2} = \dfrac{Q_M}{(0.1d_P)^2}$
ρ	ρ_P	ρ_P (same fluid)
μ	μ_P	μ_P (same fluid)
c_p	c_{pP}	c_{pP} (same fluid)
k	k_P	k_P (same fluid)
$Re = \dfrac{\rho \bar{u} d}{\mu} \propto \dfrac{\rho(Q/d^2)d}{\mu}$	$Re_P \propto \dfrac{\rho_P(Q_P/d_P^2)d_P}{\mu_P}$	$Re_M \propto \dfrac{\rho_P\left[Q_M/(0.1d_P)^2\right]0.1d_P}{\mu_P}$
		Dynamic similarity, $Re_P = Re_M$, so
		$\dfrac{\rho_P(Q_P/d_P^2)d_P}{\mu_P} = \dfrac{\rho_P\left[Q_M/(0.1d_P)^2\right]0.1d_P}{\mu_P}$
Q		$\dfrac{Q_M}{0.1} = Q_P = 300$
		Answer: $Q_M = 30 \ \text{m}^3 \ \text{h}^{-1}$

Table 5   *Solution to heat transfer problem, part (b)*

Variable	Prototype	Model
h	h_P = unknown	$h_M = 250$ W m^{-2} K^{-1}
$Pr = \dfrac{c_p\mu}{k}$	Pr_P	$Pr_M = Pr_P$ (same fluid)
$Nu = \dfrac{hd}{k}$	$Nu_P = \dfrac{h_P d_P}{k_P}$	$Nu_M = \dfrac{250 \times 0.1d_P}{k_P}$

$Nu = f(Re, Pr)$, as $Re_P = Re_M$ and $Pr_P = Pr_M$, then Nu_P must equal Nu_M

	By Nusselt number similarity, $\dfrac{h_P d_P}{k_P} = \dfrac{250 \times 0.1d_P}{k_P}$	
h	Answer: $h_P = 25$ W m^{-2} K^{-1}	

in model and prototype, $h_M d_M = h_P d_P$:

$$h_P = \frac{h_M d_M}{d_P} = \frac{h_M d_M}{10 d_M} = \frac{h_M}{10} = \frac{250}{10} = 25 \, \text{W m}^2 \, \text{K}^{-1}$$

6.7.4.7 Achievements.

- Determined the conditions under which to operate a geometrically similar model so that physical similarity is obtained with the prototype.
- Determined the heat transfer coefficient for the prototype based upon the measured heat transfer coefficient of the model.

6.7.5 Mass Transfer

Analysis of mass transfer systems is complicated by the range of units in which gas and liquid concentrations may be defined, *e.g.* mass concentration, molar concentration, mass fractions, mole fractions, partial pressures, *etc.* This leads to a wide range of units for the mass transfer coefficient.

Let us consider, as an example, the mass transfer of component "A" from the inside surface of a pipe to the bulk of a liquid flowing through the pipe. The liquid-phase mass transfer coefficient for this case, k_L, is defined by

$$N_A = k_L(c_{A_S} - c_{A_L})$$

where N_A is the molar flux of A [kmol A m^{-2} s^{-1}] and c_A is the molar concentration of A [kmol A m^{-3}] at the liquid surface (S) and in the bulk of the liquid (L). In this case, k_L has the units of m s^{-1}.

The mass transfer coefficient is expected to be a function of

- fluid flow properties, density, ρ, and viscosity, μ
- a characteristic length, the pipe diameter, d
- a characteristic velocity, mean velocity $\bar{u}$
- diffusivity of A in B, D_{AB}

giving $k_L = f(\rho, \mu, d, \bar{u}, D_{AB})$.

Dimensional analysis gives the following dimensionless groups:

$$\frac{k_L d}{D_{AB}} = f\left(\frac{\rho \bar{u} d}{\mu}, \frac{\mu}{\rho D_{AB}}\right)$$

or

$$Sh = f(Re, Sc)$$

- Schmidt number:

$$Sc = \frac{\mu/\rho}{D_{AB}} \equiv \frac{\text{momentum diffusivity}}{\text{mass diffusivity}}$$

- Sherwood number:

$$Sh = \frac{d/D_{AB}}{1/k_L} \equiv \frac{\text{diffusive resistance to mass transfer}}{\text{convective resistance to mass transfer}}$$

6.7.5.1 Example Correlation. Pipe flow, liquids:[7]

$$Sh = 0.023 \, Re^{0.83} Sc^{0.33}$$

(16)

$$2100 < Re < 35,000; \ 0.6 < Sc < 3000$$

6.7.5.2 Problem. (a) A long pipe, of 25 mm internal diameter, has a 25 mm long section of its wall made of solid naphthalene. Estimate the rate of mass transfer from the naphthalene if liquid benzene at 7°C flows through the pipe at a mean velocity of 1 m s^{-1}.

Benzene properties:[8] density, $\rho = 890$ kg m^{-3}; viscosity, $\mu = 0.00076$ Pa s. Naphthalene solubility in benzene:[9] $c_{C_{10}H_{8_S}} = 1.28$ kmol m^{-3}. Naphthalene diffusivity in benzene:[9] $D_{C_{10}H_8-C_6H_6} = 1.19 \times 10^{-9}$ m^2 s^{-1}.

Dimensionless Groups. Reynolds number:

$$Re = \frac{\rho \bar{u} d}{\mu} = \frac{890 \times 1 \times 0.025}{0.00076} = 29,276$$

Schmidt number:

$$Sc = \frac{\mu}{\rho D_{C_{10}H_8-C_6H_6}} = \frac{0.00076}{890 \times 1.19 \times 10^{-9}} = 717.6$$

Sherwood number, from Equation (16) noting that both *Re* and *Sc* are within the specified ranges:

$$Sh = 0.023 \, Re^{0.83} \, Sc^{0.33} = 0.023 \times 29,276^{0.83} \times 717.6^{0.33} = 1026.5$$

Mass Transfer Coefficient.

$$k_{\mathrm{L}} = Sh \times \frac{D_{\mathrm{C_{10}H_8\text{-}C_6H_6}}}{d} = 1026.5 \times \frac{1.19 \times 10^{-9}}{0.025}$$

$$= 4.89 \times 10^{-5}\,\mathrm{m\,s^{-1}}$$

Mass Transfer Rate. The molar flux, $N_{\mathrm{A}} = k_{\mathrm{L}}(c_{\mathrm{C_{10}H_{8_S}}} - c_{\mathrm{C_{10}H_{8_L}}})$, assuming the concentration of naphthalene in the bulk of the liquid phase is negligible, *i.e.* $c_{\mathrm{C_{10}H_{8_L}}} = 0$, becomes

$$N_{\mathrm{A}} = k_{\mathrm{L}} \times c_{\mathrm{C_{10}H_{8_S}}} = 4.89 \times 10^{-5}\,\mathrm{m\,s^{-1}} \times 1.28\,\mathrm{kmol\,m^{-3}}$$

$$= 6.26 \times 10^{-5}\,\mathrm{kmol\,m^{-2}\,s^{-1}}.$$

$$\text{Mass flux} = 6.26 \times 10^{-5}\,\mathrm{kmol\,m^{-2}\,s^{-1}} \times 128.174\,\mathrm{kg\,kmol^{-1}}$$

$$= 0.0080\,\mathrm{kg\,m^{-2}\,s^{-1}}.$$

$$\text{Mass transfer rate} = \text{flux} \times \text{area} = 0.0080\,\mathrm{kg\,m^{-2}\,s^{-1}} \times \pi 0.025^2\,\mathrm{m^2}$$

$$= 1.58 \times 10^{-5}\,\mathrm{kg\,s^{-1}}.$$

Problems with Different Units for k. If we are dealing with gas-phase mass transfer, $N_{\mathrm{A}} = k_{\mathrm{G}}(p_{\mathrm{A_i}} - p_{\mathrm{A_G}})$, where N_{A} is again the molar flux of A [kmol A m^{-2} s^{-1}] and p_{A} is the partial pressure of A in the gas phase (*e.g.* Pa) at the gas–solid interface (i) and in the bulk of the gas (G). In this case, the gas-phase mass transfer coefficient k_{G}, has the units kmol m^{-2} s^{-1} Pa^{-1}, and the Sherwood number becomes $Sh = (k_{\mathrm{G}}RTd/D_{\mathrm{AB}})$, where R has the units J kmol^{-1} K^{-1}.

6.7.5.3 Example Correlation. Pipe flow, gases:[7]

$$Sh = 0.023\,Re^{0.83}\,Sc^{0.44} \tag{17}$$

$$2000 < Re < 35,000;\ 0.6 < Sc < 2.5.$$

6.7.5.4 Problem Continued.
(b) Estimate the rate of mass transfer from the naphthalene if, instead of the liquid benzene in (a) above, air at 40°C and 1 bar flows through the pipe with a mean velocity of 10 m s^{-1}.

Air properties:[8] density, $\rho = 1.11$ kg m^{-3}; viscosity, $\mu = 0.000019$ Pa s
Naphthalene vapour pressure:[9] $p_{\mathrm{C_{10}H_{8_i}}} = 44.72$ Pa
Naphthalene diffusivity in air:[9] $D_{\mathrm{C_{10}H_8\text{-}Air}} = 9.02 \times 10^{-6}$ m^2 s^{-1}

Dimensionless Groups.   Reynolds number:

$$Re = \frac{\rho \bar{u} d}{\mu} = \frac{1.11 \times 10 \times 0.025}{0.000019} = 14,656$$

Schmidt number:

$$Sc = \frac{\mu}{\rho D_{C_{10}H_8\text{-Air}}} = \frac{0.000019}{1.11 \times 9.02 \times 10^{-6}} = 1.89$$

Sherwood No. from Equation (17), noting that both *Re* and *Sc* are within the specified ranges:

$$Sh = 0.023\, Re^{0.83}\, Sc^{0.44} = 0.023 \times 14,656^{0.83} \times 1.89^{0.44} = 87.4$$

Mass Transfer Coefficient.

$$k_L = Sh \times \frac{D_{C_{10}H_8\text{-Air}}}{RTd} = 87.4 \times \frac{9.02 \times 10^{-6}}{8314 \times 313 \times 0.025}$$

$$= 1.21 \times 10^{-8}\, \text{kmol}\, \text{m}^{-2}\, \text{s}^{-1}\, \text{Pa}^{-1}$$

Mass Transfer Rate.   The molar flux, $N_A = k_G(p_{C_{10}H_{8_i}} - p_{C_{10}H_{8_G}})$, assuming the concentration of naphthalene in the bulk of the gas phase is negligible, *i.e.* $p_{C_{10}H_{8_G}} = 0$, becomes

$$N_A = k_G \times p_{C_{10}H_{8_i}}$$

$$= 1.21 \times 10^{-8}\, \text{kmol}\, \text{m}^{-2}\, \text{s}^{-1}\, \text{Pa}^{-1} \times 44.72\, \text{Pa}$$

$$= 5.41 \times 10^{-7}\, \text{kmol}\, \text{m}^{-2}\, \text{s}^{-1}.$$

$$\text{Mass flux} = 5.41 \times 10^{-7}\, \text{kmol}\, \text{m}^{-2}\, \text{s}^{-1} \times 128.174\, \text{kg}\, \text{kmol}^{-1}$$

$$= 6.9 \times 10^{-5}\, \text{kg}\, \text{m}^{-2}\, \text{s}^{-1}.$$

$$\text{Mass transfer rate} = \text{flux} \times \text{area} = 6.9 \times 10^{-5}\, \text{kg}\, \text{m}^{-2}\, \text{s}^{-1} \times \pi 0.025^2\, \text{m}^2$$

$$= 1.36 \times 10^{-7}\, \text{kg}\, \text{s}^{-1}.$$

6.7.5.5   Achievements.   Estimated the rate of mass transfer from a pipe wall, in (a) liquid and (b) gas flows.

These two cases illustrate the use of dimensionless correlations for mass transfer and some of the problems with units in mass transfer.

Correlations for mass transfer coefficients for a wide range of geometries and flow conditions are available in the literature.[7]

6.7.6 Correlation of Experimental Data – Formation of Gas Bubbles at an Orifice

6.7.6.1 Problem. Benzing[10] reports measurements of the diameter of bubbles D, formed in a liquid at an orifice and how they varied with the orifice diameter d, and the physical properties of the liquid. An extract from the data presented is summarised in Table 6.

Carry out a dimensional analysis and correlate this data in a form that is convenient to use if it is required to predict the bubble diameter for a given system. Represent the effects of viscosity and surface tension so that they each appear in only one group.

6.7.6.2 Solution. Now, $D = f(d, \mu, \rho, \sigma, g)$; $m = 6, n = 3$ (**M, L, T**); therefore, $m - n = 3$. The term g is included as gravity (buoyancy) affects the detachment of the bubbles at the orifice. Liquid viscosity, μ, and surface tension, σ, are thought to be significant in the formation of bubbles at orifices, so the recurring set is chosen as d, ρ and g; leaving D – the independent variable – μ and σ are to appear in separate groups. Dimensional analysis gives the following:

$$\frac{D}{d} = f\left(\frac{\mu^2}{d^3 \rho^2 g}, \frac{\sigma}{d^2 \rho g}\right)$$

We have now reduced the number of variables from 6 to 3.

Table 6 *Experimental data of bubble formation at an orifice*[10]

| | | | | Bubble diameter, D (cm) | |
| | | | | Orifice diameter, $d = 0.293$ cm | Orifice diameter, $d = 0.438$ cm |
System	*Liquid viscosity μ* (P)	*Liquid density ρ* (g cm^{-1})	*Surface tension σ* (dynes cm^{-1})		
Water–air	0.0081	0.996	70.2	0.512	0.612
Water–H$_2$	0.0081	0.996	70.2	0.516	0.630
7.5% Ethanol solution–air	0.0107	0.982	51.4	0.480	0.583
45% Sugar solution–air	0.52	1.197	58.1	0.486	0.585
Oil–air	0.57	0.920	36.5	0.411	0.491

Trying a correlation of the form

$$\frac{D}{d} = a\left(\frac{\mu^2}{d^3\rho^2 g}\right)^b \left(\frac{\sigma}{d^2\rho g}\right)^c$$

least-squares analysis of the data in Table 6 yields $a = 1.83$, $b = -0.002$ and $c = 0.276$. The small value of b indicates that the effect of liquid viscosity on bubble formation is small or negligible. Removing the $(\mu^2/d^3\rho^2 g)$ group from the analysis yields

$$\frac{D}{d} = 1.86\left(\frac{\sigma}{d^2\rho g}\right)^{0.281}$$

with negligible change in the variance. Figure 6 compares the experimental values of D/d (circles) with this final correlation (solid line).

6.7.6.3   Achievements.

- Reduced the number of variables for the analysis of some experimental data.
- Produced a correlation of the experimental data in dimensionless form.

Note that Rowe[11] presents a cautionary tale of the dangers of blind usage of the results of least squares correlation of engineering data. He simulated experimental data by selecting values from a table of random numbers and obtained some seemingly reasonable correlations between them. The principal reasons for the apparent reasonableness were

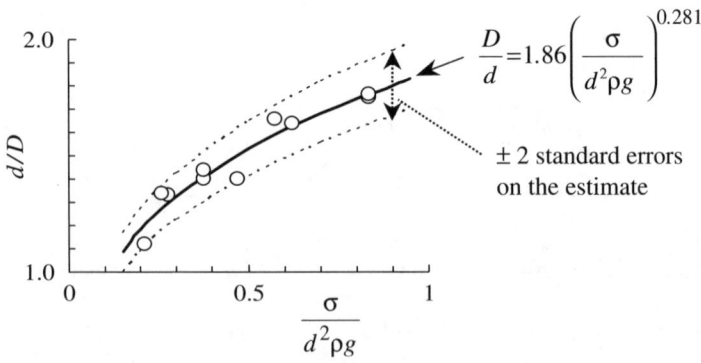

Figure 6   *Correlation of experimental data*

two-fold:

- the use of log–log plots (often used to present dimensionless data), which tend to concentrate data and make the correlation look better; and
- the inclusion of the same primary variable (in separate dimensionless groups) on each axis resulting in a plot of that variable against itself.

6.7.7 Application of Scale-Up in Stirred Vessels

During the mixing of a single-phase liquid in a stirred tank, the important parameters are the rate of power input, P, the impeller stirring speed, N, gravitational acceleration, g, the fluid density, ρ and viscosity, μ, the impeller diameter, D and the other geometric characteristic lengths of the vessel (*i.e.* baffle width and depth, impeller shape, *etc.*).[5,12]

$$P = f(D, N, \rho, \mu, g, \text{geometry}) \qquad (18)$$

Assuming geometric similarity, dimensional analysis yields the following dimensionless groups:

$$Po = f(Re_I, Fr) \qquad (19)$$

- Power number:

$$Po = \frac{P}{\rho N^3 D^5} \equiv \frac{\text{drag forces on impeller}}{\text{inertial forces}}$$

- Impeller Reynolds number:

$$Re_I = \frac{\rho N D^2}{\mu} \equiv \frac{\text{inertial forces}}{\text{viscous forces}}$$

- Froude number:

$$Fr = \frac{N^2 D}{g} \equiv \frac{\text{inertial forces}}{\text{gravitational forces}}$$

The Froude number is associated with the formation of a vortex on the liquid free surface around the impeller shaft. At low impeller speeds

($Re_I < 300$) or when baffles are present in the vessel, a vortex is not formed and the Fr can be omitted from the above relation. The formation of a vortex is best avoided, since it limits power input and can result in poor mixing.

$$Po = f(Re_I \text{ only}) \qquad (20)$$

See Chapter 3, "*Concepts of Fluid Flow*" for experimental details of *Po* versus Re_I relationships.

6.7.7.1   Scale-Up at Constant Re_I.   For a given fluid, $Re_I \propto ND^2 =$ constant, *i.e.* $N \propto D^{-2}$. Scale-up at constant Re_I implies constant *Po*, yielding, for a given fluid, $P \propto N^3D^5$. Therefore, scale-up at constant Re_I gives $P \propto N^3D^5 \propto (D^{-2})^3 \times D^5 \propto D^{-1}$. Power requirement decreases as scale increases! This is not a suitable scale-up criterion.

6.7.7.2   Scale-Up Based on Constant Power Per Unit Volume.   In this approach, the power curve is used for scale-up, similar to single-phase systems. If we assume that mixing takes place in the turbulent region, *Po* $= (P/\rho N^3D^5) =$ constant. For a given fluid, with *Po* constant, $P \propto N^3D^5$ and $(P/V) \propto (N^3D^5/D^3) \propto N^3D^2$. According to this criterion, constant power per unit volume implies that N^3D^2 should be constant between model and prototype.

Table 7 lists these and a number of other scale-up criteria for stirred tanks. Here it is assumed that the fluid is the same on scale up, *i.e.* density and viscosity are both constant, and that the vessel is operating in the fully turbulent, constant power number regime.

6.7.8   Incompatibility of Some Equations

It is decided to model a full-scale prototype, unbaffled, stirred vessel with a one-tenth scale model. The liquid in the prototype has a kinematic viscosity, v, of $10^{-7} \text{ m}^2 \text{ s}^{-1}$. As we have seen above, power number is a function of both Reynolds number and Froude number for unbaffled vessels. To ensure power number similarity, we need to ensure both Reynolds number and Froude number are similar from prototype to model.

Froude number similarity, as g is constant, $N_M^2 D_M = N_P^2 D_P$, Thus,

$$\frac{N_M}{N_P} = \sqrt{\frac{D_P}{D_M}} = \sqrt{\frac{10}{1}} = \sqrt{10}$$

Table 7 *Some scale-up criteria for geometrically similar stirred vessels in the fully turbulent regime[12]*

Scale-up criteria	Keep constant	$P \propto$	Comment
Constant Reynolds number	ND^2	D^{-1}	Unrealistic scale-up criterion, larger scales require less power!
Constant power per unit volume: implies similar turbulence micro-scales	N^3D^2	N^3D^5	May be difficult to achieve due to large power requirements
Constant impeller tip speed: (often used to limit shear damage to sensitive fluids)	ND	D^2	Industrial rule-of-thumb: keep tip speed below 10 m s^{-1} for shear sensitive products
Constant average rate of shear Constant mixing time Chemical reaction: (constant residence time distribution and constant mean residence time)[13]	N	D^5	May be impossible to achieve due to large power requirements
Constant heat transfer coefficient h: $Nu \propto Re^{2/3}$	$N^{2/3}D^{1/3}$	$D^{3.5}$	May be difficult to achieve due to large power requirements
Solids suspension: speed to "just suspend" solids[14]	$ND^{0.85}$	$D^{2.45}$	One of many correlations for particle suspension

Reynolds number similarity:

$$(ND^2/\nu)_M = (ND^2/\nu)_P$$

rearranging

$$\nu_M = \nu_P \frac{N_M}{N_P} \left(\frac{D_M}{D_P}\right)^2 = 1 \times 10^{-7} \times \sqrt{10} \left(\frac{1}{10}\right)^2$$

$$= 3 \times 10^{-9}\, \text{m}^2\, \text{s}^{-1}$$

Thus we require a model liquid with a kinematic viscosity of 3×10^{-9} m^2 s^{-1}. There is no such liquid.[9]

6.7.8.1 Notes. In many situations, it may be impossible to maintain similarity between all dimensionless groups. Therefore, one should try to maintain similarity with the most important dimensionless group and take into account the effect(s) of the other group(s) in some other way.

6.8 CONCLUSIONS

- Dimensional analysis is a powerful tool.
- Dimensional analysis reduces the number of variables to be analysed in a problem. $Q_1, Q_2, Q_3, \ldots, Q_m$ is reduced to $\Pi_1, \Pi_2, \Pi_3, \ldots, \Pi_{m-n}$.
- Dimensional analysis tells us that the new, dimensionless variables are related in some way, $\Pi_1 = f(\Pi_2, \Pi_3, \ldots, \Pi_{m-n})$.
- Dimensional analysis does not tell us the form of the function f.
- Selection of variables $Q_1, Q_2, Q_3, \ldots, Q_m$?
  - experience
  - too few, data will not correlate
  - too many, redundant dimensionless groups will appear
- Multiple forms of Πs. Selection of different recurring sets within the rules will give different dimensionless groups, although these new groups will be products of the powers of the original groups. Any number of dimensionless groups can be formed in this way.
- Similarity of dimensionless groups may be used for scale-up of physically similar systems.
- It may not be feasible to maintain similarity between all dimensionless groups on scale-up.

NOMENCLATURE

Variables

Symbol	Description	Dimensions	SI unit
a	Acceleration	$L\,T^{-2}$	$m\,s^{-2}$
a, b, c	Constants	–	–
A	Area, cross-sectional area	L^2	m^2
A,B	Component A,B	–	–
c	Molar concentration	$M\,L^{-3}$	$kmol\,m^{-3}$
c_p	Specific heat capacity	$H\,M^{-1}\,\Theta^{-1}$	$J\,kg^{-1}\,K^{-1}$
d	Diameter, pipe diameter	L	m
D	Impeller diameter	L	m
D_{AB}	Diffusivity of component A in component B	$L^2\,T^{-1}$	$m^2\,s^{-1}$

(*continued*)

f	Function	n/a	n/a
F	Force	$\mathbf{M\ L\ T^{-2}}$	N
g	Gravitational acceleration	$\mathbf{L\ T^{-2}}$	$m\ s^{-2}$
h	Heat transfer coefficient	$\mathbf{H\ L^{-2}\ T^{-1}\ \Theta^{-1}}$	$W\ m^{-2}\ K^{-1}$
k	Thermal conductivity	$\mathbf{H\ L^{-1}\ T^{-1}\ \Theta^{-1}}$	$W\ m^{-1}\ K^{-1}$
k_G	Gas-phase mass transfer coefficient	$\mathbf{L^2\ T^{-1}}$	$m^2\ s^{-1}$
k_L	Liquid-phase mass transfer coefficient	*Various*	*Various*
l	Characteristic length, pipe length	$\mathbf{L}$	m
m	Number of variables	–	–
n	Number of fundamental dimensions	–	–
N	Rate of rotation	$\mathbf{T^{-1}}$	Hz
	Molar flux	$\mathbf{M\ L^{-2}\ T^{-1}}$	$kmol\ m^{-2}\ s^{-1}$
p	Pressure, partial pressure	$\mathbf{M\ L^{-1}\ T^{-2}}$	Pa
P	Power	$\mathbf{M\ L^2\ T^{-3}}$	W
q	Heat flux	$\mathbf{H\ L^{-2}\ T^{-1}}$	$W\ m^{-2}$
Q	Quantity	n/a	n/a
Q_v	Volumetric flow rate	$\mathbf{L^3\ T^{-1}}$	$m^3\ s^{-1}$
R	Universal gas constant	$\mathbf{L^2\ T^{-2}\ \Theta^{-1}}$	$J\ kmol^{-1}\ K^{-1}$
s	Distance moved	$\mathbf{L}$	m
t	Time	$\mathbf{T}$	s
T	Absolute temperature	$\mathbf{\Theta}$	K
u	Initial velocity	$\mathbf{L\ T^{-1}}$	$m\ s^{-1}$
u_∞	Velocity relative to infinity or undisturbed flow	$\mathbf{L\ T^{-1}}$	$m\ s^{-1}$
$\bar{u}$	Average velocity across a cross-section	$\mathbf{L\ T^{-1}}$	$m\ s^{-1}$
V	Volume	$\mathbf{L^3}$	m^3
X	Length	$\mathbf{L}$	m
Y	Length	$\mathbf{L}$	m
β	Coefficient of thermal expansion	$\mathbf{\Theta^{-1}}$	K^{-1}
μ	Dynamic viscosity	$\mathbf{M\ L^{-1}\ T^{-1}}$	$Pa\ s,\ N\ s\ m^{-2}$
Π	Dimensionless group	–	–
ρ	Fluid density	$\mathbf{M\ L^{-3}}$	$kg\ m^{-3}$
σ	Surface tension	$\mathbf{M\ T^{-2}}$	$N\ m^{-1}$

Subscripts

f	Frictional
G	Gas, gas-phase
I	Impeller

(continued)

i	Interface
L	Liquid, liquid-phase
M	Model
P	Prototype
s	Sphere
S	Surface

Symbols

Δ	"The change in ..."
[=]	"Has the dimensions of ..."

Fundamental Dimensions

M	Mass
L	Length
T	Time
Θ	Temperature
H	Heat

Dimensionless Groups

Table 8 lists some common dimensionless groups used in chemical engineering. More complete lists of dimensionless groups are available in the literature.[15–17]

Table 8   *Some common dimensionless groups used in chemical engineering*

Symbol	Name	Formula	Ratio	Usage
C_D	Drag coefficient	$\dfrac{F}{A\frac{1}{2}\rho u^2}$	$\dfrac{\text{drag force on body}}{\text{inertial forces}}$	Drag forces on bodies
c_f	Friction factor	$\dfrac{\Delta p_f d}{2l\rho u^2}$	$\dfrac{\text{frictional forces}}{\text{inertial forces}}$	Fluid friction in pipes, *etc.*
Fl	Flow	$\dfrac{Q_v}{ND^3}$	$\dfrac{\text{volumetric flow rate}}{\text{rate of swept volume}}$	Flow created by impellers, fans and pumps
Fr	Froude	$\dfrac{u^2}{lg}$	$\dfrac{\text{inertial forces}}{\text{gravitational forces}}$	Free surfaces
Gr	Grashof	$\dfrac{l^3\rho^2\beta g\Delta T}{\mu^2}$	$\dfrac{\text{buoyancy} \times \text{inertial forces}}{\text{viscous forces}^2}$	Natural convection heat transfer
Le	Lewis	$\dfrac{Sc}{Pr}$	$\dfrac{\text{thermal diffusivity}}{\text{mass diffusivity}}$	Simultaneous heat and mass transfer

Table 8 (*continued*)

Symbol	Name	Formula	Ratio	Usage
Nu	Nusselt	$\dfrac{hl}{k}$	$\dfrac{\text{convective heat transfer}}{\text{conductive heat transfer}}$	Heat transfer and flow
Pe	Peclet (heat)	$Re\ Pr$		Heat transfer and flow
	Peclet (mass)	$Re\ Sc$		Mass transfer and flow
Po	Power	$\dfrac{Po}{\rho N^3 D^5}$	$\dfrac{\text{drag force on impeller}}{\text{inertial forces}}$	Power consumed by impellers, fans and pumps
Pr	Prandtl	$\dfrac{c_p \mu}{k}$	$\dfrac{\text{momentum diffusivity}}{\text{thermal diffusivity}}$	Heat transfer and flow
Re	Reynolds	$\dfrac{\rho u l}{\mu}$	$\dfrac{\text{inertial forces}}{\text{viscous forces}}$	Fluid flow
Sc	Schmidt	$\dfrac{\mu}{\rho D_{AB}}$	$\dfrac{\text{momentum diffusivity}}{\text{mass diffusivity}}$	Mass transfer and flow
Sh	Sherwood	$\dfrac{k_L l}{D_{AB}}$	$\dfrac{\text{diffusive resistance to mass transfer}}{\text{convective resistance to mass transfer}}$	Mass transfer and flow
St	Stanton	$\dfrac{Nu}{Re\ Pr}$		Heat transfer and flow
We	Weber	$\dfrac{\rho u^2 l}{\sigma}$	$\dfrac{\text{inertial forces}}{\text{surface tension forces}}$	Flow with interfaces

REFERENCES

1. J.M. Coulson and J.F. Richardson, in *Chemical Engineering*, Chapter 1, 5th edn, vol 1, J.R. Backhurst and J.H. Harker (eds), Elsevier Butterworth-Heinemann, Oxford, 1999.
2. B.S. Massey, in *Mechanics of Fluids*, Chapter 5, 8th edn, J. Ward-Smith (ed), Taylor & Francis, Abingdon, Oxford, 2006.
3. J.R. Welty, C.E. Wicks, R.E. Wilson and G. Rorrer, *Fundamentals of Momentum, Heat and Mass Transfer*, Chapter 11, 4th edn, Wiley, New York, 2001.
4. J.W. Mullin, *AIChE J.*, 1972, **18**(1), 222.
5. R.H. Perry and D. Green, *Perry's Chemical Engineers' Handbook*, Section 1, 7th edn, McGraw Hill, New York, 1997.
6. G. Astarita, *Chem. Eng. Sci.*, 1997, **54**(24), 4681.

7. R.H. Perry and D. Green, *Perry's Chemical Engineers' Handbook*, Section 5, 7th edn, McGraw Hill, New York, 1997.
8. R.H. Perry and D. Green, *Perry's Chemical Engineers' Handbook*, Section 2, 7th edn, McGraw Hill, New York, 1997.
9. DETHERM, *Thermophysical Properties Database*, Vn. 2.1.0.0, DECHEMA e.V., 2006.
10. R.J. Benzing and J.E. Myers, *Ind. Eng. Chem.*, 1955, **47**, 2087.
11. P.N. Rowe, *Trans. Inst. Chem. Eng.*, 1963, **41**, CE 70.
12. N. Harnby, M.F. Edwards and A.W. Nienow, *Mixing in the Process Industries*, 2nd edn, Butterworth-Heinemann, Oxford, 1997.
13. A.W. Nienow, *Chem. Eng. Sci.*, 1974, **29**, 1043.
14. Th.N. Zwietering, *Chem. Eng. Sci.*, 1958, **8**, 244.
15. R.H. Perry and D. Green, *Perry's Chemical Engineers' Handbook*, Section 6, 7th edn, McGraw Hill, New York, 1997.
16. D.F. Boucher and G.E. Alves, *Chem. Eng. Prog.*, 1959, **55**(9), 55.
17. J.P. Catchpole and G. Fulford, *Ind. Eng. Chem.*, 1966, **58**(3), 46.

CHAPTER 7

An Introduction to Particle Systems

PAOLA LETTIERI

7.1 INTRODUCTION

Particles and processes involving particles are of enormous importance in the chemical and allied industries. It has been estimated that particulate products, which include pharmaceuticals, detergents, agrochemicals, pigments and plastics, generate $1000M annually for the US economy and Du Pont estimates that 60% of its manufactured products rely heavily on particle science and technology. Despite this, particle technology is a relatively new subject and although the science is advancing rapidly there is still much reliance on empiricism.

The behaviour of powders is often quite different from the behaviour of liquids and gases. Engineers and scientists are used to dealing with liquids and gases whose properties can be readily measured, tabulated and even calculated. With particle systems the picture is quite different. The flow properties of certain powders may depend not only on the particle size, size distribution and shape, but also on surface properties, on the humidity of the atmosphere and the state of compaction of the powder. These variables are not easy to characterise and so their influence on the flow properties is difficult to predict. In the case of particle systems it is almost always necessary to perform appropriate measurements on the actual powder in question rather than to rely on tabulated data. The measurements made are generally measurements of bulk properties, such as bulk density and shear stress.

Given the wide range of industrial applications that involve particle systems, the influence of the operating conditions such as temperature, pressure, velocities, reactor design and any other special conditions, such as the presence of liquid in the reactor, can also significantly affect the fluid–particle contacting efficiency and the interaction between particles, thus affecting the overall flow properties of the powders.

In light of the complexity of such systems, research in particle technology has been devoted to the development of both experimental techniques and theoretical methodologies to determine the particle properties, the fluid–particle and particle–particle interactions. Substantial mathematical effort is currently being directed towards the development of computational fluid dynamics (CFD) modelling as a new tool to support engineering design and research in multiphase systems. The recent development of mathematical modelling of particle systems behaviour, together with the increased computing power, will enable us to link fundamental particle properties directly to powder behaviour and predict the interaction between particles and gaseous or liquid fluids.

The aim of this chapter is to provide a brief introductory guide to the:

(i) Characterisation of solid materials (size analysis, shape and density);
(ii) Interaction between particles and a fluid;
(iii) Fluidisation, in relation to fluidised beds (as an example of a common particle process).

A number of relevant worked examples are also presented, to provide the reader with the direct application of the theory presented.

7.2 CHARACTERISATION OF SOLID MATERIALS

The starting point of any study of particle systems is the characterisation of particles in terms of their size, shape and density. Figure 1 shows the range of sizes of some commonly encountered particulate materials compared with other quantities, such as the wavelengths of electromagnetic radiation and the sizes of molecules.

7.2.1 Size and Size Distribution

Only a minority of systems of industrial interest contain powders with a uniform particle size, *i.e. monodisperse*. Most systems generally show a distribution of sizes (*polydisperse*) and it is then necessary to define the average dimension. There are many different definitions of particle size,[1,2] the most commonly used, particularly in fluidisation, is the so called *volume-surface* mean or the *Sauter mean* diameter. This is the

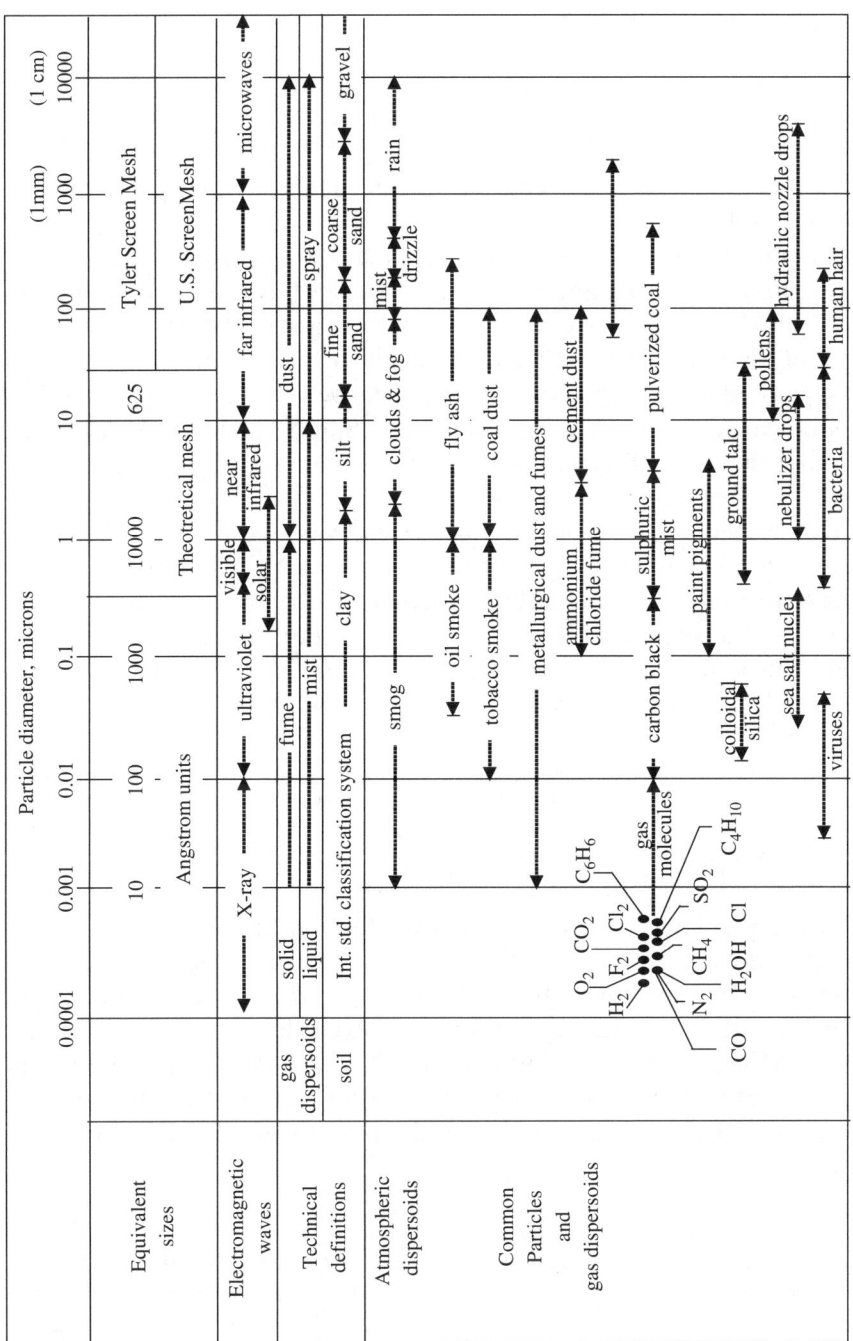

Figure 1 *Particle size ranges (Adapted from Ref. 5)*

diameter of a particle having the same external surface-to-volume ratio as a sphere and is given by:

$$d_{VS} = \frac{1}{\sum x_i / d_{pi}} \tag{1}$$

where x_i is the mass fraction of particles in each size range given by the sieve aperture d_{pi}.

A number of methods are available for measuring particle size, such as sieving, light scattering and microscopy. In sieve analysis the sample is placed on the top of a stack of sieves whose mesh size decreases with height of the stack. The stack is vibrated and as the powder falls through it is separated into fractions, which are then weighed, thereby giving the mass of particles between each sieve size. Standard sieves are in sizes such that the ratio of adjacent sieve sizes is the fourth root of two (*e.g.* 45, 53, 63, 75, 90 μm).

Other possible geometrical diameters can be used to determine the mean particle diameter of a polydisperse system. Examples are the surface average, d_s, and volume average diameters, d_v; where d_s is defined as the diameter of a sphere having the same surface area as the particle and d_v is the diameter of a sphere having the same volume as the particle. These are given by:

$$d_s = \sqrt{\sum x_i d_{pi}^2}; \quad d_v = \sqrt[3]{\sum x_i d_{pi}^3} \tag{2}$$

A comparison between the surface average and volume average diameters with the Sauter mean diameter shows that the latter is always skewed towards the finer end of the distribution and, therefore, is the smaller equivalent mean particle diameter in value. Table 1 provides a quantitative comparison of the mean diameters d_{vs}, d_s and d_v calculated using Equations (1) and (2) for two industrial fluid catalytic cracking catalysts (FCC1 and FCC2)[3] whose particle size distribution is shown in Figure 2. See worked example E1 (section 7.5.1) for the determination of the mean particle diameter and size distribution of the FCC1 catalyst sample.

Table 1 *Geometrical mean particle diameters for two FCC catalysts*

	$d_{vs}(\mu m)$	$d_s(\mu m)$	$d_v(\mu m)$
FCC1	71	91	102
FCC2	57	104	124

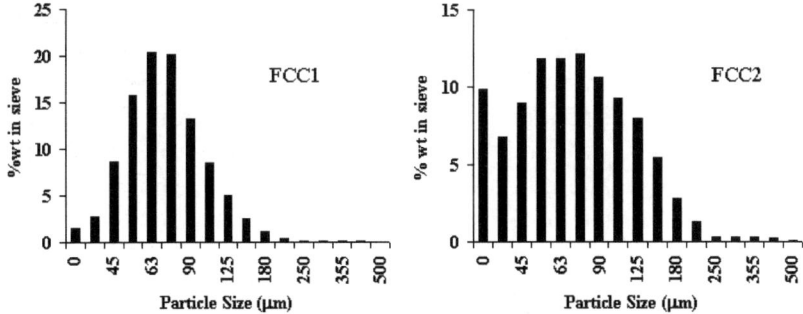

Figure 2 *Particle size distributions of two industrial FCC catalysts[3]*

Table 2 *Width of size distribution based on relative spread*

σ_d/d_{50}	Type of distribution
0	Very narrow
0.03	Narrow
0.17	Fairly narrow
0.25	Fairly wide
0.33	Fairly wide
0.41	Wide
0.48	Wide
0.6	Very wide
0.7	Very wide
>0.8	Extremely wide

The relative diameter spread σ_d/d_{50} can then be used to compare the width of size distribution of particles having different mean diameter. Plotting the cumulative size fraction versus particle size, the diameter spread is defined as:

$$\sigma_d = \frac{d_{84\%} - d_{16\%}}{2} \tag{3}$$

The median size diameter, d_{50}, is the size corresponding to the 50% value on the graph of cumulative size fraction versus particle size, see example E1. The values of σ_d/d_{50} are used to give an indication of the width of the size distribution, as summarised in Table 2.[4]

7.2.2 Particle Shape

This is a fundamental property affecting powder packing, bulk density, porosity, permeability, flowability, attrition and the interaction with fluids. It is generally considered that shape can be described in terms of two characteristics, form and proportions. Form refers to the degree to which a particle approaches some standard such as a sphere, cube or tetrahedron, while proportion distinguishes one spheroid from another of the same class. A simple form of shape factor is the sphericity, ϕ which is defined as the ratio of surface area of a sphere having the same volume as the particle to the surface area of the particle. Thus:

$$\phi = \frac{\text{surface area of equivalent volume sphere}}{\text{surface area of the particle}} = \left(\frac{d_v}{d_s}\right)^2 \qquad (4)$$

and $0 < \phi < 1$. d_v and d_s are as defined in the previous section.

The sphericity ϕ can be calculated exactly for such geometrical shapes as cuboids, rings and manufactured shapes, see Table 3. However, most particles are irregular and there is no simple generally accepted method for measuring their sphericity. Values for some common solids have been published (see Ref 4). However, these should be regarded as estimates only. Table 4 shows that ϕ is between 1 and 0.64 for most materials.

Other methods are available for quantifying shape factors and these are described in detail in Refs. 2 and 5. Using a Scanning Electron Microscope (SEM), for example, the shape and surface characteristics of

Table 3   *Sphericity for some regular solids*

Shape	Relative proportions	Sphericity ϕ
Spheroid	1:1:2	0.93
	1:2:2	0.92
	1:1:4	0.78
	1:4:4	0.70
Ellipsoid	1:2:4	0.79
Cylinder	Height = diameter	0.87
	Height = 2 × diameter	0.83
	Height = 4 × diameter	0.73
	Height = 1/2 × diameter	0.83
	Height = 1/4 × diameter	0.69
Rectangular parallelepiped	1:1:1	0.81
	1:1:2	0.77
	1:2:2	0.77
	1:1:4	0.68
	1:4:4	0.64
	1:2:4	0.68
Rectangular tetrahedron	–	0.67
Rectangular octahedron	–	0.83

Table 4 *Sphericity of some common solids*

	Sphericity ϕ
Crushed coal	0.75
Crushed sandstone	0.8–0.9
Round sand	0.92–0.98
Crushed glass	0.65
Mica flakes	0.28
Sillimanite	0.75
Common salt	0.84

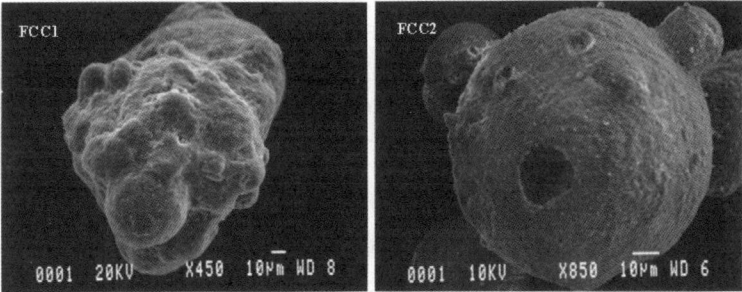

Figure 3 *SEM pictures of two FCC catalyst particles[3]*

particles can be observed. Photographs of two fluid catalytic cracking catalysts particles are reported as typical examples,[3] see Figure 3. The sphericity of the FCC2 catalyst particle is more accentuated and therefore a higher value of the shape factor is expected compared to FCC1. Using the Ergun Equation (see section 7.2.2, equation 25), a value of the particle's sphericity can be evaluated, which gives a sphericity $\phi_{FCC1} = 0.70$ for FCC1 and a value $\phi_{FCC2} = 0.92$ for FCC2.

7.2.3 Particle Density

The density of a particle immersed in a fluid is the *particle density*, defined as the mass of the particle divided by its hydrodynamic volume, V_p.[2] This is the volume "seen" by the fluid in its fluid dynamic interaction with the particle and includes the volume of all open and closed pores, see Figure 4.

$$\rho_p = \frac{\text{mass of particle}}{\text{volume the particle would displace if its surface were non-porous}} = \frac{M}{V_p} \quad (5)$$

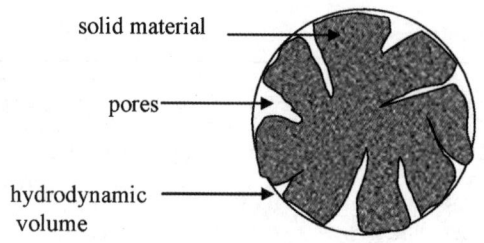

solid material

pores

hydrodynamic
volume

Figure 4   *Hydrodynamic volume of a particle (Adapted from Ref. 2)*

For non-porous solids the particle density is equal to the true, skeletal, or absolute density, ρ_{ABS}, which can be measured using either a specific gravity bottle or air pycnometer:

$$\rho_p = \frac{\text{mass of particle}}{\text{volume of solids material making up the particle}} \tag{6}$$

For porous materials $\rho_p < \rho_{ABS}$ and cannot be measured with such methods. A mercury porosimeter can be used to measure the density of coarse porous solids but is not reliable for fine materials, since the mercury cannot penetrate the voids between small particles. In this case, helium is used to obtain a more accurate value of the particle density. Methods to measure the particle density of porous solids can be found in Refs. 2 and 5.

In the case of a bed of particles immersed in a fluid, the bulk density of the bed ρ_b, which includes the voids between the particles, is defined as follows:

$$\rho_b = (1 - \varepsilon)\rho_b \tag{7}$$

where ε is the bed voidage, *i.e.* the volume fraction occupied by the fluid.

7.3   INTERACTION BETWEEN PARTICLES AND FLUIDS

The fluid–particle and particle–particle interactions dominate the fluid dynamics of particle systems and are of major importance in predicting the behaviour of complex operations, such as fluidisation, pneumatic transport and flow of slurries. In this section, the simple case study of a single particle immersed in a fluid is the starting point for the subsequent examination of the behaviour of fluidisation processes, which involve many particles in up-flowing fluids. This extends what is presented in Chapter 3, in which the concepts of Fluid Flow are introduced and parameters such as drag coefficient and Reynolds Number are defined.

7.3.1 Single Particles

We consider the motion of a single spherical particle immersed in a stationary fluid and falling at a constant velocity and we examine the forces exerted on the particle to estimate the steady velocity of the particle relative to the fluid, *the particle terminal fall velocity*, u_t. It is important to emphasise at this point that the equilibrium conditions experienced by the particle falling at u_t are equivalent to that of a motionless particle suspended in an upwardly flowing fluid with velocity u_t, as represented in Figure 5.

A stationary particle suspended in a fluid experiences a *buoyancy force* F_b, evaluated from Archimedes' principle as the weight of fluid displaced, $\rho_f g v$, where ρ_f is the fluid density, g is the acceleration due to gravity and v is the volume of the particle. If the particle begins to move the fluid will exert an additional force, the *drag force*, made up of two components; the skin friction drag, which is a direct result of the shear stress at the surface due to fluid viscosity, and the form drag due to differences of pressure over the surface of the particle.

The drag force on a single particle falling through a fluid is a function of a dimensionless drag coefficient, C_D, the projected area of the particle and the inertia of the fluid.[6] For a sphere:

$$F_d = C_D \frac{\pi d_p^2}{4} \frac{\rho_f u^2}{2} \tag{8}$$

From a balance of forces on a stationary particle, as represented in Figure 5:

Gravitation − Buoyancy − Drag = acceleration = 0

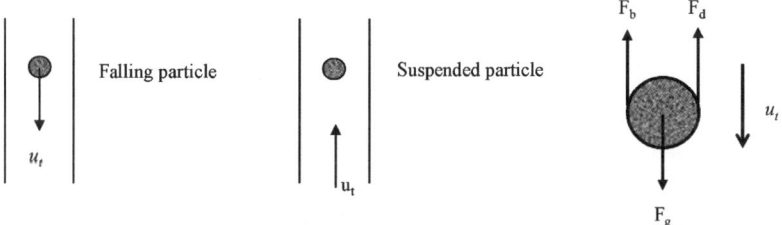

Figure 5 *Fluid–particle interaction*

Hence:

$$C_D \frac{\pi}{8} d_p^2 \rho_f u^2 = \frac{\pi d_p^2}{6} (\rho_p - \rho_f) g \tag{9}$$

or

$$C_D Re_t^2 = \frac{4}{3} Ar \tag{10}$$

where

$$Re_t = \frac{\rho_f u_t d_p}{\mu}; \quad Ar = \frac{d_p^3 \rho_f (\rho_p - \rho_f) g}{\mu^2} \tag{11}$$

From Equation (10), u_t can be written as follows:

$$u_t = \left[\frac{4 d_p (\rho_p - \rho_f) g}{3 \rho_f C_D} \right]^{1/2} \tag{12}$$

Equation (12) is implicit, as u_t is a function of C_D.
At low Reynolds numbers ($Re_p < 1$) Stokes' law applies and:

$$F_d = C_D \frac{\pi d_p^2}{4} \frac{\rho_f u^2}{2} = 3 \pi d_p \mu u \tag{13}$$

with

$$C_D = \frac{24}{Re_p} \quad \text{and} \quad Re_p = \frac{\rho_f u d_p}{\mu} \tag{14}$$

In the region where Newton's law applies ($750 < Re_p < 3.4 \times 10^5$), C_D is almost constant at 0.445.

For the intermediate flow regime, a correlation for C_D has to be chosen. A correlation that covers much of the range of interest for fluidisation is due to Dallavalle:[7]

$$C_D = \left(0.63 + \frac{4.8}{\sqrt{Re_p}} \right)^2 \tag{15}$$

The relationship between drag coefficient and Reynolds number is shown in Figure 6. Changes in drag coefficient correspond to changes in

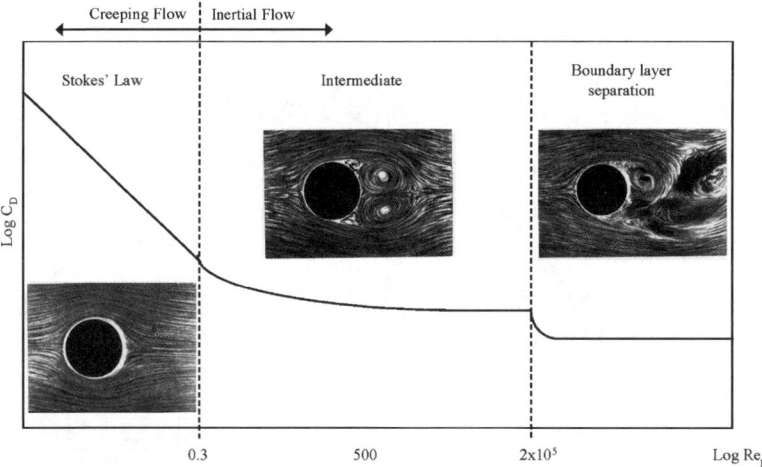

Figure 6 *Drag coefficient of a sphere and flow pattern development*

the flow pattern around a sphere, as also represented in Figure 6. At low Reynolds numbers, up to $Re_p = 1$, the flow pattern past the particle is symmetrical (*i.e.* the flow does not separate); in the intermediate regime, with values of $1 < Re_p < 500$, the flow will tend to separate, giving rise to the formation of the so-called Karman vortices. A further increase in the Reynolds number will result in fully separated flow. The higher the Reynolds number, the lower the drag coefficient, *i.e.* the resistance applied to the particle to flow.

When the flow regime is not known, the *analytical calculation* of u_t from Equation (12) requires choosing an equation for C_D, such as, for example, Equation (15), and solving a "non-trivial" system of two non-linear equations. To avoid the analytical solution, an *iterative procedure* can be applied by guessing an initial value of C_D or Re_p:

- Step 1: Guess an initial value of C_D (or Re_p).
- Step 2: Calculate value of u_t from Equation (12).
- Step 3: Calculate Reynolds number from Equation (11).
- Step 4: Check initial guess for C_D by computing it from knowledge of Re_p and using, for example, Equation (15). If C_D is equal to the initial guess, then the u_t value calculated at step 2 is correct, otherwise go to step 5.
- Step 5: Choose a new value of C_D and go back to step 2.

Alternatively, a *graphical solution* can be applied to determine u_t from an empirical plot based on the two dimensionless groups reported in

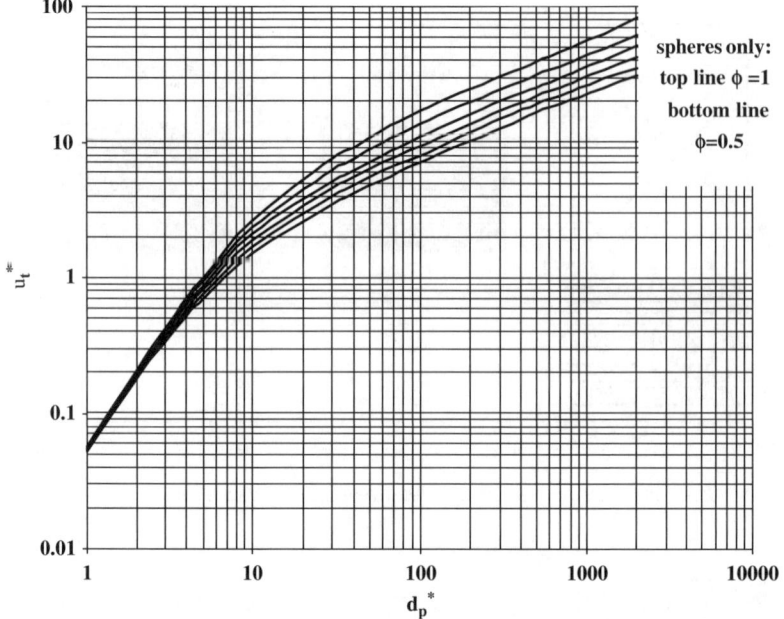

Figure 7 *Empirical diagram for the calculation of the particle terminal fall velocity (Adapted from Ref. 8)*

Equations (16) and (17), which are depicted in Figure 7.[8] Worked example E2 (section 7.5.2) shows the application of this procedure to the calculation of u_t.

$$d_p^* = d_p \left[\frac{\rho_f(\rho_p - \rho_f)g}{\mu^2} \right]^{1/3} = Ar^{1/3} \qquad (16)$$

$$u_t^* = u_t \left[\frac{\rho_f^2}{\mu(\rho_p - \rho_f)g} \right]^{1/3} = \frac{Re_t}{Ar^{1/3}} \qquad (17)$$

7.3.2 Flow Through Packed Beds

We now examine the case in which the particles are stationary and in direct contact with their neighbours, which is known as a packed bed.

In a packed bed, the fluid–particle interaction force is insufficient to support the weight of the particles. Hence, the fluid that percolates through the particles loses energy due to frictional dissipation. This results in a loss of pressure that is greater than can be accounted for by the progressive increase in gravitational potential energy. We now determine this additional energy loss, which is also the energy required to overcome the weight of the particles in a fluidised bed.

The total drop in fluid pressure across a length L of bed is given by:

$$\Delta p = \Delta P + \rho_f \, gL \tag{18}$$

where $\rho_f gL$ represents the hydrostatic contribution and ΔP is the portion lost due to the frictional interaction between the fluid and the particles. It represents the energy lost by the fluid and dissipated as heat.

In laminar flow, the flow through a packed bed of solid particles has been analysed in terms of the fluid-flow through parallel straight tubes, in light of the similarity in the expression of the pressure drop through a porous media as given by Darcy[9] and that in cylindrical tubes proposed by Hagen–Poiseuille.[6]

When a fluid passes through a bed of porous material the pressure drop per unit length of bed is given by Darcy's law as:

$$\frac{\Delta P}{L} = k \mu U \tag{19}$$

where U is the volumetric flux (volumetric flow rate per unit cross section of the empty tube), μ is the fluid viscosity and k is the permeability of the medium. The Hagen–Poiseuille equation for laminar flow through cylindrical tubes is:

$$\frac{\Delta P}{L} = \frac{32 \mu U}{D^2} \tag{20}$$

The application of the analogy involves simply replacing the fluid flux U and the bed diameter D in the Hagen–Poiseuille equation with equivalent terms relating to flow through porous media in order to take into account that a fraction of the cross-sectional area of the bed is occupied by particles. Following this reasoning, the expression for the pressure drop through a bed of spherical particles for laminar flow can be obtained:

$$\frac{\Delta P}{L} = \frac{72(1-\varepsilon)^2}{\varepsilon^3} \frac{\mu U}{d_p^2} \tag{21}$$

where the flux U in the Hagen-Poiseuille equation has been taken as indicating the superficial velocity of fluid relative to the tube wall (volume flow of gas/cross section of the bed). Using some empirical evidence, the constant in Equation (21) has been subsequently increased from 72 to 150

and Equation (21) with the new constant value has become known as the
Blake and Kozeny equation (see Ref. 10 for the full theoretical development). Equation (21) applies to the flow region where the pressure drop is
due solely to viscous losses. For beds of particles larger than about 150
μm in diameter, inertial forces become important. For the turbulent flow
regime the pressure drop in a straight tube can be written as:

$$\Delta P = 4f \frac{1}{2D} \rho_f L U^2 \tag{22}$$

where f is the friction factor. Substituting the fluid flux U and the
diameter D in Equation (22), we obtain:

$$\Delta P = \frac{3f(1-\varepsilon)}{\varepsilon^3} \frac{\rho_f L U^2}{d_p} \tag{23}$$

It was found experimentally that $3f = 1.75$. With this value, Equation
(23) becomes known as the *Burke–Plummer* equation:

$$\Delta P = \frac{1.75(1-\varepsilon)}{\varepsilon^3} \frac{\rho_f L U^2}{d_p} \tag{24}$$

Experiments have shown that equations (21) and (24) can be combined to cover for any flow condition, giving what is known as the Ergun
equation:[11]

$$\frac{\Delta P}{L} = \frac{150(1-\varepsilon)^2}{\varepsilon^3} \frac{\mu U}{d_p^2} + \frac{1.75(1-\varepsilon)\rho_f U^2}{\varepsilon^3 d_p} \tag{25}$$

where the first term is the viscous term and the second is the inertial
term.

In general, the Ergun equation for flow through a packed bed of non-
spherical and polydisperse-sized particles becomes:

$$\frac{\Delta P}{L} = \frac{150(1-\varepsilon)^2}{\varepsilon^3} \frac{\mu U}{(\phi d_{sv})^2} + \frac{1.75(1-\varepsilon)}{\varepsilon^3} \frac{\rho_f U^2}{\phi d_{sv}} \tag{26}$$

The Ergun equation may be rearranged to form dimensionless groups,
as graphically represented in Figure 8. Using Equation (25) we obtain:

$$\left(\frac{\Delta P}{L}\right)\left(\frac{d_p}{\rho_f U^2}\right)\left(\frac{\varepsilon^3}{1-\varepsilon}\right) = 150\frac{\mu(1-\varepsilon)}{d_p \rho_f U} + 1.75 \tag{27}$$

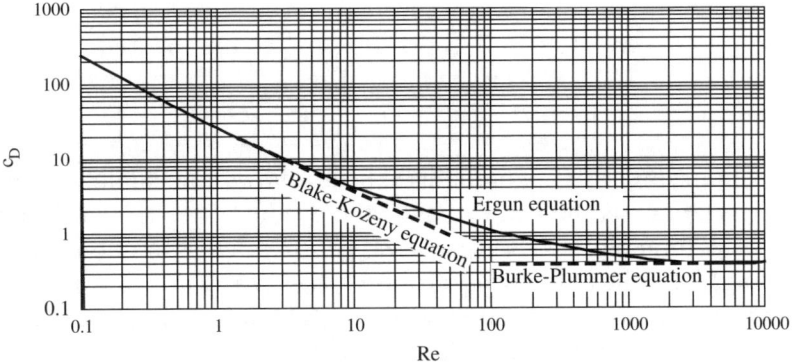

Figure 8 *The Ergun equation (Adapted from Ref. 11)*

It is worth noting the similarity between Figure 8 and the friction factor for tube flow represented in Figure 6. This highlights the physical meaning of pressure drop across the bed as energy dissipation due to friction between the particles.

7.4   FLUIDISED BEDS

As the velocity of the fluidising fluid (gas or liquid) is increased through a bed of solid particles, there comes a point where the drag force exerted by the fluid on the particles, which is proportional to the global pressure drop across the bed, is balanced by the buoyant weight of the suspension. At this point, the particles are lifted by the fluid, the separation between them increases and the bed becomes fluidised. Thus:

$$\Delta P = \frac{\text{weight of particles and fluid} - \text{upthrust}}{\text{bed cross-sectional area}} = \frac{Mg}{A} \qquad (28)$$

where M is the mass of particles in the bed and A is the cross sectional area of the bed.

For a bed of particles of density ρ_p, fluidised by a fluid of density ρ_f to form a bed of depth L and voidage ε in a vessel of cross-sectional area A, Equation (28) becomes:

$$\Delta P = \frac{[AL(1 - \varepsilon)\rho_p g + AL\varepsilon\rho_f g] - AL\rho_f g}{A} \qquad (29)$$

from which:

$$\frac{\Delta P}{L} = (\rho_p - \rho_f)(1 - \varepsilon)g \tag{30}$$

An important property of fluidised beds follows immediately from this simple relation; as the bed expands, the product $(1-\varepsilon)\,L$, which represents the total volume of particles per unit cross section, remains unchanged. As the fluid flux is increased, L increases and $(1-\varepsilon)$ decreases so as to maintain their product at a constant value. As a consequence, the pressure drop through the fluidised bed remains constant for further increases in fluid velocity, as shown in Figure 9.

7.4.1 Minimum Fluidisation Velocity

The fluid velocity at which the particles become suspended is the minimum fluidisation velocity, u_{mf}. For spherical particles, the velocity at minimum fluidisation can be predicted by combining Equations (25) and (30), given that the values of the bed voidage at minimum fluidisation, ε_{mf}, are known *a priori*:

$$(\rho_p - \rho_f)g = \frac{150\,(1 - \varepsilon_{mf})\,\mu\,u_{mf}}{\varepsilon_{mf}^3}\frac{1}{d_p^2} + \frac{1.75\,\rho_f\,u_{mf}^2}{\varepsilon_{mf}^3}\frac{1}{d_p} \tag{31}$$

which can also be written as:

$$Ar = \frac{150(1 - \varepsilon_{mf})}{\varepsilon_{mf}^3}\,Re_{mf} + \frac{1.75}{\varepsilon_{mf}^3}\,Re_{mf}^2 \tag{32}$$

From this, the minimum fluidisation velocity can be obtained:

$$u_{mf} = \frac{\varepsilon_{mf}^3\,(\rho_p - \rho_f)d_p^2 g}{150(1 - \varepsilon_{mf})\mu} \tag{33}$$

Equation (33) is strictly applicable only to conditions of viscous flows where the pressure drop is solely due to viscous energy losses.

If Equation (31) is expressed as a function of the sphericity factor, ϕ, as shown in Equation (26), then a value of ϕ can be back-calculated by knowing u_{mf} and ε_{mf}, as done for the FCC catalysts shown in Figure 3.

If ε_{mf} is not known, than we can use the equation proposed by Wen-Yu[12], which expresses u_{mf} solely as a function of the physical

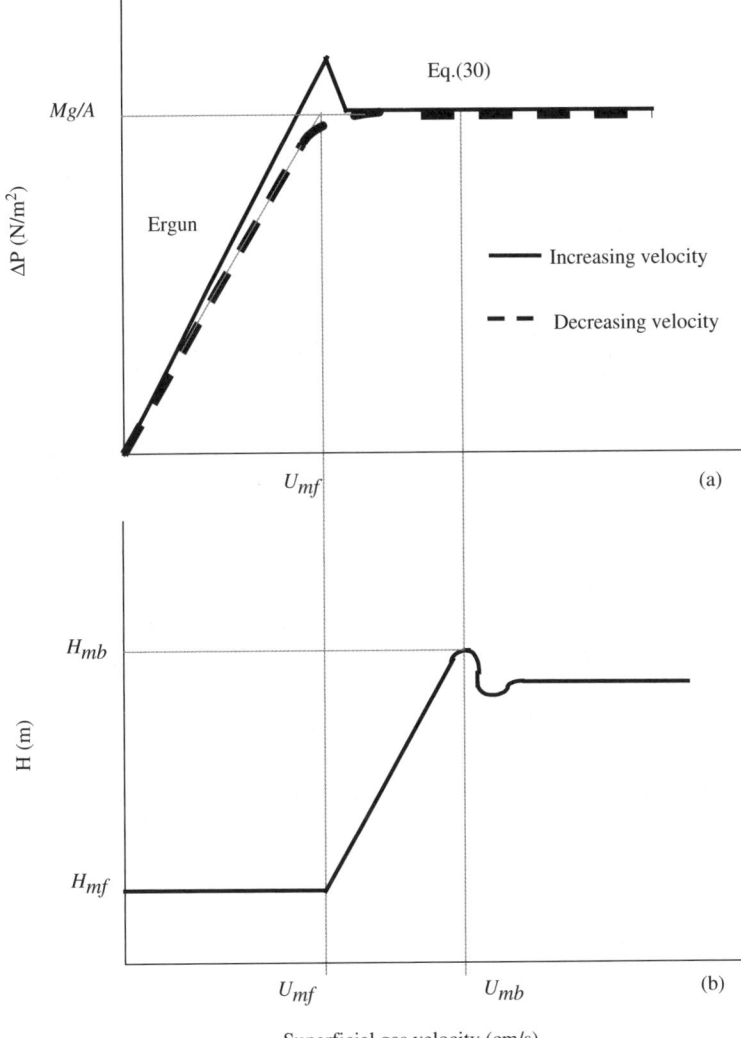

Figure 9 *Pressure drop and bed expansion profiles for gas–solid systems of fine particles*

characteristics of the fluid and the particle. Wen-Yu used experimental evidence to reformulate Equation (32) as follows:

$$Ar - 1650 \, Re_{\mathrm{mf}} \mid 24.5 \, Re_{\mathrm{mf}}^2 \tag{34}$$

This formulation assumes values of the voidage at minimum fluidisation of around 0.38.

For fine spherical particles (*below about 100 μm*) u_{mf} is obtained from the viscous term of Equation (34)

$$u_{mf} = \frac{d_p^2(\rho_p - \rho_f)g}{1650\mu} \qquad (35)$$

For larger particles, it becomes:

$$u_{mf}^2 = \frac{d_p(\rho_p - \rho_f)g}{24.5\rho_f} \qquad (36)$$

The sensitivity of u_{mf} on the bed voidage is shown in the worked example E3 (section 7.5.3), in which the calculation of u_{mf} using the Ergun and Wen-Yu equations is compared.

Most fluidisation processes are operated at high temperatures and pressures. It is important, therefore, to be able to predict changes in fluidisation with the operating conditions. Using Equations (35) and (36), the effect of temperature and pressure can be determined. With increasing temperature, *gas viscosity increases* while *gas density decreases*. For small particles, the fluid–particle interaction is dominated by the viscous effects. Equation (35) shows that u_{mf} varies with $1/\mu$, and u_{mf} should therefore decrease with temperature. For large particles, the inertial effects dominate; Equation (36) predicts that u_{mf} will vary with $(1/\rho_f)^{0.5}$, u_{mf} should therefore increase with temperature.

With increasing pressure, gas viscosity is essentially independent of pressure, while gas density increases. For small particles, u_{mf} remains almost constant. For large particles, u_{mf} decreases with pressure.

7.4.2   Types of Fluidisation Regimes

At fluid velocities higher than at minimum fluidisation, different types of fluidisation behaviour are possible:

(i) uniform expansion;
(ii) bubbling or slugging;
(iii) turbulent;
(iv) lean fluidisation.

In liquid–solid systems, and for gas–solid systems of small particles, increasing the fluidising velocity beyond u_{mf} gives rise to a uniform expansion of the bed. The homogeneous expansion of a gas–solid system is generally described using the Richardson–Zaki equation:[13]

$$u = u_t \varepsilon^n \qquad (37)$$

Equation (37) describes the relationship between fluidising velocity and voidage as a function of the particle terminal fall velocity and of a parameter n which depends on the particle Reynolds number according to the following relations:[10]

$$
\begin{aligned}
n &= 4.8 & Re_t &< 0.2 \\
n &= 4.6\, Re_t^{-0.03} & 0.2 &< Re_t < 1 \\
n &= 4.6\, Re_t^{-0.1} & 1 &< Re_t < 500 \\
n &= 2.4\, Re_t & &> 500
\end{aligned}
\tag{38}
$$

In gas–solid systems, as the gas velocity is further increased, gas bubbles form within the bed to give a bubbling fluidised bed. The velocity at which the first bubble forms is called the minimum bubbling velocity, u_{mb}. An empirical correlation typically used to predict u_{mb} at ambient conditions has been given by Abrahamsen and Geldart:[14]

$$
u_{mb} = 2.07 \exp(0.716\, F_{45}) \frac{d_p\, \rho_g^{0.06}}{\mu^{0.347}}
\tag{39}
$$

where F_{45} is the mass fraction of particles smaller than 45 μm.

The pressure drop and bed expansion profiles shown in Figure 9 represent the different stages from fixed to bubbling fluidisation for a typical gas–solid system of fine particles. Knowing the bed height, the average bed voidage can be obtained:

$$
\varepsilon = 1 - \frac{M}{\rho_p H A}
\tag{40}
$$

An increase in gas velocity beyond minimum bubbling leads to more vigorous bubbling with larger bubbles that may eventually become as large as the vessel diameter, giving a *slugging bed*. This may occur typically with high aspect ratio (H/D) beds for Geldart Group B particles. Turbulent fluidisation occurs for even higher velocities. Here the two-phase structure disappears and bed takes on a foam-like appearance. Still higher gas velocities result in particles being transported out of the bed, until only a low concentration of rapidly moving particles remains in contact with the gas, as in *pneumatic conveying systems*.[2] These types of systems are used for transporting particulate solids in a conveying fluid, where the solids are typically at least three times denser than the transporting fluid. A typical example is the transport of particles in a pipeline using gas. The distinction between pneumatic conveying systems and *hydraulic conveying systems* lies in the ratio between the solid and fluid density, with the solid density being generally of the order of magnitude of the fluid density in the latter. In pneumatic conveying systems, *two types* of flows are generally encountered. "Dilute phase" is characterised by low particle concentrations (less than 1% by volume) and relatively high velocities (greater than 20 m s^{-1}), which results in high attrition of the particles. Dilute phase conveying is carried out in systems in which the solids are fed to the pipe from hoppers at a controlled flow rate and cyclone separators are used to separate the solids from the gas stream at the end of the line. "Dense phase" conveying systems are characterised instead by high particle concentrations (greater than 30% by volume) and low velocities (1–5 m s^{-1}). In this case the solids are not entirely suspended in the gas stream; particle attrition is reduced due to the low solid velocities, reducing in turn pipeline erosion and product degradation. Both types of flows are used in horizontal and vertical transport systems. Figure 10 shows various types of fluid–particle contacting, from fixed bed, to bubbling fluidisation, slugging and pneumatic transport.

The minimum fluidisation velocity represents the lower limit of the range of operative conditions at which a fluid-bed process can be operated, while the particle terminal fall velocity represents the upper limit beyond which the particles will start leaving or elutriating from the bed. To avoid or reduce carryover of particles from a fluidised bed, gas velocity has to be kept between u_{mf} and u_t. For polydisperse systems, in calculating u_{mf}, the mean diameter d_{vs} is used for the particle size distribution present in the bed. In calculating u_t, the smallest size of solids present in appreciable quantity in the bed is used.

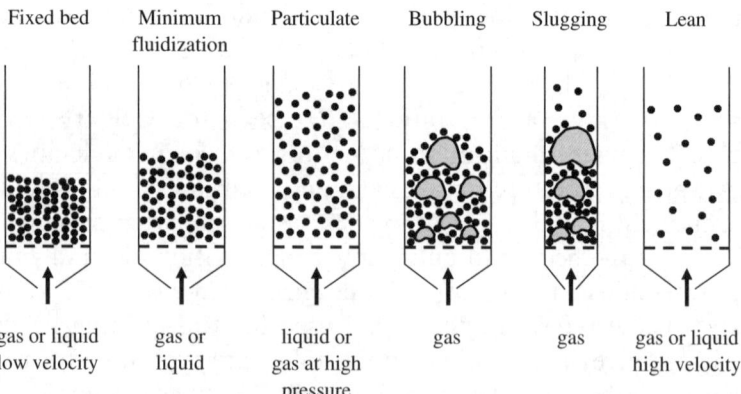

Figure 10   *Types of fluid–particle contacting*

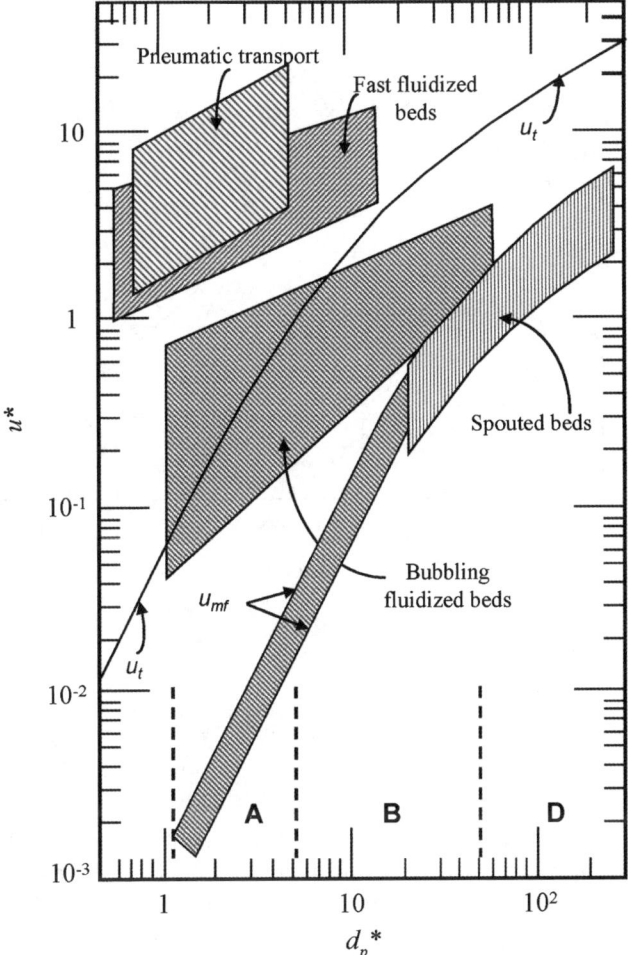

Figure 11 *Empirical diagram for the prediction of the fluidisation regime (Adapted from Ref. 15)*

Using the dimensionless parameters introduced with Equations (16) and (17) (substituting u_t^* and u_t with u^* and u), for a given particle system and a given fluidising velocity, the fluidisation regime can be determined using the diagram reported in Figure 11.[15] Similarly, the operating velocity needed to attain a certain fluidisation regime can also be determined from Figure 12, see worked example E4 (section 7.5.4).

7.4.3 Classification of Powders

In an attempt to classify the different fluidisation behaviours, Geldart[16] proposed an empirical classification, which divides fluidisation behaviour

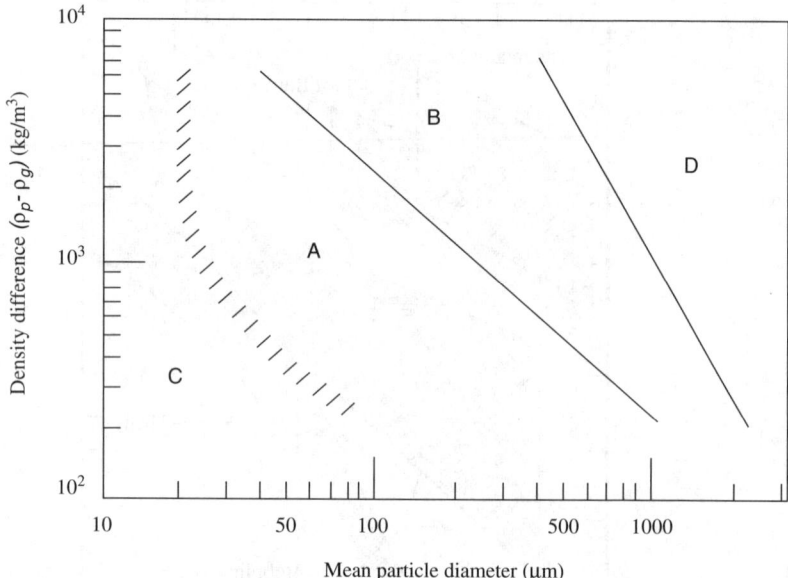

Figure 12   *Classification of fluidisation behaviour for air at ambient conditions (Adapted from Ref. 16)*

according to the mean particle size and gas and particle densities. Boundaries between these groups were proposed in the form of a dimensional plot of $(\rho_p - \rho_g)$ versus d_p, as shown in Figure 12. Geldart proposed four typical fluidisation regimes for materials fluidised with air at ambient conditions, which he indicated as A, B, C and D.

Group C materials are usually cohesive and very difficult to fluidise. The behaviour of this group of powders is strongly influenced by the interparticle forces, which are greater than those exerted by the fluidising gas. As a result, the particles are unable to achieve the separation they require to be totally supported by drag and buoyancy forces and true fluidisation does not occur, with the pressure drop across the bed being lower than the theoretical value given by Equation (30). Bubbles, as such, do not appear; instead the gas flow forms channels through the powder. An attempt to fluidise a Group C powder is shown in Figure 13. Fluidisation of Group C materials can sometimes be improved using mechanical stirring or vibrations, which break up the stable channels. In the case of plastic materials, fluidisation may be promoted by adding some sub-micron particles, which modify the contact geometry, thereby reducing the interparticle forces.

Group A powders are those which exhibit a stable region of non-bubbling expansion between u_{mf} and u_{mb}, as shown in Figure 10.

Figure 13 *Fluidisation behaviour of a group C powder: an attempt to fluidise a group C powder produces channels or a discrete plug*

Geldart[4] distinguished these powders as those for which $u_{mb}/u_{mf} > 1$. At gas velocities above u_{mb} bubbles begin to appear, which constantly split and coalesce, and a maximum stable bubble size is achieved. The flow of bubbles produces high solids and gas back-mixing, which makes the powders circulate easily, giving good bed-to-surface heat transfer.

Group B powders are characterised by having $u_{mb} = u_{mf}$. Bubbles rise faster than the interstitial gas velocity, coalescence is the dominant phenomenon and there is no evidence of a maximum bubble size, as defined for Group A materials. Bubble size increases with increasing fluidising gas velocity, see Figure 14. The interparticle forces are considered to be negligible for these powders.

Group D powders are large particles that are distinguished by their ability to produce deep spouting beds. The distinction between this and the previous groups concerns also the rise velocity of the bubbles, which is, in general, less than the interstitial gas velocity. The gas velocity in the dense phase is high and solids mixing is relatively poor, consequently, back-mixing of the dense phase is small. Segregation by size is likely when the size distribution is broad, even at high gas velocities. The flow regime around these particles may be turbulent, causing particle attrition and, therefore, elutriation of the fines produced.

7.4.4 Interparticle Forces, Measurement of Cohesiveness and Flowability of Solid Particles at Ambient Conditions

When trying to describe the fluidisation of different materials, the nature of the forces acting between adjacent particles becomes of major

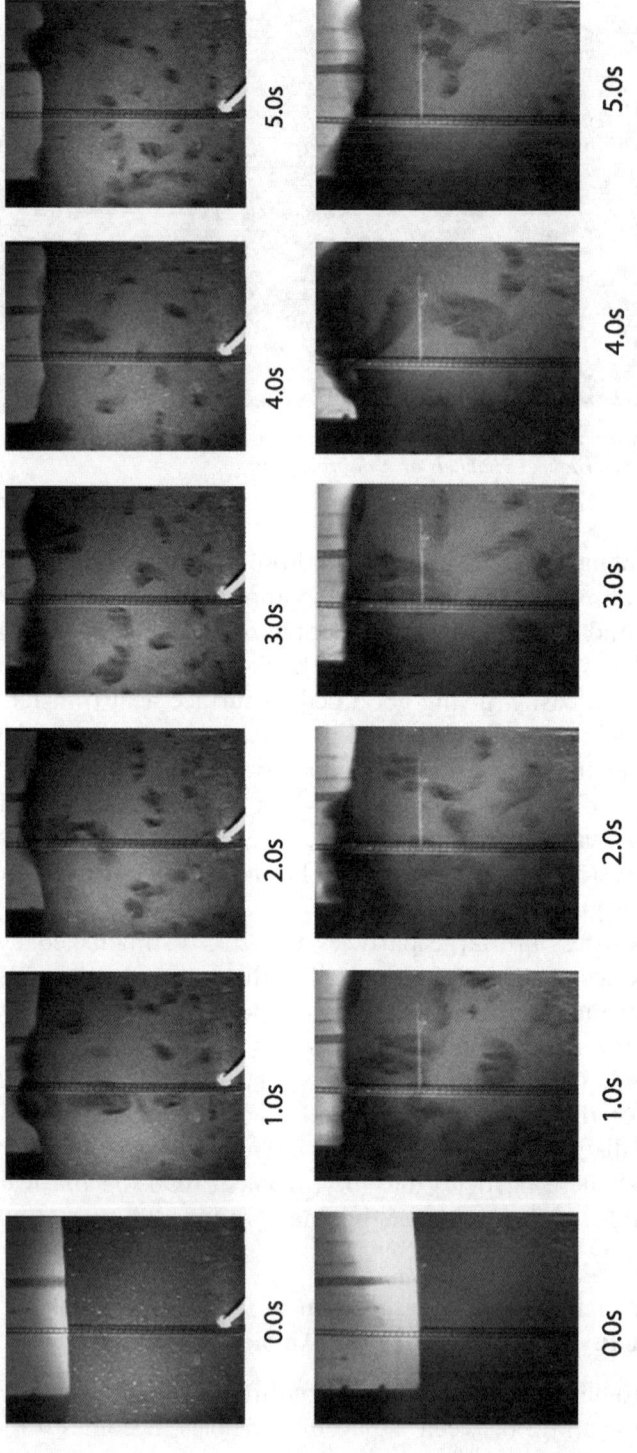

Figure 14 *Fluidisation behaviour of a Group B powder: effect of gas flow on bubbles in a two-dimensional fluidised bed, (i) lower gas velocity, (ii) higher gas velocity. A max. stable bubble size is never achieved*

Table 5 *Mechanisms of adhesion*

Without material bridges	With material bridges
Van der Waals forces	Capillary forces
Electrostatic forces	Solid bridges
Magnetic forces	Sintering
Hydrogen bonding	

Table 6 *Properties of bulk solids to determine flowability with the Carr method*

Bulk density loose, ρ_{BDL}	Angle of repose
Bulk density packed, ρ_{BDP}	Angle of fall
Hausner ratio	Angle of difference
Compressibility	Angle of spatula
Cohesion	Angle of internal friction

importance in order to predict their contribution to the fluid-bed behaviour. As described in the previous section, fluidisation of Geldart Group C materials is dominated by interparticle forces, whose effects become negligible in the fluidisation behaviour of Group B powders. The role of the interparticle forces on the fluidisation behaviour of Group A powders is, on the other hand, far from being unequivocally understood, mainly due to the difficulty in recognising the nature of the forces involved and, therefore, of quantifying their effect on the fluidisation behaviour.

Particle–particle contacting can be the result of different mechanisms of adhesion, which in turn can influence, in different ways, the fluidisation behaviour of solid materials, see Ref. 17. An extensive review on the subject is also reported in Israelachvili.[18] Table 5 lists various mechanisms of adhesion with and without material bridges.

It would be desirable to have simple tests capable of characterising the fluidisation behaviour or "flowability" of particulate materials on the basis of their *bulk* properties. To this end, Carr[19] developed a system to characterise bulk solids with respect to flowability. Table 6 summarises the properties which are determined. In Carr's method a numerical value is assigned to the results of each of these tests, and is summed to produce a relative flowability index for that particular bulk material. Given the extensive use of these empirical techniques in academia and industry, a brief review on the subject is reported here. Nevertheless, it should be emphasised that these techniques allow measurements of the flowability or cohesion of materials solely in their stationary or compressed status and at ambient conditions. A direct relationship between these measurements and the fluidisation behaviour can therefore prove very difficult, in particular when operating conditions cause the properties of the materials to change.

7.4.4.1   Loose and Packed Bulk Density.   The loose bulk density is determined by gently pouring a sample of powder into a container through a screen, and is measured before settlement takes place. The packed or tapped bulk density is determined after settling and deaeration of the powder has occurred due to tapping of the sample. Details on the techniques are reported by Geldart and Wong.[20]

7.4.4.2   Hausner Ratio and Powder Compressibility.   The ratio between the loose and packed bulk density, ρ_{BDL} and ρ_{BDP} respectively, is known as the Hausner Ratio (*HR*) and is used as an indication of the cohesiveness of the materials, see Ref. 20. In addition to the *HR*, the powder compressibility is also used to define cohesiveness. This is expressed as $100(\rho_{BDP} - \rho_{BDL})/\rho_{BDP}$.

7.4.4.3   Cohesion.   This test is used for very fine powders (below 70 μm). Material is passed through three vibrating sieves in series. The material left on each sieve is weighed and a cohesion index is determined from the relative amounts retained. Carr[19] defined cohesion as the apparent surface force acting on the surface of powders, which are composed of millions of atoms. The number of points of contact within the powder mass determines the effect of this force. Thus, cohesiveness increases with decreasing particle size, since the number of contact points increases as the particle size decreases.

7.4.4.4   Angle of Repose, Angle of Fall and Angle of Difference.   *The angle of repose* is defined as the angle between a line of repose of loose material and a horizontal plane. This is determined by pouring the powder into a conical pile from a funnel and screen assembly. Materials with good flowability are characterised by low angles of repose. *The angle of fall* is determined by dropping a small weight on the platform on which a loose or poured angle of repose has been formed. The fall causes a decrease of the angle of repose forming a new angle identified as the angle of fall. The more free-flowing the powder, the lower the angle of fall. *The angle of difference* is determined by the difference between the angle of repose and the angle of fall. The greater this angle the better the flow.

7.4.4.5   Angle of Spatula and Angle of Internal Friction.   *The angle of spatula* is a quick measurement of the angle of internal friction. This is the angle, measured from the horizontal, that a material assumes on a flat spatula that has been inserted into a pile and then withdrawn vertically. A free-flowing material will have formed one angle of repose on the spatula's blade. A cohesive material will have formed several angles of repose on the blade, the average of these is then taken. The higher the angle of spatula of a material, the less its flowability. *The angle of internal friction* has been defined as the angle at which the dynamic equilibrium between the moving particles of a material and its bulk solid is achieved. This is of particular interest for flows in hoppers and bins. The measurement of this angle is described below.

Measurements of Angles of Friction. Shear cells are generally used to determine the properties which influence powder flow and handling.[21] Some of these properties are called "failure properties". These include the effective angle of internal friction, the angle of wall friction, the failure (or flow) function, the cohesion and ultimate tensile strength. The requirement for getting powders to flow is that their strength is less than the load put on them, *i.e.* they must fail, hence the name "failure properties". From soil mechanics, a solid is characterised by a yield locus that defines the limiting shear strength under any normal stress. Plotting shear stress τ and normal stress σ_n, the yield locus for a Coulomb material intersects the τ axis at a value of τ which is defined as cohesion C at an angle ϕ_i, which is defined as the angle of internal friction of that material.

In a shear cell, such as the Jenike shear cell,[22] the test powder is consolidated in a shallow cylindrical chamber which is split horizontally. The lower half of the cell is fixed and a shear force is applied to the upper moveable part at a constant low rate (see example E5, section 7.5.5, where experiments were obtained at a constant rate of 0.03 rpm, corresponding to a linear velocity of 1 mm min^{-1}). Shearing can be carried out for each of a series of normal loads on pre-consolidated samples, so that at the end of the test the relationships between the shear stress and normal stress at various bulk densities are obtained.

Construction of the Static Internal Yield Locus. Once the shear cell is loaded with the powder specimen, the pre-consolidation load N_c is applied on the cell lid and the pre-shearing phase starts. When a steady-state value is attained for the measured shear stress, the pre-shearing is stopped. Then the shearing phase starts under a normal load N_1 lower than N_c and a peak value of the shear stress is reached corresponding to the material failure. The sequence of pre-shearing and shearing phases is then repeated, keeping the same value of the normal load for each pre-shearing phase, N_c, and using decreasing normal loads for each shearing phase, *i.e.* $N_1 > N_2 > N_3 > \ldots > N_N$.

Following the theory of Jenike,[23] the solid normal stress acting in the shear plane can, in general, be calculated by simply dividing the external normal load by the shear surface area, which is assumed to correspond to the cross-sectional area of the cell.

$$\sigma_i = \frac{N_i}{A_{\text{cell}}} \tag{41}$$

Figure 15 reports a typical shear stress chart and the derived yield locus corresponding to the applied consolidation stress. The yield locus is the interpolation of the experimental points (σ, τ) corresponding to failure. It is worth pointing out that different yield loci are obtained when applying different consolidation stresses σ_c. The choice of the normal stresses used in the shearing phases and the number of phases used in each experiment, on the other hand, are arbitrary, as all the (σ, τ) points obtained in the shearing phases will belong to the same yield locus, provided that the same consolidation stress is applied in the pre-shearing phases. The values of the static angle of internal friction φ and the cohesion C were worked out as the slope and the intercept with the τ axis of the yield locus, respectively, as shown in Figure 15. The circle passing through σ_c and tangent to the yield locus is called the Mohr Circle. The two intercepts of this Mohr circle and the σ axis are the major principal stresses σ_1 and σ_2 corresponding to the consolidation applied. By repeating the whole experiment with different consolidation loads N_c, a family of incipient yield loci can be obtained. As a result, different angles of internal friction and values of the cohesion can be obtained as a function of the major principal stress σ_1.

Jenike carried out many experimental measurements on free-flowing and cohesive materials. He found that the yield locus of a dry material would be a straight line passing through the origin, as shown in

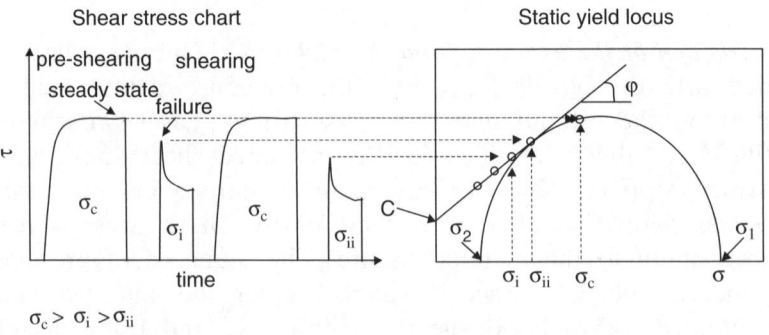

Figure 15 *Construction of the static yield locus*

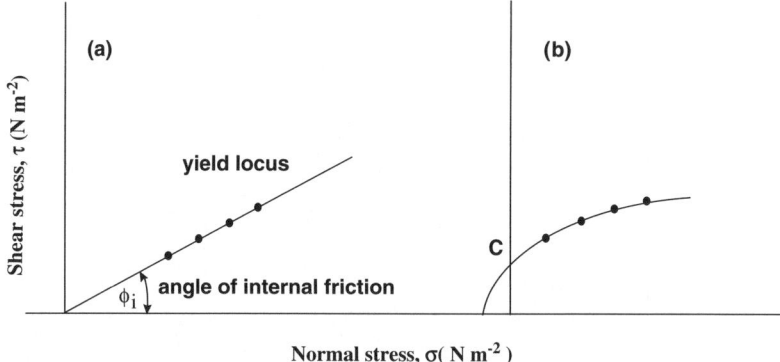

Figure 16   *Yield loci of (a) free-flowing materials; (b) cohesive materials*

Figure 16(a). The term "Cohesionless" was therefore used to describe materials which have a negligible shear strength under zero normal load ($\sigma_n = 0$). On the other hand, Jenike found that the yield loci of cohesive materials differ significantly from a straight line and have a non-zero intercept, indicated by C. Moreover, the position of the locus for a cohesive powder is a strong function of the interstitial voidage of the material. Fig 16(b) shows the typical yield locus for cohesive materials.

Construction of the Dynamic Internal Yield Locus.   The dynamic yield locus represents the steady state deformation, as opposed to the static yield locus which represents the incipient failure. The dynamic yield locus is constructed by plotting on a (σ, τ) plane the principal Mohr circles obtained for various consolidation stresses. The dynamic yield locus will be the curve or straight line tangent to all circles, as shown in Figure 17. The dynamic angle of internal friction δ and cohesion C_δ are independent of the consolidation stress. δ and C_δ are obtained as the slope and the intercept at $\sigma=0$ of the dynamic yield locus of the powder.

An approximation of the dynamic yield locus is given by the effective yield locus, provided that $C_\delta = 0$. The effective yield locus, shown in Figure 17, is the straight line drawn through the origin and tangentially to the Mohr's circle represented in Figure 15. Since it passes through the origin, it represents the angle of internal friction that the material would have if it were cohesionless. This angle is denoted by the effective angle of internal friction of the material, φ_e (Figure 18). The difference between the effective and the internal angle of friction can be an indication of the degree of cohesiveness of the powder. An application of the above procedure is presented in the worked example E5, which

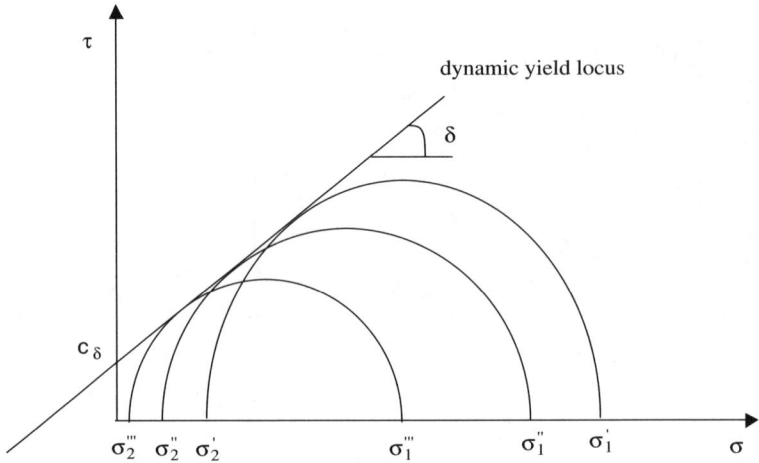

Figure 17   *Dynamic yield locus*

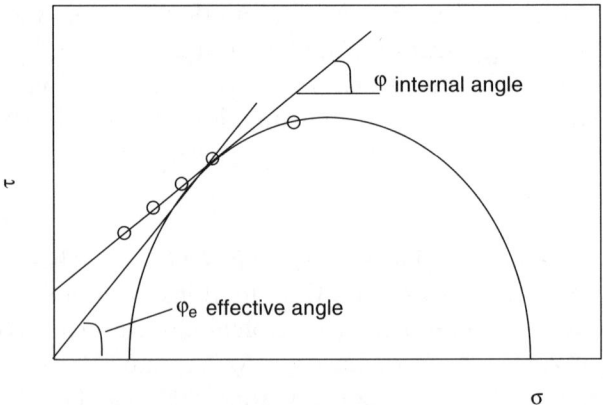

Figure 18   *Effective angle of internal friction*

illustrates the results obtained from experiments carried out on two different alumina materials and ballotini.[24]

7.4.5   Some Industrial Fluid-Bed Applications

The fluidised bed is only one of the many reactors employed in industry for gas–solid reactions, as reported by Kunii and Levenspiel.[25] Whenever a chemical reaction employing a particulate solid as a reactant or as a catalyst requires reliable temperature control, a fluidised bed reactor is often the choice for ensuring nearly isothermal conditions by a suitable selection of the operating conditions. The use of gas-solid fluidised beds

Table 7 *Some industrial applications of fluidised beds*

Physical mechanism	Application
Heat and/or mass transfer between gas/particle	Solids drying Absorption of solvents Food freezing
Heat and mass transfer between particle/particle or particle/surface	Coating of pharmaceutical tablets Granulation Mixing of solids Plastic coating of surfaces
Heat transfer between bed/surface	Heat treatment of glasses, rubber, textile fibres

Chemical mechanism	Application
Gas/gas reaction in which solid acts as catalyst	Oil cracking Manufacture of: Acrylonitrile Polyethylene
Gas/gas reaction in which solids are transformed	Coal combustion Gasification Incineration of waste Catalyst regeneration

spans across many industries, ranging from the pharmaceutical, refining and petroleum sectors to power-generation and can be classified depending on whether the physical mechanisms of heat and mass transfer are the governing factors in the process or whether chemical reactions are taking place in the fluid-bed. Examples are given in Table 7.

Examples of chemical reaction processe that employ a fluid-bed include catalytic cracking of petroleum, metallurgical processes, titanium refining, where the fluidised bed reactor is used for extracting titanium from naturally occurring ore, coal gasification, combustion and incineration, where the recent emphasis is on using fluidised bed reactors for the treatment of clean biomass and urban waste in order to produce a clean syngas from which electric energy can subsequently be obtained. Applications where mass and heat transfer are the key mechanisms include processes such as granulation, drying and mixing.

It has been shown earlier that the fluidisation properties of a powder in air at ambient conditions may be predicted by establishing in which Geldart group it lies. Table 8 shows that most fluidised bed industrial processes are operated at temperatures and pressures well above ambient. It is important to note that at operating temperatures and pressures

Table 8 *Operating conditions employed in fluidised bed processes*

Process	Example/Products	Process conditions
Drying of solids	Inorganic materials	60–110°C; 1 atm
	Pharmaceuticals	60°C; 1 atm
Calcination	Limestone	770°C; 1 atm
	Alumina	550–600°C, 1 atm
Granulation	Soap powders	5°C; 1 atm
	Food; Pharmaceuticals	20–40°C; 2–3 atm
Coating	Food; Pharmaceuticals	20–80°C; 2–3 atm
Roasting	Food industry products	200°C; 1 atm
	Sulfide ores (FeS_2)	650–1100°C; 1 atm
Synthesis Reactions	Phthalic anhydride	340–380°C; 2.7 atm
	Acrylonitrile	400–500°C; 1.5–3 atm
	Ethylene dichloride	260–310°C; 1–10 atm
	Maleic anhydride	400–500°C; 4 atm
	Polyethylene (low/high density)	75–120°C; 15–30 atm
Thermal cracking	Ethylene; propylene	750°C; 1 atm
Some other applications		
Combustion and incineration of waste solid		800–900°C; 1–10 atm
Gasification of coal and coke/ solid waste		800°C; 1–10 atm
Biofluidisation, cultivation of micro-organism		30°C; 1 atm
Semiconductor industries		300–1100°C; 1 atm

above ambient a powder may appear in a different group from that which it occupies at ambient conditions.[26] This is due to the effect of gas properties on the grouping and may have serious implications as far as the operation of the particle system is concerned.

Reliable design of commercial scale particle systems requires a good understanding of the highly complex flow phenomena and also requires detailed knowledge of how the hydrodynamics are affected by both reactor design and plant scale-up. Factors such as powder characteristics (*i.e.* density, surface properties, fines content), operating conditions (*i.e.* temperature, pressure, velocity) and reactor design (*i.e.* cooling coils, dip-legs, grid), plus any special conditions, such as the presence of liquids and other additives in the reactor, can significantly affect the gas–solid contacting efficiency. In light of this and of its wide range of industrial applications, particle science and technology remains a complex and challenging area for scientific research.

7.5 WORKED EXAMPLES

7.5.1 Example E1

Table 9 shows the results of the sieve analysis obtained for a sample of a FCC1 powder. The values reported in the first two columns of the table are the standard diameters of the sieve apertures. From these the mean diameter d_p is obtained for each two adjacent sieve sizes and the values are reported in the third column. From the mass fraction of powder in each sieve (values in the fourth column) the weight percentage is obtained and reported in the fifth column. Thus, the sum of the mass fraction over the mean diameter allows the calculation of the *volume-surface* mean particle diameter of the distribution using Equation (1):

$$d_{VS} = \frac{1}{\sum x_i/d_p} \tag{1}$$

Plotting the weight percentage versus the mean particle diameter d_p allows the *size distribution* of the powder to be obtained, as shown in Figure 2. The relative diameter spread σ/d_{50} is then calculated to obtain an indication of the width of the size distribution of the sample under analysis. The diameter spread, σ_d, is given by:

$$\sigma_d = \frac{d_{84\%} - d_{16\%}}{2} \tag{3}$$

and the median size diameter, d_{50}, are determined by plotting the cumulative size fraction, reported in Table 9, versus particle size, see also Figure 19. This gives a value for $\sigma_d/d_{50} = 0.37$, indicating a wide particle size distribution according to Table 2.

7.5.2 Example E2

A non-spherical FCC catalyst particle falls in a column of nitrogen at ambient conditions and attains a terminal fall velocity u_t. The density and viscosity of the nitrogen are $\rho_f = 1.2 \times 10^{-3}$ g cm^{-3} and $\mu = 1.8 \times 10^{-4}$ g cm^{-1} s^{-1} respectively. Calculate the terminal velocity of the particle given its physical characteristics, $d_p=70$ μm, sphericity $\phi = 0.7$

Table 9   *Worked example E1: calculation of mean particle diameter*

Min particle diameter	Max particle diameter	Mean particle diameter	Mass fraction	% Mass fraction in sieve	Mass fraction/ mean diameter	Cumulative mass fraction
0	38	19	0.015	1.5	7.89E-04	1.5
38	45	41.5	0.027	2.7	6.51E-04	4.2
45	53	49	0.086	8.6	1.76E-03	12.8
53	63	58	0.157	15.7	2.71E-03	28.5
63	75	69	0.204	20.4	2.96E-03	48.9
75	90	82.5	0.201	19.8	2.44E-03	69.0
90	106	98	0.132	13.2	1.35E-03	82.2
106	125	115.5	0.085	8.5	7.36E-04	90.7
125	150	137.5	0.050	5.0	3.64E-04	95.7
150	180	165	0.025	2.5	1.52E-04	98.2
180	212	196	0.011	1.1	5.61E-05	99.3
212	250	231	0.004	0.4	1.73E-05	99.7
250	300	275	0.001	0.1	3.64E-06	99.8
300	355	327.5	0.001	0.1	3.05E-06	99.9
355	425	390	0.001	0.1	2.56E-06	100
425	500	462.5	0.001	0.1	2.16E-06	100
500	600	550	0.000	0.0	0.00E-00	100
Total				100.0	1.40E-02	

Mean particle diameter (μm)   71.54

Mean particle diameter at 16%   84% of cumulative mass fraction
                                51   102

Diameter spread (μm)   26

Median particle diameter 69
(μm)
(at 50% of the cumulative
mass fraction)

Particle size distribution   0.37

and $\rho_p = 1.40$ g cm^{-3}, using a graphical solution method based on Equations (16) and (17).

Solution:
    First, calculate d_p^* from Equation (16):

$$d_p^* = d_p \left[\frac{\rho_f(\rho_p - \rho_f)g}{\mu^2} \right]^{1/3} = 0.0070 \left[\frac{0.0012(1.4 - 0.0012)980}{(0.00018)^2} \right]^{1/3} = 2.6$$

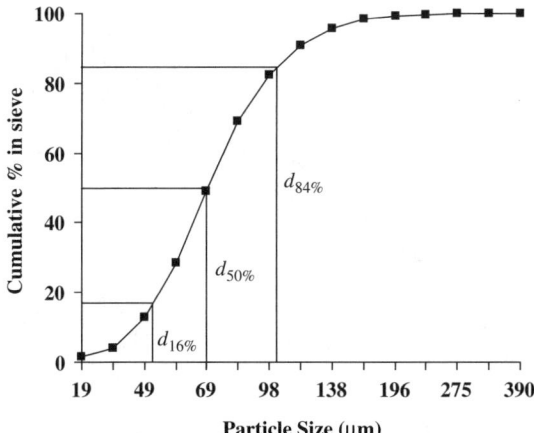

Figure 19 *Worked example E1: cumulative percentage under size versus particle size*

Next, we find u_t^* from the empirical diagram reported in Figure 7, shown again here as Figure 20, choosing the curve that corresponds to $\phi = 0.7$. This gives:

$$u_t^* = 0.28$$

Finally, from Equation (17), we obtain u_t:

$$u_t = u_t^* \left[\frac{\mu(\rho_p - \rho_f)g}{\rho_f^2} \right]^{1/3} = 0.28 \left[\frac{0.00018(1.4 - 0.0012)980}{(0.0012)^2} \right]^{1/3} = 15.5 \, \text{cm s}^{-1}$$

7.5.3 Example E3

A *fine* spherical silica-based catalyst with the physical characteristics reported below is being investigated for the development of a new fluidised bed process. Fundamental parameters such as the bed voidage and minimum fluidisation velocity, u_{mf}, have to be determined.

d_p = particle diameter = 75 μm
ρ_p = particle density = 1770 kg m^{-3}
ρ_f = fluid density = 1.22 kg m^{-3}
μ = fluid viscosity = 1.8 × 10^{-5} kg m^{-1} s^{-1}

(i) Calculate the voidage of the bed, ε, (volume fraction occupied by the voids) when the packed bed of solid particles occupies a depth $L = 0.5$ m in a vessel of diameter $D = 0.14$ m. The mass of

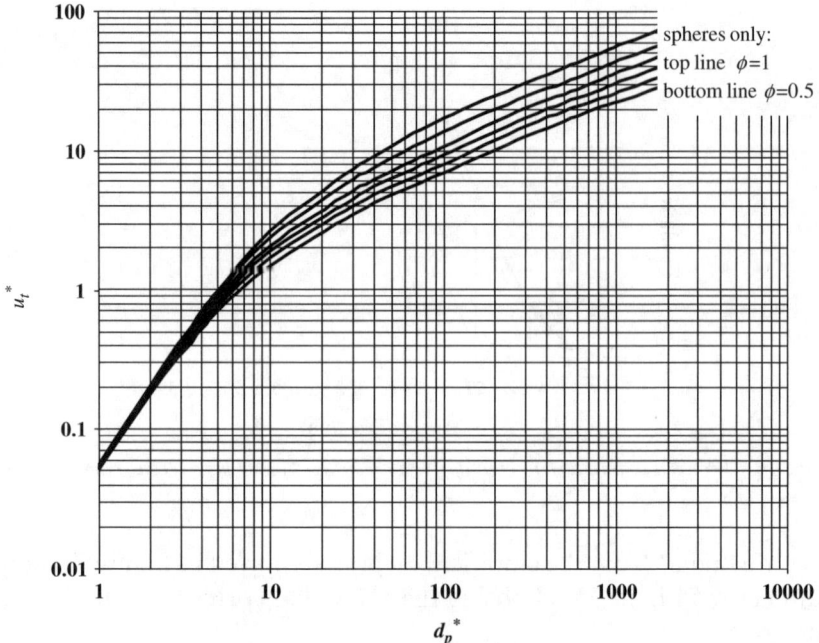

Figure 20 Worked example E2: empirical diagram for the calculation of the particle terminal fall velocity

particle is 8.5 kg. Knowing the voidage, determine the number of particles making up the bed, assuming that they are perfectly spherical.

(ii) Determine u_{mf} using the Ergun equation for pressure drop through packed beds of spherical particles. Different values of the voidage at minimum fluidisation ($\varepsilon_{mf} = 0.38$, 0.42, 0.45) are given in order to estimate the sensitivity of the calculation of u_{mf} on ε_{mf} when using the Ergun equation. Determine u_{mf} also using the Wen-Yu equation.

Solution:

(i)

$$\varepsilon = \text{volume occupied by the gas/total volume of bed}$$
$$\varepsilon = (V_{bed} - V_{particle})/V_{bed} = 1 - V_{particle}/V_{bed}$$
$$\varepsilon = 1 - M/(\rho_p\, L A) = 1 - M/(\rho_p\, L \pi D^2/4)$$

Substituting the values we obtain:

$$\varepsilon = 0.38$$

Knowing the voidage, the number of particles can be determined. The total volume occupied by the particles in the bed is equal to:

$$V = AL(1 - \varepsilon)$$

which is also equal to the number of particles multiplied by the volume of a single particle, thus:

$$AL(1 - \varepsilon) = N_p \pi \frac{d_p^3}{6}$$

from which:

$$N_p = 2.2 \, E + 10$$

(ii) The Ergun equation for pressure drop through packed beds of spherical particles (Equation 32) may be applied to give u_{mf} as follows:

$$Ar = \frac{150(1 - \varepsilon_{mf})}{\varepsilon_{mf}^3} Re_{mf} + \frac{1.75}{\varepsilon_{mf}^3} Re_{mf}^2 \qquad (32)$$

where:

$$Re_{mf} = \frac{\rho_f \, u_{mf} \, d_p}{\mu} ; \qquad Ar = \frac{d_p^3 \, \rho_f (\rho_p - \rho_f) g}{\mu^2}$$

The viscous effects dominate the fluid–particle interaction of small particles (below 100 μm), thus the inertial term of the Ergun equation can be neglected. Hence, the minimum fluidisation velocity can be obtained from:

$$u_{mf} = \frac{\varepsilon_{mf}^3 (\rho_p - \rho_f) d_p^2 g}{150 (1 - \varepsilon_{mf}) \mu} \qquad (33)$$

By substituting the given numerical values, u_{mf} can be obtained for the different values of the voidage:

ε_{mf}	u_{mf} (cm s^{-1})
0.38	0.32
0.42	0.46
0.45	0.6

For fine spherical particles, when using the equation proposed by Wen–Yu, u_{mf} is obtained from the viscous term of Equation (35):

$$u_{mf} = \frac{d_p^2(\rho_p - \rho_f)g}{1650\mu} = 0.33\,\text{cm s}^{-1}$$

The results show that the values calculated using the Ergun equation are very sensitive to the value used for the voidage at minimum fluidisation. The value obtained from the Wen–Yu equation corresponds to that obtained from the Ergun Equation (25) with a voidage of approximately 0.38, as reported in Section 7.3.2.

7.5.4   Example E4

As part of the fluid-bed process development described in E3, the operative conditions and the corresponding fluidisation regime have to be determined.

Calculate the superficial gas velocity which is needed to expand the bed and obtain an average fluid-bed voidage $\varepsilon=0.62$. Consider viscous flow regime. Use Figure 11 to check what fluidisation regime corresponds to the calculated superficial gas velocity.

Find the flow regime at which the fluid-bed would be operated if fluidised at 0.25 m s^{-1} and 1.5 m s^{-1}.

Solution:

The Richardson–Zaki equation (Equation 37) is used in fluidisation to describe the homogeneous expansion. It relates the superficial gas velocity to the fluid bed voidage and particle terminal fall velocity:

$$u = u_t \varepsilon^n \tag{37}$$

where for viscous flow regime, $n=4.8$.

Knowing the particle terminal fall velocity, u_t, the superficial gas velocity which is needed to expand the bed and obtain an average fluid-bed voidage $\varepsilon = 0.62$ can be determined. Using a similar procedure to that adopted in E2, we can use Figure 11 (repeated again here as Figure 21) to obtain the value for u_t, as follows:

From the data given, we can calculate d_p^*

d_p = particle diameter = 75 μm
ρ_p = particle density = 1770 kg m^{-3}

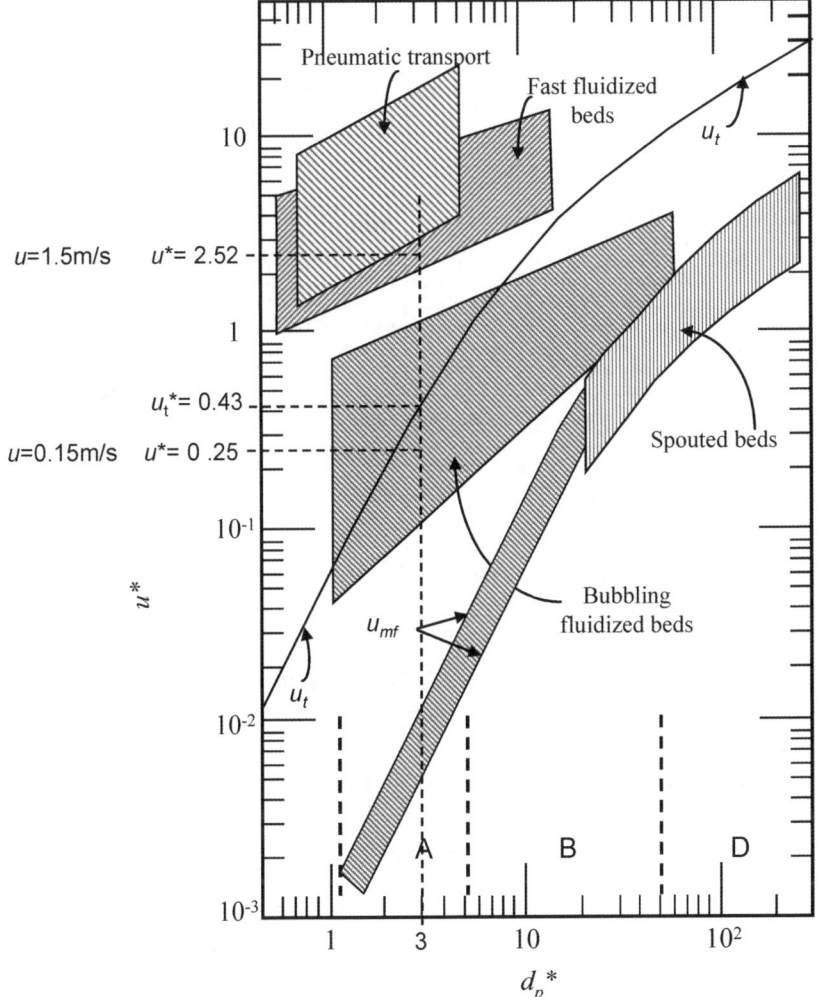

Figure 21 *Worked example E4: empirical diagram for the prediction of the fluidisation regime*

ρ_f = fluid density = 1.22 kg m^{-3}
μ = fluid viscosity = 1.8 × 10^{-5} kg m^{-1} s^{-1}

$$d_p^* = Ar^{1/3} = d_p \left[\frac{\rho_f(\rho_p - \rho_f)g}{\mu^2} \right]^{1/3} = 3$$

Using the diagram in Figure 22, the intersect of the u_t curve with the coordinate corresponding to the calculated d_p^* gives a value for $u_t^* = 0.43$.

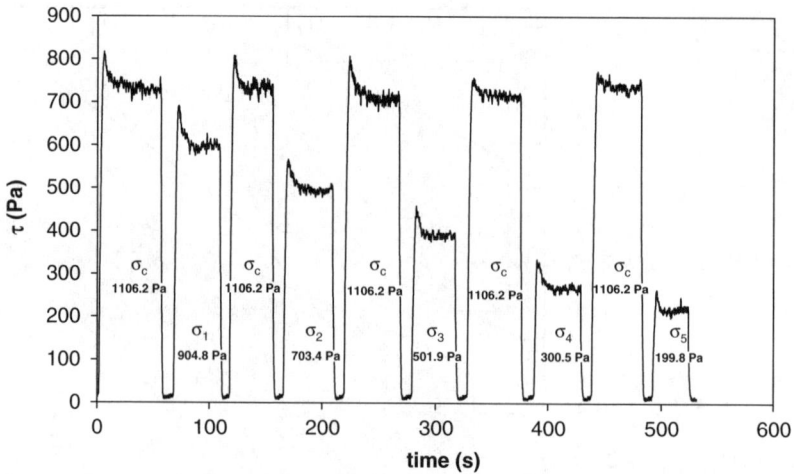

Figure 22   *Worked example E5: experimental shear stress chart ($\sigma_c = 1106.2$ Pa)*

From u_t^* we can therefore back-calculate a value for the terminal fall velocity, u_t:

$$u_t = u_t^* \left[\frac{\rho_f^2}{\mu(\rho_p - \rho_f)g} \right]^{-1/3} = 0.26\,\mathrm{m\,s^{-1}}$$

Thus, the superficial gas velocity needed to expand the bed and obtain a voidage equal to 0.62 is:

$$u = u_t\,\varepsilon^{4.8} = 0.026 \text{ m s}^{-1}$$

We can also check that the regime that corresponds to the calculated superficial gas velocity is homogeneous fluidisation, for which the Richardson–Zaki equation is valid. To do so, we calculate the value of u^* corresponding to $u=0.026$ m s^{-1}, giving $u^* = 0.015$, and using Figure 21 we can verify that the corresponding fluidisation regime is homogeneous fluidisation.

Using Figure 21, the fluidisation regime corresponding to 0.30 m s^{-1} and 1.5 m s^{-1} can also be determined:

$u = 0.15$ m s^{-1} $u^* = 0.25$ *i.e.* Bubbling fluidisation
$u = 1.5$ m s^{-1} $u^* = 2.5$ *i.e.* Fast-turbulent fluidisation

7.5.5   Example E5

The following example reports the experimental results obtained for the effective angle of internal friction and cohesion for three samples of powders.

Table 10 *Worked example E5: material physical characteristics*

	d_p (μm)	F_{45} (%)	ρ_p (kg m^{-3})
A1	75	3.2	1730
A2	49	30	1730
Ballotini	300	—	2500

Table 11 *Worked example E5: Consolidation stresses in internal yield loci experiments*

Normal load (N:)	Normal stress σ_c (Pa)
300	1106.2
210	804.1
150	602.6
90	401.2

Two alumina materials, A1 and A2 belonging to Geldart Group A, and ballotini belonging to Geldart group B (see section 7.4.3) are examined. The physical characteristics of the three materials are summarised in Table 10. A1 is virtually free of fines (particles having size below 45 μm), while A2 contains 30%wt of fine particles. In light of this, the alumina sample A1 is expected to be more cohesive than A2. As reported in section 7.4.3, the ballotini are instead expected to be free from any dominant effect of the interparticle forces.

For each powder, four static yield loci are obtained by applying the normal loads reported in Table 11, to which the corresponding consolidation stresses, σ_c, are calculated from Equation (41). Following the procedure described in section 7.4.4.7, the values of the normal stresses σ_i applied for each consolidation stress σ_c, are reported in Table 12 and also shown in Figure 22.

Figure 22 shows a typical experimental shear stress chart, obtained for A2, using a compaction with normal stress equal to 1106.2 Pa. Figure 23 shows the corresponding yield locus from which the static angle of internal friction, φ, was worked out from the slope of the yield locus. The cohesion, C, is obtained from the intercept with the shear stress axis.

As outlined in section 7.4.4.5, the static angle of internal friction and the cohesion of a granular material are a function of the consolidation stress. Therefore, they can be expressed also as a function of the major principal stress σ_1. Table 13 reports the values obtained for the effective angle of internal friction and the cohesion for all the powders

Table 12 *Worked example E5: Shearing stresses in internal yield loci experiments*

$\sigma_c = 1106.2$ Pa	$\sigma_c = 804.1$ Pa	$\sigma_c = 602.6$ Pa	$\sigma_c = 401.2$ Pa
σ_i (Pa)	σ_i (Pa)	σ_i (Pa)	σ_i (Pa)
904.8	602.6	401.2	300.5
703.4	401.2	300.5	250.1
501.9	300.5	199.8	199.8
300.5	199.8	149.4	149.4
199.8	149.4	—	—

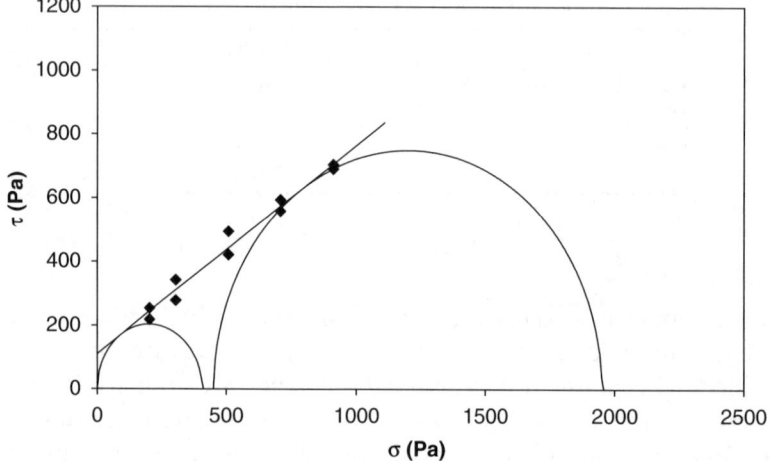

Figure 23 *Worked example E5: experimental static yield locus – Alumina A2, $\sigma_c = 1106.2$ Pa*

Table 13 *Worked example E5: failure properties of the powders*

	Alumina A1	Alumina A2	Ballotini
C (Pa) ($\sigma_1 = 700$ Pa)	98	160	24
$\varphi_e - \varphi$ (°) ($\sigma_1 = 700$ Pa)	22	25	6

investigated. Sample A2, which contains a large amount of fines exhibits the highest values for cohesion and effective angle of internal friction. The ballotini in contrast, are less cohesive, as reflected in the lowest values for both C and $\varphi_e - \varphi$.

7.6 CONCLUDING REMARKS

In this chapter we have introduced some of the fundamental parameters which characterise solid materials, including particle size analysis, shape, density and shear stress, and how they influence the flow properties of powders. The description of the fluid–particle interaction, for both the case of a single particle suspended in a fluid and when flow passes through a packed bed, has also been presented. We have introduced the concept of fluidisation and the principle forces acting on a particle under equilibrium conditions. We have also shown how particle characteristics influence the fluidisation behaviour and presented some industrial applications. A number of relevant worked examples have also been presented to provide the reader with the direct application of the theory.

NOMENCLATURE

A	Cross-sectional area of the bed	m^2
Ar	Archimedes number	—
C	Cohesion	Pa
C_δ	Cohesion from dynamic yield locus	Pa
C_D	Drag coefficient	—
d_p	Mean particle diameter	m
d_{pi}	Sieve aperture	m
d_s	Surface diameter	m
d_v	Volume diameter	m
d_{sv}	Surface/volume diameter	m
d_p^*	Dimensionless particle diameter Equation (16)	—
D	Bed diameter	m
F	Friction factor	—
F_b	Buoyancy force	N
F_d	Drag force on a single particle	N
F_{45}	Fraction of fine particles below 45 μm	
g	Acceleration due to gravity	$m\ s^{-2}$
K	Permeability	$Pa\ s^{-1}$
H	Bed height	m
L	Length of bed	m
m	Mass of a single particle	kg
n	Richardson–Zaki parameter Equation (38)	—
N_c	Pre-consolidation load	kg
N_i	Normal load in shearing phase i	kg
N_p	Number of particles in the bed	—

(*continued*)

Re	Reynolds number	—
Re_{mf}	Reynolds number at minimum fluidisation	—
Re_p	Particle Reynolds number	—
Re_t	Particle terminal Reynolds number	—
U	volumetric flux	$m^3\ s^{-1}\ m^{-2}$
u	Superficial gas velocity	$m\ s^{-1}$
u_f	Fluid phase velocity	$m\ s^{-1}$
u_{mf}	Minimum fluidisation velocity	$m\ s^{-1}$
u_{mb}	Minimum bubbling velocity	$m\ s^{-1}$
u_t	Particle terminal fall velocity	$m\ s^{-1}$
u_t^*	Dimensionless particle terminal fall velocity Equation (17)	—
V_p	Volume of particle	m^3
x_i	Mass fraction of particles in each size range	—

Greek symbols

δ	Dynamic angle of internal friction	Deg
ε	Bed voidage	—
ε_{mf}	Minimum fluidisation voidage	—
ϕ	Particle sphericity	—
ϕ_i	Angle of internal friction	Deg
φ_e	Effective angle of internal friction	Deg
μ	Gas viscosity	$Ns\ m^{-3}$
ρ_{BDL}	Loose bulk density	$kg\ m^{-3}$
ρ_{BDP}	Packed bulk density	$kg\ m^{-3}$
ρ_{ABS}	Absolute density	$kg\ m^{-3}$
ρ_b	Bulk density	$kg\ m^{-3}$
ρ_g	Gas density	$kg\ m^{-3}$
ρ_f	Fluid density	$kg\ m^{-3}$
ρ_p	Particle solid density	$kg\ m^{-3}$
ΔP	Bed pressure drop	Pa
σ	Normal stress	$N\ m^{-2}$
σ_c	Consolidation stress	$N\ m^{-2}$
σ_d	Mean deviation	m
σ_d/d_{50}	Relative diameter spread	—
$\sigma_1,\ \sigma_2$	Major stress	$N\ m^{-2}$
$\sigma_i,\ \sigma_{ii}$	Shearing stress	$N\ m^{-2}$
τ	Shear stress	$N\ m^{-2}$

REFERENCES

1. T. Allen, *Particle Size Measurements*, Chapman and Hall, London, 1996.
2. M. Rhodes, *Introduction to Particle Technology*, Wiley, London, 1998.
3. P. Lettieri, J.G. Yates and D. Newton, Homogeneous bed expansion of FCC catalysts, influence of temperature on the Richardson–Zaki index n and on the particle terminal fall velocity, *Powder Technol.*, 2002, **123**, 221–231.
4. D. Geldart, *Gas Fluidization Technology*, Wiley, London, 1986.
5. J.P.K. Seville, U. Tuzun and R. Clift, *Processing of Particulate Solids*, Chapman & Hall, London, 1997.
6. B.S. Massey, *Mechanics of Fluids*, 6th edn, Van Nostrand Reinhold, London, 1989.
7. J.M. Dallavalle, *Micromeritics*, Pitman, New York, 1948.
8. A. Haider and O. Levenspiel, Drag coefficient and terminal velocity of spherical and nonspherical particles, *Powder Technol.*, 1989, **58**, 63–70.
9. H.P.G. D'Arcy, *Les Fontaines Publiques de la Ville de Dijon*, Victor Dalmont, Paris, 1856.
10. L.G. Gibilaro, *Fluidization Dynamics*, Butterworth Heinemann, Oxford, 2001.
11. S. Ergun, Fluid flow through packed columns, *Chem. Eng. Prog.*, 1952, **48**(2), 89–94.
12. C.Y. Wen and Y.H. Yu, Mechanics of fluidization, *Chem. Eng. Prog. Symp. Ser.*, 1966, **62**, 100–111.
13. J.F. Richardson and W.N. Zaki, Sedimentation and Fluidization: Part I, *Trans. Instn. Chem. Engrs.*, 1954, **32**, 35–53.
14. A.R. Abrahamsen and D. Geldart, Behaviour of gas-fluidized beds of fine powders Part I. Homogeneous expansion, *Powder Technol.*, 1980, **26**, 35–46.
15. J.R. Grace, Contacting modes and behaviour classification of gas–solid and other two-phase suspensions, *Can. J. Chem. Eng.*, 1986, **64**, 353–363.
16. D. Geldart, Types of fluidization, *Powder Technol.*, 1973, **7**, 285.
17. J.P.K. Seville, C.D. Willett and P.C. Knight, Interparticle forces in fluidization: a review, *Powder Technol.*, 2000, **113**, 261–268.
18. J. Israelachvili, *Intermolecular and Surface Forces*, Academic Press, London, 1991.
19. R.L. Carr, Particle behaviour storage and flow, *Br. Chem. Eng.*, 1970, **15**(12), 1541–1549.

20. D. Geldart and A.C.Y. Wong, Fluidization of powders showing degrees of cohesiveness – I. Bed expansion, *Chem. Eng. Sci.*, 1984, **39**(10), 1481–1488.
21. J. Schwedes, Consolidation and flow of cohesive bulk solids, *Chem. Eng. Sci.*, 2002, **57**(2), 287–294.
22. A.E. Jenike, *Gravity Flow of Bulk Solids*, Utah Engineering Experimental Station, University of Utah, Bulletin 108, 1961.
23. A.E. Jenike, *Storage and Flow of Solids*, Utah Engineering Experimental Station, University of Utah, Bulletin 123, 1964.
24. G. Bruni, D. Barletta, M. Poletto, P. Lettieri, A rheological model for the flowability of fine aerated powders, *Chem. Eng. Sci.*, 2006, in press.
25. D. Kunii and O. Levenspiel, *Fluidization Engineering*, Butterworths, London, 1991.
26. J.G. Yates, Effects of temperature and pressure on gas–solid fluidization, *Chem. Eng. Sci.*, 1996, **51**(2), 167–205.

An Introduction to Process Control

EVA SORENSEN

8.1 INTRODUCTION

All chemical plants need process control to ensure that they are operated safely and profitably, while at the same time satisfying product quality and environmental requirements. Furthermore, modern plants are becoming more difficult to operate due to the trend towards complex and highly integrated processes. For such plants, it is difficult to prevent disturbances from propagating from one unit to other, interconnected, units. The subject of process control has therefore become increasingly important in recent years and it is vital for anyone working on a chemical plant to have an understanding of both the basic theory and practice of process control.

This chapter will give a brief summary of basic control theory as applied in most chemical processing plants. It will start with a discussion of process dynamics and why an understanding of the dynamics, or *the transient behaviour*, of a process is essential in order to achieve satisfactory control. Standard feedback controllers will then be discussed and different strategies for tuning such controllers will be presented. A brief overview of more advanced control strategies, such as feed-forward control, cascade control, inferential control and adaptive control, and for which processes these may be suitable, will be given next. Finally, an introduction will be given to plant-wide control issues.

In most process control courses for chemical engineers, the first part of the course normally deals with the development of dynamic process models from first principles (mass and energy balances), since these are used in the analysis of process dynamics and often also for controller tuning. In this chapter, however, the focus will be on process *control* and modelling will not be considered. Neither will the chapter consider frequency response techniques for process analysis and controller tuning, as these are considered beyond the scope of this book. The emphasis will therefore be on control *concepts* and the reader is referred to standard control textbooks for more information.[1-4] In addition, although instrumentation is essential for control, it is a complete subject area in itself and is therefore not considered here.

8.2 PROCESS DYNAMICS

The primary objective of process control is to maintain a process at the desired operating conditions, safely and efficiently, while satisfying environmental and product quality requirements. The subject of process control is concerned with how to achieve these goals. Luyben[3] gives the following process control laws:

First law: The simplest control system that will do the job is the best. Bigger is definitely not better in process control.
Second law: You must understand the process before you can control it. Ignorance about the process fundamentals cannot be overcome by sophisticated controllers.

The process control system should ensure that the process is maintained at its specified operating conditions at all times. To be able to do this, we must first understand the process dynamics as advised by Luyben in his Second Law.[3] The dynamics of a process tells us how the process behaves as a result of the changes. Without an understanding of the dynamics, it is not possible to design an appropriate control system for it.

In the following text, important aspects in the study of process dynamics are outlined. An example of a dynamic process is given first. *Stability* of a process is defined next, followed by a discussion of typical uncontrolled, or *open loop*, responses.

8.2.1 Process Dynamics Example

As an example of a dynamic process, consider the process in Figure 1, which is a tank into which an incompressible (constant density) liquid is pumped at a variable feed rate F_i (m^3 s^{-1}). This inlet flow rate can vary with time because of changes in operations upstream of the tank. The height in the tank is h (m) and the outlet flow rate is F (m^3 s^{-1}). Liquid leaves the tank at the base via a long horizontal pipe and discharges into another tank. Both tanks are open to the atmosphere. F_i, h and F can all vary with time and are therefore *functions of time t*.

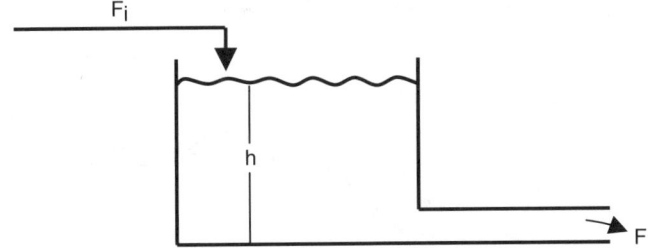

Figure 1 *Gravity-flow tank*

Let us now consider the dynamics of this tank, starting with the *steady state* conditions. By steady state we mean the conditions when nothing changes with time. The value of a variable at steady state is normally denoted by a subscript *s*. (Note that some control textbooks use a different notation, for instance, by placing a bar over the variables.) At steady state, the flow out of the tank, F_s, must equal the flow into the tank, $F_{i,s}$, and the height of the liquid in the tank is constant, h_s. At steady state we therefore have the steady-state equation:

$$F_{i,s} = F_s$$

The value of h_s is that height which provides enough hydraulic pressure head at the inlet of the pipe to overcome the frictional losses of liquid flowing out and down the pipe. The higher the flow rate $F_{i,s}$, the higher the height h_s.

Now consider what would happen dynamically if the inlet flow rate F_i changed. How will the height *h* and the outlet flow rate *F* vary with time? Figure 2 is a sketch of the problem.

The problem is to determine which curves (1 or 2) represent the actual paths that *F* and *h* will follow after a step change in F_i. The curves marked "1" show gradual increases in *h* and *F* to their new steady-state values. However, the paths could follow the curves marked "2", where the liquid height rises above its final steady-state value. This is called *overshoot*. Clearly, if the peak of the overshoot in *h* is above the top of the tank, the tank would overflow. The steady-state calculations give no information about what the dynamic response of the system will be. Before a controller can be designed to control the tank height, the designer must determine which of the curves (1 or 2) the tank height *h* and the flow rate *F* are most likely to follow, *i.e. what are the dynamics of the process?*

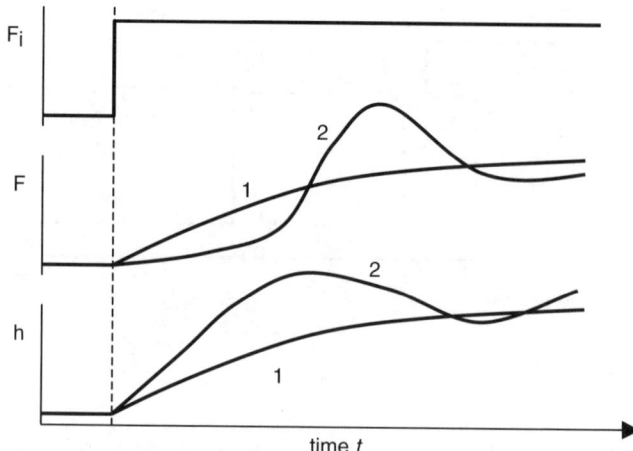

Figure 2   *Dynamic responses of gravity-flow tank*

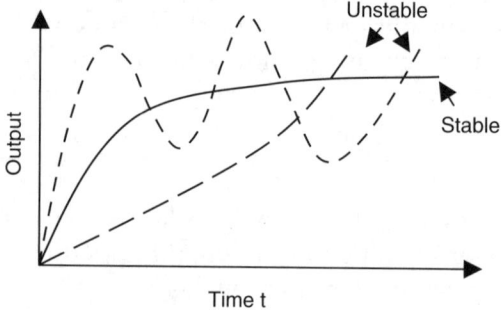

Figure 3   *Stability of a process*

8.2.2   Stability

A process is said to be unstable if its output becomes larger and larger
(either positively or negatively) as time increases, as illustrated in Figure
3. Note that no real system really does this, as some constraint is usually
encountered, for example, the level in a tank may overflow, a valve on a
flow stream may completely shut or completely open or a safety valve
may blow.

   Most processes are *open loop stable*, also called *self-regulating, i.e.*
stable without any controllers on the system. This means that after a
change in the system caused by either a disturbance or by a deliberate
change in a manipulated variable, the process will come to a new stable
operating condition, *i.e.* a new steady state. This is not necessarily the

desired operating condition, and in most cases it is not, but nevertheless, the new operating condition is stable. If, however, the process does not settle out to a new steady state, it is called non-self-regulating and will require control in order to ensure that the system remains safe and stable. In the tank example, as F is free to vary with the liquid level in the tank, the process is self-regulating since if F_i increases, the level will rise causing F to increase, thereby bringing the system to a new steady state where F rises to the new value of F_i. If, however, there is a pump on the outlet line, *i.e.* F remains constant whatever the conditions within the tank, the system will be non-self-regulating as the tank would overflow if F_i increased and F remained constant.

What makes controller design challenging is that all real processes can be made *closed loop unstable* when a controller is implemented to steer the process to specified operating conditions. In other words, a process which is open loop stable and therefore will come to a new, although not the desired, steady state after a disturbance may become unstable when a controller is implemented to steer the process towards the desired steady state. Stability is therefore of vital concern in all control systems.

8.2.3 Typical Open Loop Responses

The term *process response* is used to describe the shape of the plot one would obtain if one plotted the controlled variable as a function of time after a disturbance or set point change. In the tank example, the response would typically be given by tank height h as a function of time and would be referred to as "the response in h". Almost all chemical processes fall within a few standard response shapes, which makes analysing them easier. These standard responses are illustrated in Figure 4. The *order* of a process refers to the order of the differential equations describing the system, and refers loosely to the number of parts of the process in which material and energy can be contained. Simple processes can often be assumed to be of first order, while more complex processes are of second order or higher. In Figure 4, y refers to the variable that is observed. For instance, in the tank example, y is h.

Dead time is the length of time, t_d, after a disturbance occurs before any change in the output is noticed and is illustrated for a first order process in Figure 4(c). It occurs particularly when material or energy is being transported some distance. Clearly it means that disturbance detection is slow and control actions are made on the basis of old measurements. Dead time is the most troublesome characteristic of a

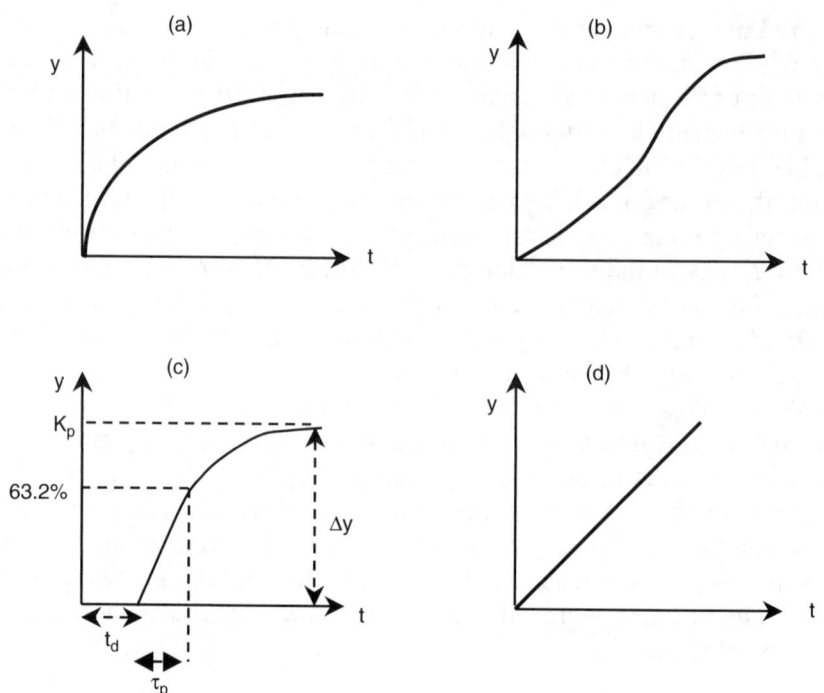

Figure 4   *Typical open-loop unit step responses. (a) First order, (b) Second (or higher) order, (c) first order with dead time and (d) non-self-regulating process*

system to control and can cause otherwise stable systems to become unstable when the control loop is closed.

The process gain, K_p, of the system is the value that the response will go to at steady state after a unit step change in the input, *i.e.* it relates the input to the output at steady state.

The *time constant*, τ_p, of a process is, for a first-order system, defined as the time it takes for the response to reach 63.2% of the final response value (see Figure 4(c)). Theoretically, the process never reaches the new steady-state value, given by the process gain K_p, except at $t \rightarrow \infty$, although it reaches 99.3% of the final steady-state value when $t = 5\tau_p$. The difference between dead time and time constant must be emphasised; dead time is the time it takes before anything at all happens, while time constant is a measure of how slowly or quickly a response settles to its new steady-state value once the change has started. A slow process that has a large time constant does not necessarily have any dead time. Time constants in chemical engineering processes can range from less than a second to a couple of hours, or even days for biochemical applications.

8.3   FEEDBACK CONTROL SYSTEMS

Most plant control systems are very simple and are normally standard feedback controllers, either P-, PI- or PID controllers. These will be described in more detail in this section, together with two techniques that can be used to tune these controllers. First, however, we need to define the *control objective, i.e.* what do we want the controller to do, and define what we mean by *feedback control*.

8.3.1   Disturbance Rejection and Set Point Tracking

The operation of a process may deviate from its desired operating conditions for two different reasons:

Disturbances. These are changes in flow rates, compositions, temperatures, levels or pressures in the process which we can not control because they are either given by the feed conditions (assumed controlled by an upstream unit), ambient conditions, such as the weather, or utilities, such as steam or cooling water.
Set point changes. These are deliberate changes in the operating conditions, such as a change in polymer grade for a polymerisation reactor, change in distillate composition for a distillation column, *etc.*

The control objective is different in the two cases. In *disturbance rejection*, the control objective is to reject the disturbance as quickly as possibly, *i.e.* to bring the process back to the original steady state by counteracting the effect of the disturbance (see Figure 5(a)). As an example, consider a person in a shower who wants to maintain the water temperature at a constant value. To achieve this, the person may have to turn the cold water tap up to increase the flow rate of the cold water if there is a sudden increase in temperature as a result of someone else in the house using cold water, *i.e.* to fill a kettle in the kitchen. (This is an example of a disturbance caused by a shared utility, here the cold water.)

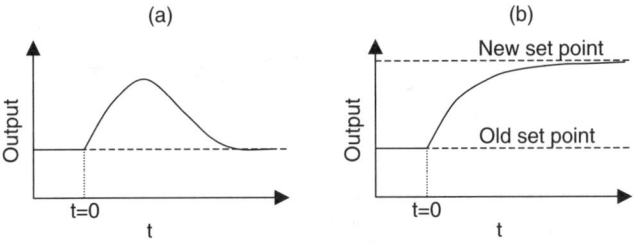

Figure 5   *Process responses: (a) disturbance rejection and (b) set point tracking*

In *set point tracking*, the set point for the controller is changed and the control objective is to bring the process to the new set point as quickly as possible (see Figure 5(b)). For a person in the shower who wants to reduce the temperature at the end of the shower time, this means turning the hot water tap down to reduce the flow rate of hot water, thereby bringing the temperature down to the new desired level.

8.3.2   Feedback Control Loop

A feedback control loop is generally illustrated as shown in Figure 6. The *Process* refers to the chemical or physical process (the tank in the example earlier). The *Measuring Device* is used to measure the value of the variable that is to be controlled (the level indicator in the tank example or a thermometer in a shower). The *Control Element* (usually a flow control valve or the setting on a heater or cooler) changes the value of the *manipulated variable*. The manipulated variable is the variable used to change the *controlled variable* (in the tank example, the output flow rate is the manipulated variable used to change the level in the tank, which is the controlled variable).

The input to the *Feedback Controller* (P-, PI-, PID- or PD-controller, discussed later) is the difference (also called error) between the set point and the measured variable; hence the minus sign in the control loop. The controller has its name from the fact that the comparison between the controlled variable and the set point is *fed back* to the controller. The Feedback Controller determines the change required in the manipulated variable to bring the error back to zero, *i.e.* the controlled variable to its set point value, and sends a signal to the control element that executes this change. The effect of this change on the process, together with the effect of any disturbances, is then measured and compared to the set point and the loop started again.

Feedback control systems can be either *analogue*, where the controller is a mechanical device, or *digital*, where the controller is a computer or

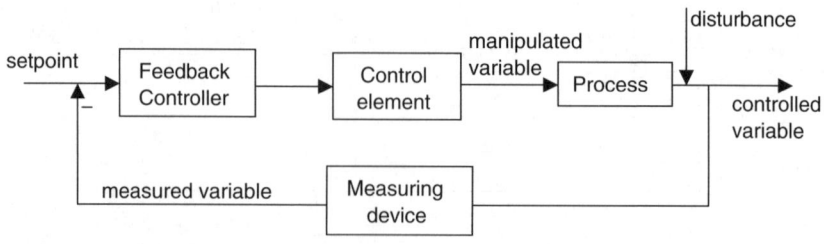

Figure 6   *Feedback control loop*

microprocessor. A digital control system will have additional elements in the control loop to convert the various signals from analogue to digital or *vice versa.*[1-11]

How often the control loop is executed is determined by the *sampling interval*, which is how often the measurement is taken. A critical pressure on an exothermic reactor may be sampled several times per second, while the level of a buffer tank may only be sampled a few times an hour. The sampling time must be chosen with care to ensure that possible changes in the process are detected early enough for the controller to take appropriate action. However, sampling too often is undesirable as it may upset the process and because a large number of data points would have to be sampled and stored.

8.3.3 P, PI and PID Controllers

The most commonly used controller in the process industries is the three term or PID controller. This controller is a feedback controller and adjusts the manipulated variable in proportion to the change in its output signal, c, from its steady state value (bias), c_s, on the basis of a measurement of the error in the controlled variable, ε, which is given by

$$\varepsilon(t) = y_{sp}(t) - y(t) \tag{1}$$

where ε is the error signal, $y_{SP}(t)$ is the set point and $y(t)$ is the measured value of the controlled variable, *e.g.* the height of the liquid in the tank example. The change required in the control signal c is obtained from the error (calculated numerically in a digital controller or obtained by a mechanism inside an analogue controller) by the following relationship:

$$c = c_s + K_c \left(\varepsilon + \frac{1}{\tau_I} \int_0^t \varepsilon \, dt + \tau_D \frac{d\varepsilon}{dt} \right) \tag{2}$$

where P is the Proportional, I is the Integral (Reset) and D is the Derivative (Preact).

The relationship involves three terms and three adjustable parameters, the *controller gain*, K_c, the *integral time*, τ_I, and the *derivative time*, τ_D (hence the name). Finding the right values of these parameters for the best possible control action is called *tuning*. There are several techniques available for controller tuning as will be discussed later.

8.3.3.1   Proportional Action.   Control action is proportional to the size of the error, and Equation 2 becomes:

$$c - c_s = K_c \, \varepsilon$$

That is, the change required in the manipulated variable is proportional to the error in the controlled variable. Some control literature refer to the *proportional band* which is defined as $100/K_c$. Some *offset* is associated with this control action, as will be shown later.

8.3.3.2   Integral Action.   Control action is proportional to the sum, or integral, of all previous errors. This controller eliminates offset. Some control textbooks refer to the *reset rate*, which is defined as $1/\tau_I$.

8.3.3.3   Derivative Action.   Control action is proportional to the rate of change of the error and so anticipates what the error will be in the immediate future.

8.3.3.4   Composite Control Action.   Composite control actions are also possible, *i.e.* P, PI or PD, as well as PID. The most common is PI as P alone will have an offset and the derivative action in PID or PD can introduce unnecessary instability in the response.

8.3.4   Closed Loop Responses

When a controller is implemented on a process, the response after either a disturbance or a set point change is called the *closed loop response* as the control loop in Figure 6 is now closed (recall that open loop means without controller). Figure 7 shows typical closed loop responses of a first-order process to a unit step change in the input. Figure 7(a) and (b) show how Proportional control introduces some offset from the final desired steady state for both set point tracking and disturbance rejection after a change at $t = 0$. The size of the offset depends on the gain of the process K_p and on the proportional gain of the controller K_c and is, for a first-order process, given by:

$$\text{Offset} = \frac{1}{1 + K_p K_c}$$

The offset decreases as K_c becomes larger and should theoretically go towards zero for infinitely large controller gains. However, using very large controller gains should generally be avoided as it may lead to unstable control if any deadtime, however small, is present in the system.

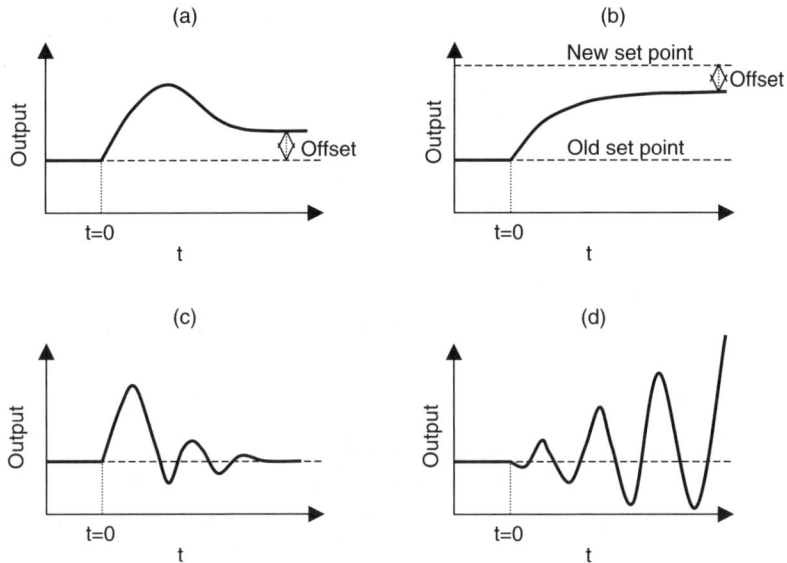

Figure 7 *Closed loop responses of first-order process. (a) P only (set point tracking), (b) P only (disturbance rejection), (c) stable PI or PID (disturbance rejection) and (d) unstable PI or PID (disturbance rejection).*

Figure 7(c) shows that Integral action removes offset but at the expense of introducing some oscillation. The level of oscillation depends on the control parameters. Systems with significant amounts of dead-time will always result in some oscillation in the closed loop response. Figure 7(d) shows how incorrect controller tuning may lead to an unstable response whereby the changes made in the manipulated variable by the controller are too large, causing the measured variable to fluctuate more and more (for the example with a person in the shower, if the person makes a large change in cold water flow rate to compensate for the loss of cold water when the kettle is being filled in the kitchen, the water temperature may become too low. The person may then compensate with a large decrease in cold water flow rate causing the water to become too hot, *etc.*).

8.3.5 Controller Tuning

How do we choose the values of the controller parameters K_c, τ_I and τ_D? They must be chosen to ensure that the response of the controlled variable remains stable and returns to its steady-state value (disturbance rejection), or moves to a new desired value (set point tracking), quickly. However, the action of the controller tends to introduce oscillations.

How quickly the controller responds and with how little oscillation, depends on the application. The following are some of the available tuning methods, of which we will consider the last two in detail (see textbooks[1-11] for numerous other methods):

(i) trial and error (not recommended);
(ii) theoretical methods (frequency response methods, see text books);
(iii) continuous cycling (or Ziegler–Nichols) method;
(iv) process reaction curve (or Cohen–Coon) method.

8.3.5.1 Continuous Cycling Method (Ziegler–Nichols Tuning). In the *continuous cycling method,* the system (process with controller) is brought to the edge of stability under Proportional control only. Suitable values of the parameters can then be determined from the proportional gain K_c found at that condition. The procedure is as follows:

(i) close the feedback loop;
(ii) turn on proportional action only (equivalent to setting $\tau_I = \infty$, $\tau_D = 0$ for a PID controller);
(iii) increase the controller gain, K_c, until the process starts to oscillate. Continue to slowly increase the gain until the cycles continue with constant amplitude, as shown in Figure 8 (called *standing oscillations*);
(iv) note the period of these cycles P_u (distance in time between two peaks) and the value of K_c at which they were obtained (called K_u);
(v) determine the controller settings according to the tuning (Ziegler–Nichols) rules in Table 1.

The settings obtained by this method are good initial estimates but are not optimal and some retuning may be necessary. Note that this method

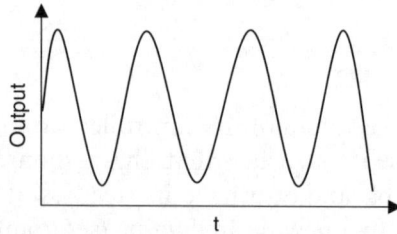

Figure 8 *Standing oscillations in Continuous Cycling (Ziegler–Nichols) method*

Table 1 *Continuous cycling (Ziegler–Nichols) tuning parameters*

	K_c	τ_I	τ_D
P	0.5 K_u	–	–
PI	0.45 K_u	$P_u/1.2$	–
PID	0.6 K_u	$P_u/2$	$P_u/8$

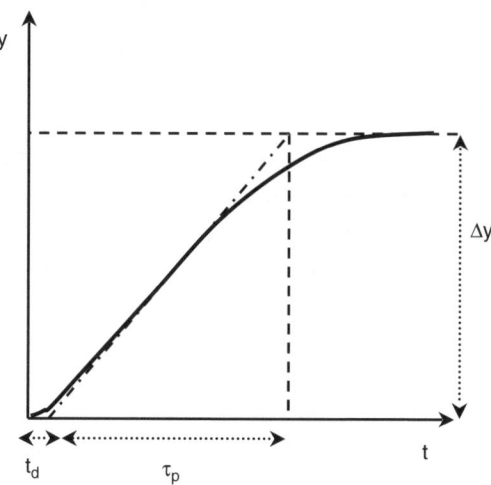

Figure 9 *Process reaction curve (Cohen–Coon) method*

operates the system at the brink of instability and if the controller gain K_c is chosen too high during the tuning procedure, the system will become unstable. The method is therefore not recommended for processes in which instability may lead to dangerous situations (*e.g.* runaway for a reactor).

8.3.5.2 Process Reaction Curve Method (Cohen–Coon Tuning). For some processes, it may be difficult or hazardous to operate with continuous cycling, even for short periods. The *process reaction curve method* obtains settings based on the *open loop* response and thereby avoids the potential problem of closed loop instability. The procedure is as follows:

(i) disconnect the control loop between the controller and the manipulated variable (valve);
(ii) make a step change in the manipulated variable (valve opening);
(iii) record the response of the process (should be similar to Figure 9);

Table 2   *Process reaction curve (Cohen–Coon) tuning parameters*

	K_c	τ_I	τ_D
P	$\tau/Kt_d(1 + t_d/3\tau)$	–	–
PI	$\tau/Kt_d(0.9 + t_d/12\tau)$	$t_d/(30 + 3t_d/\tau)(9 + 20t_d/\tau)$	–
PID	$\tau/Kt_d(1.33 + t_d/4\tau)$	$t_d/(32 + 6t_d/\tau), (13 + 8t_d/\tau)$	$4t_d/(11 + 2t_d/\tau)$

(iv) determine the controller settings according to the tuning (Cohen–Coon) rules in Table 2. The parameter K is defined as

$$K = \frac{\Delta y}{\Delta c}$$

where Δy is the final change in the output (the controlled variable) and Δc is the initial change in the input (the manipulated variable). The tangent line is drawn as a tangent to the point of inflection. The dead time t_d and process time constant τ_p are as shown in Figure 9.

The settings obtained by this method are good initial estimates but are not optimal and some retuning may be necessary as for the Continuous Cycling method.

8.3.6   Advantages and Disadvantages of Feedback Controllers

Feedback control is the most important control technique and is widely used in the process industries. Its main advantages are as follows:

(i) Corrective action occurs as soon as the controlled variable deviates from the set point, regardless of the source and type of disturbance.

(ii) Feedback control requires minimal knowledge about the process to be controlled, in particular, a mathematical model of the process is not required, although having a model is very useful for control system design.

(iii) The PID controller is both versatile and robust. If process conditions change, re-tuning the controller usually produces satisfactory control.

However, feedback control also has certain inherent disadvantages:

(i) No corrective action is taken until a deviation in the control variable occurs. Thus, perfect control, where the controlled variable does not deviate from the set point during disturbance or set point changes, is theoretically impossible.

(ii) It does not provide predictive control action to compensate for the effects of known or measurable disturbances.

(iii) It may not be satisfactory for processes with large time constants and/or long time delays. If large and frequent disturbances occur, the process may be operated continuously in a transient state and never attain the desired steady state.

(iv) In some situations, the controlled variable cannot be measured on-line, and consequently, feedback control is not feasible.

8.4 ADVANCED CONTROL SYSTEMS

Although widely used in the process industries, there are situations where feedback controllers are not sufficient to achieve good control. In this section, several more specialised strategies will be introduced that provide enhanced process control beyond that which can be obtained with conventional single-loop feedback controllers. As processing plants become more and more complex in order to increase efficiency or reduce costs, there are incentives for using such enhancements, which fall under the general classification of *advanced control*. This section introduces several different strategies that are widely used industrially and which in many cases utilise the principles of single-loop feedback controller design but with enhancements:

(i) Feed-forward control
(ii) Ratio control
(iii) Cascade control
(iv) Inferential control
(v) Adaptive control

8.4.1 Feedforward Control

In the previous section, it was emphasised that feedback control, though widely used, has certain disadvantages. For situations in which feedback control by itself is not satisfactory, significant improvements can be achieved by adding *feedforward control*. The basic concept of feedforward control is to measure (or estimate) the disturbances on-line and take corrective action *before* they upset the process. In contrast, feedback control does not take corrective action until *after* the disturbance has upset the process and generated a non-zero error signal. A diagram of a general feedforward controller is given in Figure 10. A feedforward controller consists of a model of the process which is used to predict the

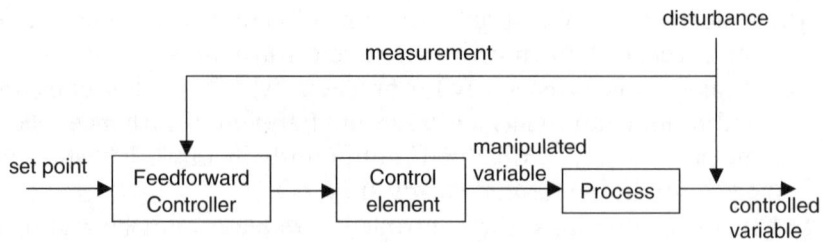

Figure 10   *Feedforward control*

performance of the process and this controller is therefore *not* of P-, PI-
or PID-type. The model determines the value of the manipulated variable which is required to keep the controlled variable at its set point
value given the value of the disturbance.

A feedforward controller has the potential of perfect control, however, in reality it has several disadvantages:

(i)   The disturbance variables must be measured on-line. In many
      applications, this is not feasible.
(ii) To make effective use of feedforward control, at least an ap-
      proximate process model must be available. In particular, we
      need to know how the controlled variable responds to changes in
      both the disturbance and the manipulated variable and the
      quality of feedforward control depends on the accuracy of the
      process model.
(iii) Ideal feedforward controllers that are theoretically capable of
      achieving perfect control may not be physically realisable. For-
      tunately, practical approximations of these ideal controllers often
      provide effective control.

8.4.1.1   Feed-Forward – Feedback Control.   In practical applications,
feedforward control is normally used in combination with feedback
control. The feedforward part is used to reduce the effects of measurable
disturbances, while the feedback part compensates for inaccuracies in
the process model, measurement errors and unmeasured disturbances.
The feedforward and feedback controllers can be combined in several
different ways, as discussed in most standard control text books.

8.4.2   Ratio Control

Ratio control is a special type of feedforward control that has had
widespread application in the process industries. Its objective is to

maintain the ratio of two process variables at a specified value. The two variables are usually flow rates, where one is a manipulated variable and the other a disturbance. The ratio controller will change the manipulated variable as the disturbance changes to maintain the specified ratio between the two variables. Typical applications include: (1) specifying the relative amounts of components in blending operations, (2) maintaining a stoichiometric ratio of reactants into a reactor and (3) keeping a specified reflux ratio in a distillation column.

8.4.3 Cascade Control

A disadvantage of feedback controllers is that corrective action is not taken until *after* the controlled variable deviates from the set point. Cascade control can significantly improve the response to disturbances by employing a second measurement point and a second feedback controller. The secondary measurement is located so that it recognises the upset condition sooner than the controlled variable. Note that the disturbance is not necessarily measured.

As an example of when cascade control may be advantageous, consider the jacketed reactor in Figure 11. The reaction is exothermic and the heat generated by the reactor is removed by the cooling water in the cooling jacket. The control objective is to keep the temperature of the reacting mixture constant. Possible disturbances are the temperature of the feed and the temperature of the cooling water. The only manipulated variable available for temperature control is the cooling water flow rate.

- *Single Feedback Control.* The conventional single loop control in Figure 11(a) will respond much faster to changes in the feed temperature than to changes in the cooling water temperature.
- *Cascade Control.* The response of the simple feedback control to changes in the cooling water temperature can be improved by measuring the cooling water temperature and taking control action *before its effect has been felt by the reacting mixture* as shown

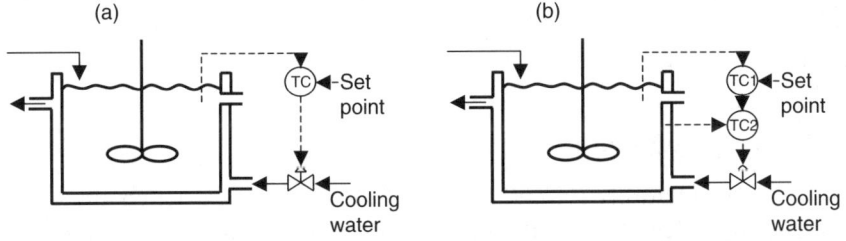

Figure 11 *Jacketed reactor. (a) single-loop control and (b) cascade control*

in Figure 11(b). Thus if the cooling water temperature goes up, the flow rate can be increased to remove the same amount of heat, and the reaction mixture is not affected.

In cascade control, we therefore have two control loops using two different measurements but sharing a common manipulated variable. The loop that measures the controlled variable (in the example, the reacting mixture temperature) is the dominant, or *primary* control loop (also referred to as the *master* loop) and uses a set point supplied by the operator, while the loop that measures the second variable (in the example, the cooling water temperature) is called the *secondary* (or *slave*) loop and uses the output from the primary controller as its set point. Cascade control is very common in chemical processes and the major benefit to be gained is that *disturbances arising within the secondary loop are corrected by the secondary controller before they can affect the value of the primary controlled output.*

8.4.4 Inferential Control

In all the control systems considered so far, it has been assumed that measurements of the controlled variable were available. In some control applications, however, the process variable that is to be controlled cannot be conveniently measured on-line. For example, product composition measurements may require that a sample be sent to the plant laboratory from time to time. In this situation, measurement of the controlled variable may not be available frequently enough or quickly enough to be used for feedback control.

One solution to this problem is to employ *inferential control*, where process measurements that can be obtained more rapidly are used with a mathematical model to *infer* the value of the controlled variable, as illustrated in Figure 12. For example, if the overhead product stream in a distillation column cannot be analysed on-line, measurement of a selected tray temperature may be used to infer the actual composition. If necessary, the parameters in the model may be updated, if composition measurement become available, as illustrated by the second measuring device in Figure 12 (dashed lines).

8.4.5 Adaptive Control

Controllers inevitably require tuning of the controller settings to achieve a satisfactory degree of control if the process operating conditions or the environment changes significantly, *e.g.* because of heat exchanger

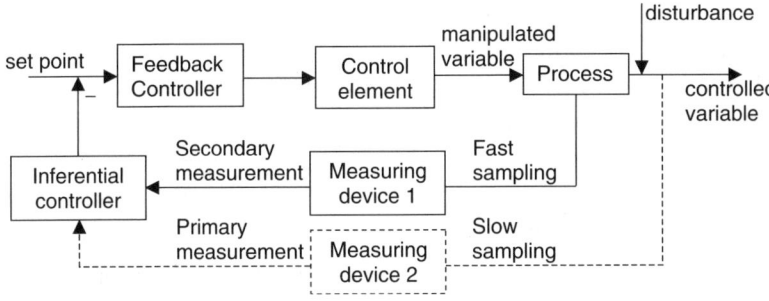

Figure 12 *Inferential control*

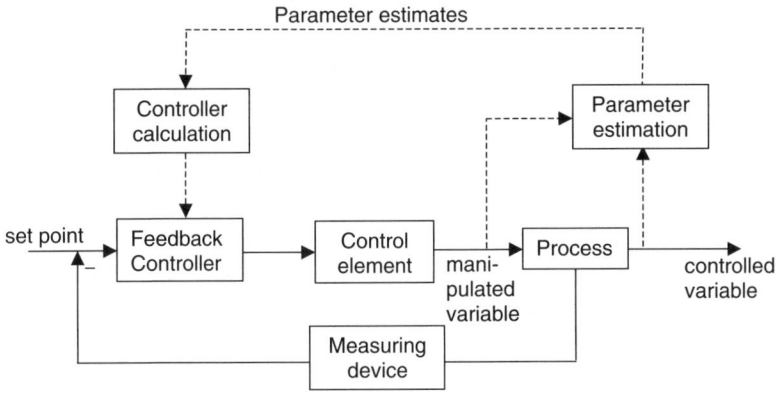

Figure 13 *Adaptive (self-tuning) control*

fouling, changes in feed composition or in product specifications or because of ambient variations, such as rain storms. If these changes occur frequently, then adaptive control techniques should be considered. An *adaptive control system* is one in which the controller parameters are adjusted automatically to compensate for changing process conditions.

If the process changes can be anticipated or measured directly and the process is reasonably well understood, then *gain scheduling* may be used.[5,11] In gain scheduling, the controller gain is changed based on the measurement of a scheduling variable, usually the controlled variable or the set point. A common example of gain scheduling is pH neutralisation, where one value of the controller gain is used at low pH and another at high pH. The adaptation from one value to the other depends on the value of the pH, *i.e.* the controlled variable.

When the process changes cannot be measured or predicted, the adaptive control strategy must be implemented in a feedback manner. Many such controllers are referred to as *self-tuning controllers*, or *self-adaptive controllers*, and a typical block diagram is shown in Figure 13.

Measurements of the controlled and the manipulated variables are used to estimate the parameters of a simple process model. This process model is then used to calculate the new control parameters based on a pre-selected tuning method.

8.5   BATCH CONTROL

Control systems for batch plants differ significantly from those of continuous plants discussed so far. In batch processing, there is a much greater emphasis on production scheduling of equipment to match available production equipment and raw materials with the demands for the products. Batch control systems, in contrast to continuous process control, involve *binary logic* and *discrete event analysis* applied to the sequencing of different processing steps in the same vessel, usually requiring the application of programmable logic controllers (PLCs). Feedback controllers are still used to handle set-point changes and disturbance rejection but they may require certain enhancements, such as adaptive control, to cope with the wide operating ranges because there is no steady-state operating point. A good discussion of batch control is given by Seborg et al.[1]

8.6   PLANT-WIDE CONTROL ISSUES

Plant-wide control is concerned with designing control systems for a large number of individual process units that may be highly interacting. A typical plant-wide control system will consist of many single-loop controllers as well as multi-variable controllers such as Model Predictive Control (MPC),[1,10] and may involve thousands of measurements, hundreds to thousands of manipulated variables and hundreds of disturbance variables. Fortunately, a plant with a large number of processing units can be analysed as smaller clusters of units.

Several additional issues arise from unit interactions which further distinguishes plant-wide control from the control of single units. Most modern plants will have significant heat integration between units in order to reduce energy costs. Recycling of un-reacted material is also used to improve the efficiency of the plant. Although both heat integration and material recycle can significantly reduce plant capital and operating costs, these techniques inevitably increase the amount of interaction among operating units and reduce the control degrees of freedom. Nevertheless, appropriate control strategies can deal with such undesirable consequences.[10]

It should be noted that interactions between control loops is not just limited to interacting units but will also occur within single units. A typical example is an exothermic reactor where a change in the control loop which controls the level in the reactor will have an effect on the amount of material in the reactor. This in turn will affect the heat removal requirements and, therefore, the cooling water control loop. Many strategies for reducing loop interactions, and for selecting control loops so as to minimise interactions, can be found in most standard control textbooks.[1-11]

8.7 WORKED EXAMPLE

An exothermic reaction A → B is taking place in a CSTR which has a cooling jacket with cooling water. The input stream is coming from an upstream unit. The main disturbances, controlled variables and manipulated variables are:

disturbances: feed temperature T_i, cooling water temperature T_{CW}
controlled variables: reactor temperature T, product concentration C_B
manipulated variables: outlet flowrate F_o, cooling water flowrate F_{cw}

In order to ensure safe operation and a satisfactory quality of the reactor product, the outlet concentration C_B is to be controlled. Due to possible side-reactions, the reactor temperature T must also be controlled. The temperature can be measured easily but a measurement of the concentration of the product B is only available from a laboratory every hour.

Propose a control system based on feed-forward – feedback control, cascade control and inferential control to achieve these control objectives.

8.7.1 Solution

The control system consists of five measurements, two manipulated variables and five control loops. In the first control loop, loop 1, a temperature measurement TT_1 is made of the reactor temperature T (where the first T refers to Temperature and the second T to Transmitter) which is sent to a feedback controller TC_1 (where the T refers to Temperature and the C to Controller) which is normally a PI-controller in order to avoid offset.

Loop 1 is cascaded with loop 2 to improve the response to disturbances in the cooling water temperature T_{CW}. The cooling water temperature is measured (TT_2) and the signal sent to a second feedback controller TC_2 which is normally a *P*-controller (see Figure 14).

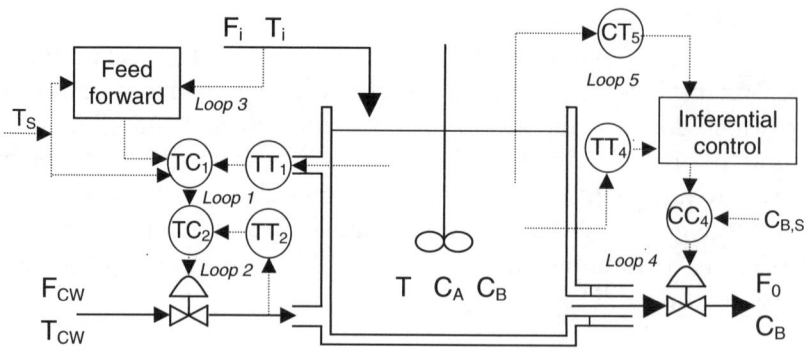

Figure 14 *Control system of a CSTR*

   The feedback control in loops 1 and 2 is combined with a feed-forward controller in loop 3 which measures the inlet temperature T_i, calculates the change in cooling water flow rate F_{CW} which is required to bring the reactor temperature T back to its set point T_S and sends this signal to the feedback controller (the feed-forward controller consists of a *model* of the process and is therefore *not* of P-, PI- or PID-type). The feed-forward control loop will therefore theoretically eliminate any disturbances in inlet temperature T_i. The feedback part of the control system, loop 1, will compensate for any inaccuracies in the feed-forward control model as well as eliminate the effect of other, unmeasured disturbances, *e.g.* in inlet flow rate F_i.

   The concentration of the product B, C_B, is not measured on-line and a measurement is only available hourly from a lab. The control of the concentration is therefore based on inferential control in loop 4 using the reactor temperature T. The inferential controller will then, from a model of the process, *infer* what the concentration C_B is and use this inferred measurement as the signal to the controller CC_3 (where the first C refers to Concentration).

   The model in the inferential controller is updated hourly with the actual measurement of the concentration C_B from the lab in loop 5 (CT_5).

   Note that the control system outlined above is not complete. Additional control loops to ensure safe operation would also be included, such as level control to avoid the reactor overfilling, *etc.*

   Had the reactor been a batch reactor instead of a CSTR, the controllers would be combined with an adaptive control system which would have updated the controllers, *i.e.* the control parameters, as the batch progressed. The batch reactor would also have had a separate logical control system to take the reactor through the batch schedule.

REFERENCES

1. D.E. Seborg, T.F. Edgar and D.A. Mellichamp, *Process Dynamics and Control*, 2nd edn, Wiley, Danvers, 2004.
2. G. Stephanopoulos, *Chemical Process Control*, Prentice Hall, New Jersey, 1984.
3. W.L. Luyben, *Process Modeling, Simulation, and Control for Chemical Engineers*, 2nd edn, McGraw-Hill, Singapore, 1990.
4. C.A. Smith and A.B. Corripio, *Principles and Practice of Automatic Process Control*, 2nd edn, Wiley, Danvers, 1997.
5. F.G. Shinskey, *Process Control Systems*, McGraw-Hill, Singapore, 1979.
6. D.R. Coughanowr, *Process Systems Analysis and Control*, 2nd edn, McGraw-Hill, Singapore, 1991.
7. M.L. Luyben and W.L. Luyben, *Essentials of Process Control*, McGraw-Hill, New York, 1997.
8. T.E. Marlin, *Process Control: Designing Processes and Control Systems for Dynamic Performance*, McGraw-Hill, Singapore, 1995.
9. B.A. Ogunnaike and W.H. Ray, *Process Dynamics, Modeling and Control*, OUP, New York, 1994.
10. W.L. Luyben, B.D. Tyreus and M.L. Luyben, *Plantwide Process Control*, McGraw-Hill, New York, 1998.
11. B.W. Bequette, *Process Dynamics: Modeling, Simulation and Control*, Prentice Hall, New Jersey, 1998.

CHAPTER 9

Economic Appraisal of Large Projects

KEN SUTHERLAND

9.1 INTRODUCTION

Much of what the chemical engineer does is connected, directly or indirectly, with large plant projects – either the building of a new one or the improvement of an old one. Assuming that it is properly designed and constructed, it should work satisfactorily – and the other chapters of this book give a good understanding of how that design process is conducted.

However, the aim of the new or improved plant is normally to make a profit for the company owning it, and profitable operation is by no means assured, unless it is as carefully planned for as in the design process. This chapter will show how the planning for a profitable operation is undertaken.

The economic appraisal process is usually concerned with a *large* project, not necessarily physically large (although that is normally the case), but certainly large in respect to the sums of money required for its design and construction – sums of money that may exceed those readily available within the company for that investment. The project is also long in terms of its construction period and operating lifetime, and it is this extended lifetime that is the prime reason for the appraisal.

A small plant, which can be designed and built in a year, will normally be financed out of annual revenue; a large plant, taking several years to build and costing perhaps more than the profit generated in those years, has to be considered quite differently, and special techniques have to be used to determine its profitability.

9.2 TIME VALUE OF MONEY

The project to be considered, then, involves an investment that is large with respect to the available funds (costing perhaps tens or hundreds of

272

millions of pounds), and long with respect to a company's normal yearly assessment of its financial position. It would be quite normal for such a project, say, a new oil refinery or pharmaceutical plant, to take 3 or 4 years to build, from inception to start-up, which will then operate for a long period, 10, 20 years or more. This leads to a situation where the investment in the project is made over an initial period and the revenues earned by the project do not start to appear until the investment is largely complete, but then they continue to arise over a long period after that.

It is, of course, possible to add up all of the revenues over the project's lifetime and to compare this sum with the total investments. If the revenues exceed the costs, then the project is apparently profitable, but if the reverse is true, it is not profitable and no manipulative techniques will be able to make it so. The comparison of costs and revenues, in an appropriate fashion, is the target of economic appraisal, and its first problem is to deal with the time value of money.

A pound today is not worth the same as a pound in a year's time: "worth" meaning its ability to buy an unchanging article (such as a loaf of bread). The change in value will be negative if nothing is done with the pound during the year, because inflation will inexorably decrease its real value. The change will, however, be positive if the pound is put to work, and if the return on that work (interest) is greater than the loss due to inflation.

The relation between the present value, P, of a sum of money, and its future value, F, is determined by the standard compound interest formula:

$$F = P\left(1 + \frac{i}{100}\right)^n$$

where i = interest rate in percent for the period concerned (usually a year), and n = the number of periods over which the calculation is being made.

In this calculation, the interest rate will be negative if inflation is working (*i.e.* $P > F$), or positive for an investment where interest is being earned (*i.e.* $F > P$).

It is this change in the value of money with time that complicates an economic appraisal – how is a pound spent at the start of construction to be compared fairly with a pound earned perhaps 15 years later? How this is done is the key to the appraisal process, and is explained later in this chapter.

9.3 THE COMPANY

First, some comments need to be made about the company making the investment, because the project does not exist in isolation. It will be part of a working company (even if that company is set up just to implement the project), which will be operating in accordance with the rules appropriate to the country in which it is based. Primarily, the company will exist for the purpose of making an adequate return for its owners (usually its shareholders), *i.e.* it should make an annual profit, which should increase each year in a well-managed company. Hence, it is necessary for the investment project to make its contribution to the profits.

The company, therefore, has responsibilities, not just to its stake-holders (its owners, its customers, its employees, past and present, and, increasingly these days, to its neighbours), but it also has legal responsibilities, mainly to issue an annual report, containing a set of accounts, especially a balance sheet (for a particular point in time, usually the last day of the financial year for the company) and a profit/loss account, describing the company's performance during that financial year.

From the balance sheet, it is possible to tell how "rich" the company is, and from the profit/loss account, how profitable it is (in the year in question, at least). The two documents provide very useful information as to the state of the company's capital, and hence its investment potential.

9.3.1 The Balance Sheet

The balance sheet, as its name implies, shows the equivalence between the assets of the company (*i.e.* what it owns), and the liabilities (*i.e.* what it owes), all calculated on the day of the balance. The total assets must equal the total liabilities.

The assets held by the company are usually grouped under two headings:

- fixed assets, which cannot easily be turned into liquid form (*i.e.* cash or its equivalent), and
- current assets, which include cash holdings and items that will yield cash during the next year.

The fixed assets include not only the property and equipment that are normally thought of as assets, but also the intangible assets, such as patent rights and any strategic investments in other companies. The current assets cover not only cash, as already stated, but also unpaid invoices for goods sold, and inventories of unused raw materials, semi-finished and finished products.

The company's liabilities are listed under three headings:

- owners' funds (*i.e.* the funds in use by the company that belong to the owners),
- long-term loans (*i.e.* funds belonging to others, not due for repayment within a year of the balance sheet date), and
- current liabilities, due for payment within a year.

The owners' funds, or equity, include the issued capital (*i.e.* the face value of the company's shares), capital reserves (funds received in ways distinct from operating profits, such as premiums on the sale of shares) and revenue reserves, which are the accumulated annual profits earned. The long-term loans include not only the sums borrowed from banks and finance houses as well as bonds issued by the company, but also the provisions for future potential (but as yet unknown) costs – for example the sums a company involved in asbestos liability litigation would prudently set aside against possible penalties and costs. Current liabilities include all bills received but not yet paid (for goods and services), and any short-term loans, due within a year, such as bank overdrafts.

From the point of view of project investment planning, the items of most interest are the level of tangible investments, the holdings of cash and the balance of owners' funds and borrowed capital. These may have a controlling effect upon the way in which a new project may be funded, or even upon its practicability.

9.3.2 The Profit/Loss Account

The profit (or loss) account, which accompanies the balance sheet in the annual accounts, is a summary of the year's activities for the company in financial terms. Its main components are given in Table 1, which shows the progression from total revenue to net profit, and how each item is calculated.

A real profit/loss account will normally have more items in it than the skeleton of Table 1, but the format given here shows the most important items. This account is less useful in the investment planning decision, since it does not concern itself with financial resources, but the calculation of profit, as shown in Table 1, is a vital component of the appraisal process, as will be seen later in this chapter.

From the profit/loss account and the appropriate parts of the balance sheet, the key performance ratios for the company can be calculated, which can be used for comparison with the company's past performance, and with those of its competitors. These include return on sales

Table 1 *Profit/loss account*

Account item	Calculation
Total sales revenues	SR
Selling costs	SC
Factory netback	FNB = SR – SC
Operating costs (variable and fixed)	OC
Overheads	OH
Depreciation	D
Production costs	PC = OC + OH + D
Gross profit (before interest and tax)	PBIT = FNB – PC
Interest paid	I
Profit before tax	PBT = PBIT – I
Corporation tax	T
Net profit (profit after tax)	PAT = PBT – T
Dividends paid to shareholders	DP
Retained earnings	RE = PAT – DP

(*i.e.* the profit as a percentage of the total sales revenue), the return on total assets, the return on capital employed and the return on equity. (Care must be exercised as to which profit term is used in these ratios.)

9.4 SOURCES OF MONEY

A major part of the early stages of any investment project will involve the raising of the necessary finance for it, from internal or external sources, and at the lowest cost. A correspondingly important part of the appraisal process is therefore to examine the various sources of funds, and to try to put some relative value or cost on them.

The money required for a new project will come from one or more of the following:

- retained profits (which actually belong to the owners);
- retained depreciation funds (nominally retained to replace the depreciating assets);
- bank loan – from the company's own bankers, from an independent finance house, including venture capital sources from an international lending agency (when an overseas aid project is being considered);
- new issue of shares (ordinary/preference);
- sale of bonds or other "convertibles"; and
- government grants or long-term loans (for a project in a designated development area).

Each of these major sources, even the company's own funds, come with some kind of cost attached. The calculation of such costs is dealt with later in this chapter.

There are other ways of raising money, but they are much less likely to be of the scale necessary to finance a large project. A joint venture with another company is an increasingly important way of doing this – at the cost of losing control of the whole project. A sale of an existing asset may be a good source – especially if the company owns easily separable assets that are no longer key parts of the business.

On the smaller scale, equipment can be leased, and this concept is expanding into BOO (build, operate and own) contracts where another company – usually a plant contractor – will supply the plant, but retain ownership of it.

9.4.1 Owners' Funds

The funds belonging specifically to the owners of the company are increased each year by the addition of the retained earnings from the year's operations. They may also decrease, if the company makes a loss, or if some are invested in new assets, or if the company chooses to pay dividends greater than the net profit (which, given good business reasons for so doing, the company is perfectly entitled to do).

The owners' funds are all invested in the company as part of the total assets, and, as a result, are a major liability of the company. They are, of course, owned by the shareholders in direct ratio to individual shareholdings.

9.4.2 Shares and Bonds

Ownership in a company is conferred by issuing a share of the company in the form of a transferable certificate, and then selling that share to a person or an institution, who thus becomes an owner of that proportion of the company represented by that share. A registered company is legally bound to issue a minimum number of shares, which must be sold to its forming members. Once this minimum number is sold, a company can then issue as many shares as it likes, and sell some or all of them to whomsoever wishes to buy them.

The money raised from the sale of these shares is kept by the company (as "owners' funds") and is used to pursue the company's business. Once sold by the company, a share can be traded (on the appropriate stock market), and the company can buy back its own shares in this way if it wants to do so, at any time. The value of the share on the stock market will fluctuate daily, according to the strength of the company and the state of the world, and may fall below the face value, or rise to many times that figure, according to demand.

In order to raise new funds, a company can, at any stage in its lifetime, issue a new batch of shares and sell some or all of them. If the company is doing well and obviously expanding, then it will probably be able to sell these shares at a premium above the face value. If, on the other hand, the company is doing badly and needs the money to stave off disaster, then it will almost certainly have to offer the shares at a discount below the face value.

This is an important way of raising money, but a new share issue dilutes the existing share ownership, and may be unpopular with existing shareholders. It is normal, therefore, to offer the new issue first to existing owners.

Shares give their owners the rights to (but no guarantee of) an annual dividend, and also rights to a share of the assets if a company is wound up, but the key characteristic of a share is its ownership of a part, however small, of the company.

Companies in need of more capital can also raise it by selling bonds. A bond is a certificate, with a face value usually much larger than that of a share, which has a fixed lifetime, usually 10 years at least, at the end of which time the bond "matures" and the company pays back its face value (by then, of course, much decreased in real value). However, the company also guarantees to pay interest at a fixed percentage of the bond's face value for this lifetime, usually twice a year, and this rate is usually quite an attractive one.

Bonds can be bought and sold, like shares, on a special bond trading market, and their trading value will vary around the face value (usually without the spectacular changes that can occur with shares). The key characteristics of a bond are the guaranteed repayment of capital and annual interest payment (so long as the company survives to the maturity of the bonds), but, unlike a share, no ownership rights are conveyed with a bond.

The sale of both shares and bonds can be keyed to their intended use for a specific project, and if the project idea is a good one, whose purpose is easily conveyed to the potential investors, then the sale is correspondingly eased.

9.5 VALUING A COMPANY

As part of an investment planning exercise, it is frequently necessary to put a value on the totality of an operating company. The calculation of this value can be quite difficult, but may be vital. The need arises either when the sale or purchase of the company as a whole is being considered, or when one wants to compare several companies as to size – in the present context this may be a way of determining the likely sums available from a sale for re-investment.

The simplest method of valuation starts from the obvious fact that to buy a company, one just has to buy all the outstanding shares. In principle, therefore, the value of the company is the number of shares multiplied by their individual value at any moment (the "capitalization"). Almost certainly the offered price will have to exceed the current share price to persuade people to sell their shares (unless the market is falling, in which case they may be glad to off load them). There may also have to be a premium on the capitalization to allow for intangibles, such as "goodwill".

This valuation is sound for the particular sale/purchase situation, but is very dependent upon the immediately current share price. It has no direct link to any measure of an intrinsic company value, certainly not one that can be used to compare the sizes of two or more companies.

So there must be other measures of value. The most obvious is gross sales volume – the more a company sells in any market, the bigger it must be. This is generally true, but what may be more important for comparative purposes is what lies behind these sales – are there enormous quantities of assets or capital required to achieve them, or is it a very lean company? The ratio of gross sales to asset value or to capital employed will therefore be a better measure.

Since companies exist primarily to provide income for their owners, *i.e.* to make a profit, their gross profit could be a more interesting figure than sales volume. The most profitable companies may be the best – but, again, it is the ratio of profit to assets or capital employed that may be more important.

To a large extent a company is its assets, so the purchase of a company implies the purchase of its assets. This is apparently a very real figure – fixed assets of plant and buildings, technology, people, spare cash and so on. Asset value, however, is a little nebulous – people can easily leave, for example – and some companies make very large sales off very small assets.

A key figure in valuing a company is the ratio of its current share price (as an indicator of what the world thinks of it) to the total net profit (earnings) divided by the number of shares (earnings per share). This is called the *P/E* ratio (price/earnings), and is a major tool in comparing the performance of a company within its class of similar companies.

While the individual valuations may not be all that difficult, the method to be used for determining the value of a company depends very much on the purpose of the valuation exercise.

9.6   THE COST OF CAPITAL

The funds that might be obtained from the various sources given in Section 9.4 come at very different costs – both real and virtual. If there are few sources available, then a simple calculation may be possible to determine an annual cost of the finance raised for the project, but it is more likely that a detailed analysis will have to be made.

The capital needed for a new major project will come from two sources: inside the company and outside it. The internal resources are the owners' funds, and this money exists in the current assets part of the balance sheet, from which it can be drawn. New capital can be raised by the sale of already issued shares held by the company, or by the sale of bonds, or by the issuance of new shares.

The normal external sources are loans from banks and other lending institutions at commercial rates of interest, and grants from governments or loans (also from overseas credit agencies) at lower than commercial rates of interest. The cost of a single source of borrowed money is easily determined – it is the annual rate of interest being charged on the loan. Where several different sources are used, a composite cost can be calculated (see below).

The cost of using the company's own funds (or, actually, the owners' funds) is less easily derived, because it contains an element of owners' future expectations and an element of risk. It is most definitely not "free", because the money can be used for a number of plans other than the one under consideration, and its cost should be set at least to equal the expected return on the most attractive of these alternatives, known as the "opportunity cost".

9.6.1   Cost of Company's Own Funds

The normal method of calculation for company funds is to use the *capital asset pricing model (CAPM)*. This was developed by share analysts keen to have a defence against accusations of negligence in selecting shares for clients as a means of assessing the real value of any share, in the form of risk and desirability. It essentially demonstrates one version of the direct proportionality between risk and return.

In its graphical form, risk is measured along the horizontal axis by a factor "beta", related to the volatility of the price of the share, resulting from any general disturbance of the market, relative to the average of all quoted shares on that market. This is plotted against share "yield" (dividends as a percentage of current share price). If all shares available on one market are plotted on this basis, it can be shown that they lie roughly on a straight line, the "market line", rising with increasing values of beta, as shown in Figure 1. Beta is lowest for long-term government bonds, and its value is set to zero for the base rate of interest. The average value for the whole market is set to 1.0, so that high risk shares have a beta value approaching 2.0.

The justification for this model is the fact that any share that sits well above the line (*i.e.* having a relatively high yield) for any reason will appear attractive to the market, investors will buy it, its price will rise as a result of this demand and its yield will thus fall back to the market line. Conversely, any share that is well below the line will appear unattractive, investors will sell, the price will fall and the yield will rise.

Beta values will vary from industry to industry, with "blue chip" companies near to zero, and high-risk telecom or biotech companies, for example, at the opposite end. Over the years, various types of companies have become grouped together in different regions of the market line, with different values of risk. An individual company can position itself among similar types of company (type of business being the most important characteristic), to derive its cost of capital from the resultant beta value, as exemplified in Figure 2. Here a company with a beta figure of 1.4 positions itself on a market line deriving from a base (beta = 0) of 5% with an average yield (beta = 1) of 11.2%. Extrapolation of the market line to the beta of 1.4 gives a yield of 13.7%, which is taken to be the "cost" of the company's own capital.

This model is so successful that large companies can use a similar method to decide which parts of the company are doing well, and which should be sold.

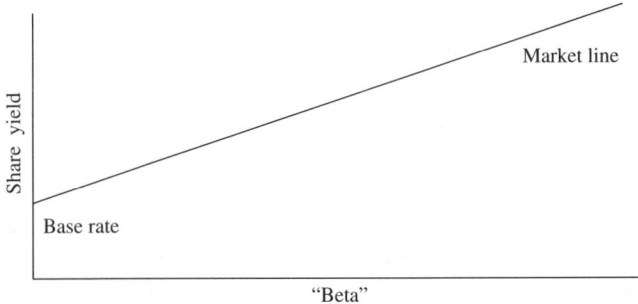

Figure 1 *CAPM market line*

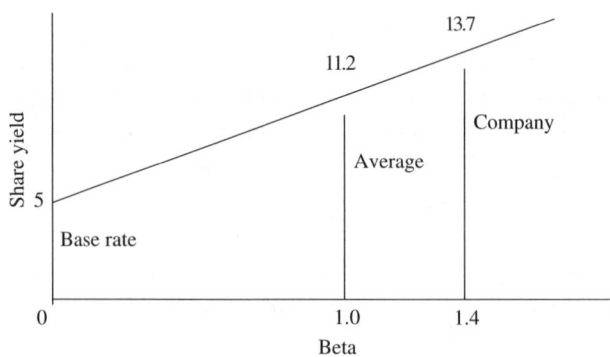

Figure 2 *Calculation of company's cost of capital*

9.6.2   Average Cost of Several Sources

The cost of the company's own funds can thus be calculated, by means of the CAPM. However, it is an unusual company that does not have other sources of finance, each with its own cost (*i.e.* interest rates for different loans). The average cost of all of its sources of capital is expressed by a single figure, the *weighted annual cost of capital (WACC)*, calculated according to the relative amounts of each source of funds.

Each source of finance is tabulated both with its effective "cost" and its proportion of the total amount of money in use. The product of the percentage cost and the (decimal) fraction share of the total funds gives the component of the final weighted cost for each source, as shown by the example in Table 2.

Table 2 shows a company using its own funds (at its CAPM cost) and three other sources of capital: a cheap government loan and two commercial bank loans, one at a very high rate of interest (because the loan was made to finance what the lender considered to be a high-risk venture). In the case of bank loan 1, the entries show that the rate of interest charged on the loan is 8.5%, and that it represents 30% of the total funds. Its contribution to the weighted average is then 30% of 8.5%, or 2.55%. The total of the four contributions in the last column gives the WACC of 12.45%.

The calculated WACC is a figure that changes with time, as a loan is paid off, or another taken out, or the proportion of company funds increases. Despite this changeability, the WACC provides a good bench-mark against which to measure the rate of return on a new project, and it is precisely in this way that it is used later in this chapter.

Table 2 *Calculation of weighted annual cost of capital*

Source of funds	Rate = cost (%)	Proportion of total (%)	Product = cost component
Company funds @ CAPM cost	13.7	25	3.425
Government loan	3.5	10	0.35
Bank loan 1	8.5	30	2.55
Bank loan 2	17.5	35	6.125
WACC			12.45%

9.7 COST OF CONSTRUCTION

A major part of the investment appraisal exercise is the calculation of the total amount of capital needed:

- to design and engineer the plant;
- to buy the land (if necessary), and all the components of the plant;
- to construct and install the plant;
- to prepare the plant for operation, including staff recruitment and training;
- to provide working funds for the first few months of operation; and
- to pay interest on any money borrowed during the construction period.

The financial appraisal is normally undertaken quite early in the life of the project, well before the final designs are complete, so a great deal of the cost calculation is based on the skill of the estimator, and much of it is done by factoring on known figures (such as how much a similar plant cost in another place or at another time). As it is an estimate, the final figure will prudently include a contingency allowance – the better the estimate is believed to be, the lower this contingency needs to be.

Most of the work in the estimating process relies on the application of well-established factors to a basic figure of the cost of all of the main items of the plant, as delivered to the site entrance. This figure has thus to be calculated as accurately as possible, based upon the plant design as available at the time of the estimate. It depends upon the identification and rough sizing of the key process plant items (such as reactors, distillation columns, heat exchangers, and so on), and the determination of approximate delivered prices for these items, from charts and other information published in textbooks, journals and on the internet.

A typical set of figures is shown in Table 3, which includes a range of factor values for each of the installation and engineering items. Which particular figure is chosen from the range shown is based on the experience of the estimator, although the quoted mean figure is the most likely. It should be noted that some ranges are much wider than others, a situation dictated by the nature of process plant in general.

To the basic purchased equipment cost (including delivery to the site) of 100 are added each of the subsequent factors to make up the final total. The contractor's fee is necessary if the plant design and construction work is given to a contractor to undertake (which is very often the case, because the company needing the plant is unlikely to have the necessary engineers on its payroll).

The interest accrued on any necessary loan taken out during the construction period will obviously depend markedly on the loan draw down and conditions, but 10% of the fixed capital investment (FCI) is a good approximation. The working capital is more sensibly calculated in

Table 3 *Total capital employed*

Expenditure item	Range (%)	Mean
Purchased equipment	100	100
Installation of purchased equipment	40–50	45
Instrumentation and controls	15–30	23
Piping and valves	30–75	52
Electrical services	15	15
Plant buildings	20–30	25
Service facilities	50–70	60
Land	5–10	7
Site improvement	10–15	13
Total installed plant		340
Engineering and supervision	30–35	33
Construction costs	35–40	37
Contractor's fee	15–25	20
Total engineering		90
Contingency (10% of installed plant + engineering)		45
Fixed Capital Investment (FCI)		475
Interest accrued on loan during construction period (10% FCI)		50
Start-up costs (10% FCI)		50
Total installed cost (TIC)		575
Working capital (15% FCI)		75
Total capital employed (TCE)		650

terms of production and selling costs – but until these are available, 15% of FCI is another good approximation.

The calculation embodied in Table 3 shows that the total amount of capital required for the project is somewhere around six times the cost of the purchased and delivered equipment (a ratio of 5–7 can be expected, depending upon the complexity of the plant). However, it is equally important that the way in which this capital is to be spent be determined – in order that the demands on the company's funds and borrowing capacity can be foreseen well ahead of their requirement, and also so that the way in which any loan has to be drawn upon can be seen, and hence any accrued interest be calculated.

It is normally sufficient that the capital needs be calculated on a yearly basis, and they are then assumed all to be paid out on the first day of that year (except for the start-up costs and the working capital, which are assumed to be paid on the day the plant starts to operate).

9.8 REVENUE CALCULATION AND NET CASH FLOW

With the capital needs calculated, it is now necessary to determine what profit (if any) the plant will make. It is assumed that the project is to be implemented as a result of some sound market research that might have shown an appropriate need in the marketplace. This research might also have shown what the plant capacity should be, and should have given guidance as to the likely sales price for the product – although this will normally be a management decision based upon the market entry strategy. An established product will probably have to go in at a discount on the present market price; while a new product may be able to be priced on a cost-plus-reasonable-profit basis.

Given, then, the production rate and sales price, the total revenue for the project can be calculated (perhaps with allowance made for under-running in the first year or two as the faults are dealt with). It will cost money to get the product to market – a sales force, advertising, cost of shipment and distribution, and so on. There may also be a royalty on sales payable to the owner of the technology used in the plant. These costs, generally called selling costs, are subtracted from the revenues to derive the sum that actually gets back to the project – the "factory netback".

9.8.1 Operating Cost

The major annual cost for the project will probably be the cost of operating the plant, the production cost. This is normally thought of as having four components:

- variable operating costs;
- fixed operating costs;
- overheads; and
- depreciation.

The variable costs are those directly resulting from the operation of the plant and dependent upon how much product is made. They will include the costs of all raw materials and other materials consumed by the process, of all utility services (electricity, gas, water, compressed air, steam, *etc.*), plus the costs of packaging, and maybe a waste disposal cost, if that is related to production rate. These costs will vary if, for example, an underrun in the first year or two is being allowed.

The fixed operating costs are those that do not change with production level. The most important is the cost of operating, maintenance and service labour, plus its management, but also included are such items as insurances of all kinds, local taxes, waste disposal (which may be a fixed cost rather than a running cost) and energy tax (carbon levy). Apart from labour, the cost of which is based upon the nature of the plant (its complexity, its level of automation), these other fixed costs are usually obtained by factoring from the plant's capital cost.

If the project and the company are the same, then these are all the costs to be considered. However, it is usual for the project to be situated among other plants belonging to the company on a larger site, and then there are general costs for the whole site (such as a fire station, the security guard system, a canteen) to be shared among all the plants. This overhead charge (direct if it is for the same site, plus indirect for the company's other activities) is usually derived on a capital investment proportional basis.

9.8.2 Depreciation

This leaves just one major production cost item to deal with, that of depreciation.

Fixed assets (excluding land) used by a company usually decrease in value over time, even when properly maintained. This can be due to product obsolescence or simply wearing out (with an ultimate need to be replaced). If an asset life is less than a year, the expenditure on it is included in the profit/loss account as a cost. If, however, its life is greater than 1 year, then a way has to be found to adjust both profit/loss account and balance sheet to give recognition to the actual life of the asset.

The initial cost of the asset is shown as a fixed asset in the balance sheet. In the profit/loss account, the decrease in value is shown each year by allowing a proportion of the original value as a notional operating cost, the "depreciation allowance". A correspondingly reduced asset value is then shown in the end-of-year balance sheet, with an equivalent increase in the "cash" assets (so that the total assets are not changed).

Typically, a chemical company will depreciate its plant and equipment (on-sites) at 10–15%, and its buildings (off-sites) at 5–8% (a lower rate because they may be reusable for some other purpose when the plant's life is over). The actual values used will be what the company and its auditors think appropriate and prudent. The depreciation allowance is usually linear ("straight-line", *i.e.* same amount each year) but other methods are used, *e.g.* "declining value" (same percentage of value at start of each year), which leaves a terminal value at the end of the plant's life, which has then to be written off at one go.

Depreciation is allowed as an operating expense, and hence as part of the production cost, as far as calculation of corporation tax liability is concerned. The rates and timing of such allowance against tax can vary, but the most likely method is to count it annually against an annual tax assessment.

The things to remember about depreciation, however it is calculated, are that it is a part of the production cost, but that it does *not* represent a cash outflow, merely an internal transfer.

A complication of this picture lies in the use of the word "amortisation", which also means the progressive writing down of the value of an asset, and thus could include "depreciation" or be used instead. It is, confusingly, separated from depreciation in the currently used profit measure "EBITDA" or earnings before interest, tax, depreciation and amortisation, where amortisation refers to the writing down of a financial asset, with depreciation limited to fixed assets.

(A further complication occurs for assets that increase in value with time, such as land, or mineral resources, or even buildings. Revaluation of assets requires the efforts of skilled accountants, with matching balance sheet entries under fixed assets and capital reserves.)

9.8.3 Annual Profit

The total production cost (variable plus fixed operating costs, overheads and depreciation) when subtracted from the factory netback (total revenues less selling costs) leads to an operating profit, often called "gross profit", but more correctly profit before interest and tax (PBIT). From this, then, is subtracted the annual interest paid on the loans directly involved in financing the project, to give the profit before tax (PBT).

The corporation tax payment is usually a percentage of the PBT figure, and is frequently actually paid in the year following that in which the profits were earned. However, it is customary to account for it in the same year, as far as feasibility calculations are concerned.

With the tax deducted from the PBT, the net profit, net earnings or profit after tax (PAT) is determined, to be the prime source of real income for the project, but not the only one.

9.8.4 Net Cash Flow

The net cash flow into the project each year is made up of two items: the PAT and the depreciation allowance for that year. The depreciation is added to the PAT because, unlike all the other costs, it does not actually leave the balance sheet.

In order to carry out a proper financial appraisal, the net cash flow figures must be calculated as inputs to the project's funds for each year of its operating life. Although these sums actually arise throughout the year, they are conventionally taken as arising on the last day of the year.

9.9   FINANCIAL REQUIREMENT OF NEW PROJECT

The main object of a financial appraisal is to answer the question relating to the project, "is it worth undertaking it?". This is not the same as "will it be profitable?", but rather "will it be *adequately* profitable?".

An alternative way of deciding this is to ask, "is this the best way of spending the total required capital, or are there better ways of investing the money (*i.e.* ways that will show a greater return)?".

The company must decide if the return that is forecast to be earnable from the project is equal to or greater than some target figure of earnings that meets the company's growth goals. It is up to the finance director of the company to set this target, and then up to the project's designers to see if they can achieve it.

The financial appraisal process must be able to answer both of these basic questions: what can this project earn, and which of several alternative projects is the best, economically?

(It is recognised that greatest return may not be the only criterion, or even the most important, in making a decision on a new plant: it may be necessary to fight competitors or to take a share in a new geographical market. But greatest return is the usual wish.)

How, then, should a target figure for the return on the investment be established? The simplest is to look at current interest rates – but these are at historically low levels, and it should not be difficult for a project to exceed these rates. Next, one looks at relevant business ratios (as percentages) – say the ratio of sales to capital employed, both historically in the same company, and among competitors in the same general business group – as a guiding figure.

The most important assessment is the comparison with the company's present level of cost of capital, as measured by the CAPM for a single source, or the WACC for capital derived from several sources. As interest rates have fallen quite considerably, it is likely that these models, being historical, are higher than would be the case for capital raised now. However, to set either of these as the target rate would probably be safe, were it not for the risks involved in any new project.

In fact, the target rate, often called the hurdle rate of return (HRR – "hurdle" in the sense of needing to jump over it), is usually set at a premium above the WACC, this premium being designed to reflect the uncertainties surrounding the project. These include:

- the extra risk over the company's usual operations, arising from
    (i) factors inherent in the project,
    (ii) uncertain status of the company in the (new) business area,
    (iii) specific uncertainties relating to the business area itself, and
    (iv) exceptional risks relating to the country in which the plant is to be built (if different from the country where the company is based);
- a contingency factor to cover omissions and bias in the project proposal; and
- allowances for exceptional inflation rates or currency fluctuations in the country where the plant is to be located, if this is far from the company's main operations.

Values for this risk factor, and therefore for the HRR, vary widely with the industry and company, but, for the bulk chemicals and petrochemicals sector, it is unlikely to be less than 15% and may be over 20%.

Whatever rates are used for comparative purposes, it is absolutely vital to ensure that they are calculated on the same basis: before or after taxes have been paid, before or after interest, with or without depreciation as a cost.

9.10 THE APPRAISAL PROCESS

The project financial appraisal process is now a question of taking the calculated investments, made in the assumed way, and comparing them with the calculated cash flows, arising in the assumed pattern over the plant's operating lifetime, in such a way as to show the effective rate of return on the investment.

There are two simple factors that can be used for comparison, either with established targets or among several competing projects, which make no allowance for the ways in which money is spent or received, and hence do not allow for the time value of money:

- return on investment (RoI) = average PBIT/total capital employed, and
- payback time = point in time at which cumulative net cash flows just equals total employed capital.

These are useful factors, despite their limitations.

A simple graph of net cash flow out of, or in to, the project's account illustrates the way in which the cumulative net cash flow moves over the lifetime of the project, and enables the payback time to be easily seen, as in Figure 3. The payback time may be expressed as "less than 3 years", or more precisely, such as 2.4 years or 2 years 5 months.

In general, the lower the payback time for any project, the more profitable it is likely to be, but as a complete indicator, this method fails to allow for the different patterns of payments and revenues.

9.10.1 The Discounted Cash Flow Process

The prime task in project appraisal is to compare the total income from the project over its productive lifetime, with the total investment needed

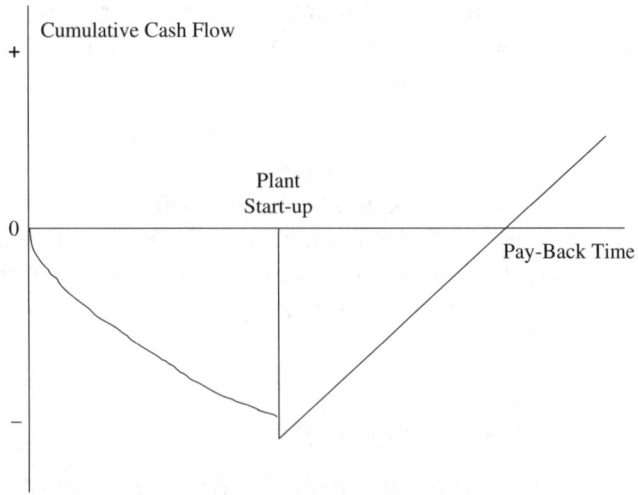

Figure 3 *Payback time*

to build it and get it running. To do this properly requires some measure that will enable the analyst to

- make a fair comparison between inputs and outputs;
- give an index by which to show the value of a new project to a company;
- enable fair comparisons among several competing projects; and/or
- enable sensitivity analyses to be made easily for different project parameters.

The discounted cash flow (DCF) method, to be described here, does enable the making of a fair comparison between inputs and outputs, and provides both an effective rate of return and an easy means of assessing the effects of different approximations.

The essence of the DCF process is that it does take into account the passage of time during the life of the project. It works by modifying all sums of money, costs or revenues, to give their nominal values at a particular point in time, so that a balance can be struck between the sum of all costs and the sum of all revenues at that point. The method works equally well at any point in time over the project's life, and the first decision to be taken is to select the most meaningful point to use.

The two balance points in time that are most commonly used are

- the commencement of work on the project, and
- the day of start-up of the plant.

There is some sense in selecting the date of the appraisal as a balance point, but this cannot easily allow for the potential delays between conception and start of construction (although it can give a good incentive to reducing delays between idea and implementation for a deserving project).

In fact, the best balance point is the date on which operation starts, because at that time, all expenditures should be complete, and no revenues have yet started to accrue. All sums of money, input or output, are then discounted, forwards or backwards, to this comparison point in time.

9.10.2 Present Value

The discounting process enables the conversion of any sum of money involved in the project to its "present value" (PV) at the point of comparison. (Confusingly, "present" does not necessarily mean "now", and, in fact, in the most sensible way of doing DCF analyses, it means on the day of start-up, which may be several years ahead of "now". Even the different choice of the comparison point as the day of construction commencement does not prevent confusion about the term "present", because commencement of the project may still be well ahead of the "now" of doing the appraisal.)

Choice of the day of start-up as the comparison thus requires the acceptance that the "present" is actually some time in the future, and it employs two kinds of discounting (although one is only the reciprocal of the other). The fact that at start-up all expenditures are past, and that all income is in the future, eliminates the need to use negative cash flows, and is a slightly easier concept to grasp. It is also easier to manage construction periods that are not a whole number of years in this way.

The method of calculation is based on transferring the value of money forward or backwards in time, by discounting.

The *present value PV* of a sum *S* invested *in the past*, at an interest rate $i/100$ per period, after *n* such periods is

$$PV = S\left(1 + \frac{i}{100}\right)^n$$

forward discounting

Correspondingly, the *present value PV* of a sum *S* arising *in the future*, after *n* periods, at an interest rate of $i/100$ per period is

$$PV = S\left(1 + \frac{i}{100}\right)^{-n}$$

backward discounting

Note that forward discounting *increases* the sums concerned, and that backward discounting *decreases* the appropriate amounts.

In these two discounting equations, the interest rate, *i*, measured in percentage, is called the discount factor (DF), and its choice is a complex matter, to be discussed a little later.

The calculation of "present" values requires the observance of two main conventions, that

- all cash flows are assumed to occur at one point in the year (even if they actually arise uniformly throughout it), with expenditures aggregated at the beginning of the year, and income at the end of the year; and
- all discounting is done annually, forward to the start-up date for expenditures, and backwards for all incomes.

A further convention is that working capital is paid in one sum at start-up (together with those costs that are specific to the start-up process itself), and working capital (only) is repaid to the revenue account at the end of the project's life (together with any scrap value of the plant).

Tables of DFs are available in various places, but calculations are now much simplified by use of spreadsheet functions.

9.10.3 Net Present Value

The *net present value* (NPV) is one of the key terms in the appraisal technique – it refers to the difference between the present values of all costs and that of all revenues:

$$NPV = PV(\text{revenues}) - PV(\text{costs})$$

and may be negative or positive. Negative NPV figures usually indicate an unattractive situation.

The NPV is calculated at a particular value of the DF, but the value corresponding to $DF = 0$, *i.e.* the base case, just looking at the simple, undiscounted sums of costs and revenues, is a useful figure also.

All other things being equal, the project showing the highest value of NPV at a particular value of the DF (including zero) will be the most attractive, financially.

(If cash flow patterns are very irregular, then discounting may have to be done other than annually, but this is only needed when requiring a fairly accurate analysis, or when comparing different projects with widely varying cash flow patterns.)

The value of the DF to be used in the discounting formulae is chosen according to the purpose of the calculation. If the purpose is to demonstrate a positive NPV, or to compare NPVs for different projects, then a value will be chosen to match the company's own finances, and the value could then be

- the company's cost of equity capital (by the CAPM calculation);
- the overall cost of capital to the company (by the WACC calculation);
- the opportunity cost of capital, chosen by the company to match other development opportunities; or
- the hurdle rate of return, chosen by the company to cover all known risks.

Obviously, the last of these, $DF = HRR$, is the most sensible.

9.10.4   Project Lifetime

The probable life of the project will depend very much on the market place to be served by the product, but a 5-year life should be achievable, and over 10 years would normally be expected for a project of the size being considered here.

It is necessary to select a value for the effective lifetime of a project, so that the undiscounted NPV is positive (*i.e.* a lifetime longer than the payback time). A value of 10 years is conventionally chosen

- because it is a sensible real figure; and
- because the PVs of almost any practical sum and DF combination become very small for 11 years or more, and so add very little to the PV(revenues), and hence to the NPV.

An important consideration is whether or not to assign a scrap (or still-operating sale) value to the plant at the end of its economic life. Unless the plant is to be "run into the ground", there is usually some scrap value, even if only from the plant buildings, which is usually calculated as a percentage of the FCI at the start, and is then included as an additional cash flow at the end of the last year of the analysis (together with the working capital).

9.11   THE INTERNAL RATE OF RETURN

There is thus an inverse relationship between NPV and DF – the higher the latter, the smaller the value of NPV. The highest value of NPV is the base figure, when the sums are not discounted ($DF = 0$), and, eventually, for increasing values of DF, NPV becomes negative.

There is then a single value of the DF for which, uniquely, the NPV is zero. This value of the DF is called the *internal rate of return* (IRR), or sometimes, the DCF rate of return. The IRR is the other key term in the DCF process

$$NPV = 0 \quad \text{for } DF = IRR$$

Because the calculation of the NPVs takes into account the ways in which the relevant sums of money have been invested or recovered, as well as their size, the IRR is characteristic of the project and its cash flows: the IRR is the rate of return achieved by investing the total capital employed in the project in the same pattern as in building the project, to produce income in the pattern of the net cash flows of the project.

In this definition, the pattern of payments or earnings is the way in which payments are made during the construction period and the way in which earnings are received once production has started. In most financial investments, especially by the private investor, the payments are of regular amounts and usually are at regular time intervals, while receipts will either be a single sum at the end of the savings period, or an annuity. In a large plant project, however, the payments will certainly be of unequal amounts, probably paid irregularly at various times during the construction period, and the receipts may well be irregular too, if only (the investors may hope) increasing every year. The IRR then determines the exact rate of return on such an irregular investment, and if the payment or receipt pattern changes, then so will the IRR – which can thus be used to compare different investment patterns, for example.

9.12 THE DCF APPRAISAL PROCESS

To determine the intrinsic value of the project (as well as an alternative means of comparing several different projects, or even several different versions of the same project), the IRR is calculated as that unique discount rate that makes the NPV equal to zero. The IRR cannot be calculated from a single equation, but must be determined iteratively, and the final stage may be graphical.

The IRR will be the ultimate goal of the DCF process, but it is usually a good idea also to determine the NPV of the project for one of the company rates of return – and the hurdle rate is the most obvious.

It is unlikely that an IRR of 15% or less will be of much interest to a company, certainly for a new development, while it is also unlikely that a rate of over 35% is feasible, unless it is for a very exotic and expensive product. The discount rates used in appraisal calculations, therefore, usually lie in the 15–35% range. In this range, the DFs for 11 years or more are so large as to make any calculation beyond 10 years rather pointless, so confirming the choice of 10 years as an economic lifetime.

The basic steps of a DCF appraisal are shown in Table 4, which describes a project whose total capital employment is £124 million, spent as shown over a 3-year construction period (including expenditure at year 0, *i.e.* the point of start-up). Revenues rise during the first 2 years of operation to a maximum figure from year 3 onwards.

Calculations of NPV are included in Table 4 for the base case (DF = 0), for the company's hurdle rate (15%), and for four other DF values, chosen to enable a graph to be plotted between the values of DF and NPV (Figure 4).

Table 4   *A typical appraisal calculation*

| Period | Base | Discount rate (%) | | | | |
		15	10	25	17.5	20
Year −3	25.00	38.02	33.28	48.83	40.56	43.20
Year −2	45.00	59.51	54.45	70.31	62.13	64.80
Year −1	30.00	34.50	33.00	37.50	35.25	36.00
Year 0	24.00	24.00	24.00	24.00	24.00	24.00
Total	124.00	156.04	144.73	180.64	161.93	168.00
Year 1	29.50	25.65	26.82	23.60	25.11	24.58
Year 2	34.10	25.78	28.18	21.82	24.70	23.68
Year 3	38.70	25.45	29.08	19.81	23.86	22.40
Year 4	38.70	22.13	26.43	15.85	20.30	18.66
Year 5	38.70	19.24	24.03	12.68	17.28	15.55
Year 6	38.70	16.73	21.85	10.15	14.71	12.96
Year 7	38.70	14.55	19.86	8.12	12.52	10.80
Year 8	38.70	12.65	18.05	6.49	10.65	9.00
Year 9	38.70	11.00	16.41	5.19	9.07	7.50
Year 10	62.70	15.50	24.17	6.73	12.50	10.13
Total	397.20	188.68	234.88	130.45	170.68	155.26
NPV	273.20	32.64	90.16	−50.19	8.75	−12.74

NOTES
1. The project takes three years to build, numbered −3, −2, and −1 respectively.
2. "Year 0" applies to the moment of start-up, and is of zero length in time.
3. The column headed "Base" contains the undiscounted costs (Years −3 to 0) and revenues (Years 1 to 10), expressed in £ millions.
4. The remaining columns show these base sums discounted, forwards and backwards at the discount rates denominating the columns.
5. Working capital is paid out in Year 0, and recovered in Year 10.

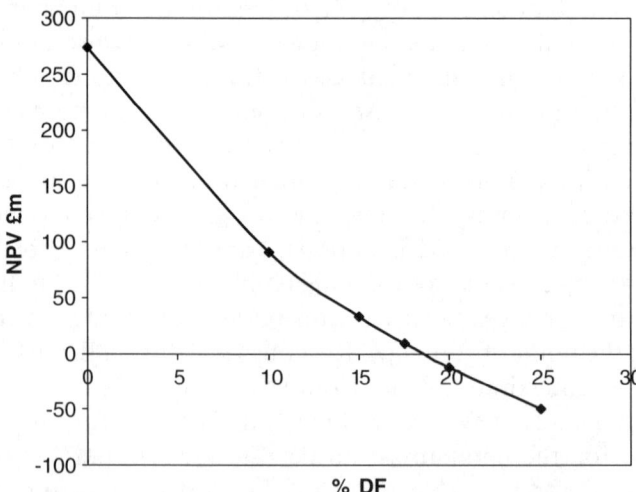

Figure 4   *Relationship between NPV and DF*

The NPV figure for the base case is more than twice the capital investment, so in the simplest terms the project is profitable. This means that the returns are comfortably large enough to repay any loan taken out to pay for the project.

It can be seen that the NPVs become negative between DF = 17.5% and DF = 20%, so that is where the IRR will lie, and from the graph of Figure 4 the value can be seen to be close to 18.5%.

9.12.1 Graphical Interpolation

For DCF calculations being done by hand, the final stage will most easily be a graphical interpolation – the alternative being a long-winded iteration process to find the actual value (which is not normally sought to better than one place of decimals).

The curve relating NPV to DF is not linear, as can be seen in Figure 4, although it can be taken as linear for very short intervals of DF. On the other hand, the NPV/DF curve is a *smooth* one, so any point not lying on a smooth curve must have been calculated or plotted incorrectly.

The aim in selecting values of the DF to use in NPV calculations is to find two values, preferably no more than 1% apart, one of which gives a positive NPV, and the other a negative value. For this interval of DF, the NPV/DF plot can be taken as linear, and the actual value of the IRR is then calculated by triangulation (using the theory of similar triangles), as illustrated in Figure 5.

In the case illustrated in the previous example, the NPV figures would be calculated for DFs of 18% and 19%, and, in the example of Figure 5, *A* would be 18%, and *x* would be 0.5%, giving an IRR of 18.5%.

9.12.2 Computer Calculation

The whole repetitive process of discounting calculations is avoidable by use of a modern spreadsheet, such as Microsoft Excel, which can

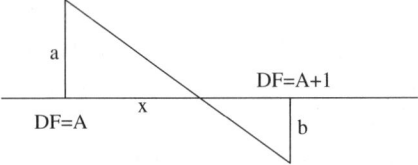

if a and b are two known values of NPV at discount factors A% and (A+1)%, then the IRR is found at the point (A+x)%, where:

$$a/x = b/(1-x) \quad \text{or} \quad x = a/(a + b)$$

Figure 5 *Graphical interpolation*

undertake the iteration process automatically and very rapidly. The basic annual cost and cash flow data, of course, still have to be calculated by hand (even if by a properly constructed spreadsheet), but then Excel's "GoalSeek" tool for example can produce the value of IRR directly.

However, Excel has to be treated with care when it comes to plotting points calculated by Excel onto a graph – for demonstration or interpolation purposes.

9.12.3 Interest during Construction

If a sum is borrowed to build any or all of the plant, then it is very likely that this loan will be called upon (or "drawn down") during the construction period. The moment any part of it is borrowed, then interest becomes due from that moment.

In all cases of such borrowing, however, the company does not expect the interest to be paid by the project as it comes due, because the project has, as yet, no revenues from which to pay the interest. The lender will be paid its interest from other funds belonging to the company, and the interest due during construction is then turned into a capital sum (capitalised), which is then added to all of the other construction costs. The loan will normally be arranged at a fixed interest rate, so the extra sum can be calculated quite easily.

9.12.4 Repayment of Loans

All loans have to be repaid to the lender, but it is usual for the loan terms to be such as to allow repayment over several years: at least 5, and maybe 10. In this case, the outstanding loan will decrease with each year of the plant's lifetime, and the interest payment for each year will fall accordingly.

Conventionally, the loan interest, paid in any year, is assumed due on the outstanding sum at the start of that year, while any repayment does not take effect until the end of the year.

For any project, the repayment of the loan must effectively come out of its proceeds over the years, *i.e.* the total net revenues (cash flow) must be at least as large as the amount borrowed, on top of the other costs of construction, *etc.* However, the repayment element of the loan is *not* included in a financial appraisal, because its equivalent in plant cost has already been included in the balance as part of that cost. (An alternative way of looking at this is that the company, in paying back the loan, is just replacing that part of the original cost with an equal but differently sourced sum of money.)

9.12.5   Effect of Plant Size

A calculated IRR is a unique index for a particular set of costs and revenues, and enables easy comparison with the corresponding index for other situations. However, it will be apparent that if all of the figures in the above example had been multiplied by two, then the same IRR would have resulted. This implies that the two plants, differing by a factor of two in size, are equally attractive financially.

This is true – and so other factors have to be brought into the decision process. Can the company afford to build a bigger plant? Can the market accept twice as much product?

It is, of course, also true that a plant of twice the output would probably cost significantly less than twice as much as the smaller one (a price factor of $2^{0.6}$, if the six-tenths rule is used), and so would be more attractive financially – but the same questions have to be asked.

9.13   OTHER ASPECTS

There are some other aspects of financial appraisal that need consideration.

9.13.1   Non-Profit Projects

This chapter has been aimed at the revenue-earning project, the amounts of revenue leading to an annual cash flow, from which profitability, and hence economic feasibility, can be determined. But what if there are no revenues? A new public service investment, such as a major (non-toll) road, or an overseas aid project, such as a water pipeline system for a sub-Saharan country, generate no revenues – yet may have huge investment costs.

These kinds of project must still undergo a stringent appraisal process, to ensure that each is the best way of spending the money required. It follows that some kind of "revenue" must be calculated, and this can be done on the basis of savings on other expenditures, or benefits to the community. Such calculations are not easy, but they can be done, and "cost–benefit analysis" is a well-practised technique.

9.13.2   Sensitivity Analyses

A major benefit of the DCF process, especially when used in combination with Excel's calculating powers, is that it enables rapid calculation of alternative circumstances, of the "what if" variety. This allows the determination of the sensitivity of the appraisal to, say, changes in sales price (the most important variation), or raw material cost, or electricity cost, or an increase in the labour needs, or an underestimate of construction costs. If the spreadsheets are properly designed, then it may take only a few key presses on the computer to derive IRR values for different sales prices or construction costs, for example.

9.13.3 Inflation

It is normal, in undertaking an economic appraisal, to ignore the effects of inflation – which applies both to costs and revenues (assuming that prices rise roughly to keep pace with inflation). If inflation is included in the calculation, then a slightly higher value will normally found for the IRR, so that its exclusion presents a slightly less favourable case.

9.13.4 Post-Construction Audit

The total expenditure on the plant, after construction and start-up, should be carefully developed and recorded for the company's files, against a need on a future project – the best possible guide to an estimate of the cost of a future plant is the detailed cost of an earlier example.

The operational costs, sales costs and resultant profitability should be recalculated once the plant is running to its full capacity, and the revenue should be checked against the forecast figure. Maintenance costs should also be carefully checked for future use.

APPENDIX – WORKED EXAMPLE

Consider the production of acetone from i-propyl alcohol, at a full annual rate of 100,000 tonnes, on an existing site (*i.e.* no land purchase cost). From previous experience, the delivered equipment cost can be expected to be £14.5 million.

As this is a fluid-based process, the only solid material being the catalyst, the following equipment scale factors can be used.

A. Total Capital Employed

		£ Millions
Purchased and delivered equipment (P&DE)		14.500
Land		0
Site improvement	15% P&DE	2.175
P&DE installation	40% P&DE	5.800

Pipework and valves	40% P&DE	5.800
Instrumentation and controls	17% P&DE	2.465
Electrical services	12% P&DE	1.740
Buildings for plant)	23% P&DE	3.335
Other site services) = "Buildings"*	55% P&DE	7.975
Vehicles and distribution	7% P&DE	1.015

Installed plant			44.805

Process, design and detailed engineering	30% P&DE	4.350
Construction	36% P&DE	5.220
Contractor's fee	17% P&DE	2.465
Royalties		0

Engineering		12.035

Contingency	10% of IP + E	5.684

Fixed capital investment (FCI)	62.524

Loan interest accrued during construction (2 years)
$(25 \times 2 \times 0.08) + (25 \times 0.08)$ 6.000

Start-up costs	10% of FCI	6.252

Total installed cost	74.776

Working capital	15% of FCI	9.379

Total capital employed	84.155

(Funded as £50 millions loan, paid in two equal instalments, and £34.155 millions company equity.)

*Depreciation is calculated on FCI only, at different rates for "plant and vehicles" and "buildings". Of the installed plant total:

Plant and vehicles	= 33.495 or 74.76%, so proportion of FCI	= 46.743
Buildings	= 11.310 or 25.24%, so	= 15.781

and depreciation will then be calculated on these amounts.

B. Production Cost

From mass balance and energy balance calculations, the raw material and energy costs are

		£ Millions	
Raw materials and catalyst	£175/tonne product	17.500	
Utilities	£ 21.7/tonne	2.170	
Packaging		0	
Variable manufacturing cost			19.670

for a full year's production.

Fixed manufacturing cost (*i.e.* expenditure on operations that is independent of amount produced).

Process labour	5 shifts of 5 × 52 weeks	0.572	
Laboratory and maintenance labour	25% of shift	0.143	
Other staff	6 × 52 weeks	0.129	
Supervision	20% total labour	0.169	
Payroll costs	27% payroll	0.274	
Total labour			1.287

Waste disposal	Fixed fee 1%	PC*	1% PC
Maintenance material costs	4% PC		4% PC
Communications	0.5% PC		0.5% PC

(*PC = full year's production cost, yet to be calculated)

Insurances	2.5% FCI	1.563
Local taxes	2% FCI	1.250

Fixed manufacturing costs		4.100 + 5.5% PC

Overheads	3% FCI	1.876

Depreciation (straight line)
$$12.5\% \text{ of P \& V } (46.743) = 5.843 \text{ (for 8 years only)}$$
$$8\% \text{ of buildings } (15.781) = 1.262$$

Total depreciation		7.105

Contingency	5% PC	5% PC

Production cost = variable cost + fixed costs + overheads + depreciation + contingency = 32.751 + 10.5% PC

$$\text{or} \quad PC = 32.751 + 0.105 \, PC$$

$$0.895 \, PC = 32.751$$

or   Production cost = £36.593 millions per full year

It is prudent to allow some production shortfall in the first year of production, say 10%.

All that then changes is the variable cost, which reduces by 10%, or £1.967 millions

So, production cost in Year 1 is 36.593 − 1.967 = £34.626 millions

[NB In later operating years, the depreciation allowance decreases, because Plant and vehicle depreciation operates at 12.5% and therefore is not levied after Year 8. The PC thus decreases after Year 8.]

C. Sales Price Calculation

If the sales price is £SP per tonne of product, then gross sales revenue (in a year of full production)	100,000 SP
The cost of sales = 5% sales revenue =	5000 SP
Factory netback = gross revenue − selling costs =	95,000 SP

Production cost in full year $=$ 36.593

PBIT $=$ FNB $-$ PC $=$ 95,000 SP $-$ 36.593

Annual interest on loan $=$ 8% of 50.000   4.000

Profit before tax $=$ PBIT $-$ Interest   95,000 SP $-$ 40.593

PAT $=$ PBT less 30% tax $=$ 0.7 (95,000 SP $-$ 40.593)

Good performance will have net profit (PAT) equal to 12.5% of gross sales revenue:

PAT $=$ 12.5% gross sales $=$ 0.125 $\times$
100,000 SP $=$ 12,500 SP

Equating these two derivations of PAT:

$$12,500 \, SP = 0.7 \, (95,000 \, SP - 40.593)$$
$$(95,000 - 12,500/0.7) \, SP = 40.593 \, \text{millions}$$
$$SP = \text{Selling price} = £526.21$$

A selling price of £550.00/tonne will be used.

D. Net Cash Flow

The annual net cash flow is the annual revenue for the project that is to be balanced against the investments. It is calculated as the net profit (PAT) in each year *plus* the depreciation allowance for that year:

$$NCF = PAT + D$$

In the final year of operation, usually (and here) taken as year 10 after start-up, the NCF is increased by the recovery of the working capital (and also by any scrap value for the plant, which is ignored here).

Component	Year 1	Years 2–8	Year 9	Year 10
Gross sales	49.50	55.000	55.000	55.000
Factory netback	47.025	52.250	52.250	52.250
Production cost	34.626	36.593	30.75	30.75
Gross profit (PBIT)	12.399	15.657	21.50	21.50
Loan interest	4.00	4.00	4.00	4.00
Profit before tax	8.399	11.657	17.50	17.50

(*continued*)

Component	Year 1	Years 2–8	Year 9	Year 10
Profit after tax	5.879	8.160	12.250	12.250
Depreciation	7.105	7.105	1.262	1.262
Working capital				9.379
Net cash flow	12.984	15.265	13.512	22.891

These net cash flow figures are now used as the bases for the discounted revenues.

E. Investment Schedule

The investments are made as company equity and loan, to cover the fixed capital investment, the interest accrued on the loan during the 2-year construction period, the start-up costs (testing and commissioning, and staff recruitment), and the working capital.

These sums are conventionally considered to have been paid at the start of each of the 2 years of construction (years −2 and −1) and at the point of start-up (year 0 – a "year" of zero length). The investment schedule is thus:

At start of	Equity	Loan	Total
Year −2	8.262	25.0	33.262
Year −1	10.262	25.0	35.262
Year 0*	15.631		15.631
Total	34.155	50.0	84.155

*Payment for start-up costs and working capital

F. Discounting Process

The investments for the construction period and the net cash flows for the 10-year operating lifetime can now be discounted to determine the net present values at various discount factors. The undiscounted payments and revenues can be seen in the left hand column:

Disc factor	0	0.05	0.08	0.09	0.10	0.15
Year −2	33.26	36.67	38.80	39.52	40.25	43.99
Year −1	35.26	37.03	38.08	38.44	38.79	40.55
Year 0	15.63	15.63	15.63	15.63	15.63	15.63

(*continued*)

PV costs	84.16	89.33	92.51	93.59	94.67	100.17
Year 1	12.98	12.37	12.02	11.91	11.80	11.29
Year 2	15.26	13.85	13.09	12.85	12.62	11.54
Year 3	15.26	13.19	12.12	11.79	11.47	10.04
Year 4	15.26	12.56	11.22	10.81	10.43	8.73
Year 5	15.26	11.96	10.39	9.92	9.48	7.59
Year 6	15.26	11.39	9.62	9.10	8.62	6.60
Year 7	15.26	10.85	8.91	8.35	7.83	5.74
Year 8	15.26	10.33	8.25	7.66	7.12	4.99
Year 9	13.51	8.71	6.76	6.22	5.73	3.84
Year 10	22.89	14.05	10.60	9.67	8.83	5.66
PV revs	156.24	119.25	102.97	98.29	93.92	76.01
NPV	72.09	29.92	10.46	4.70	−0.75	−24.16

From this table a number of key discoveries can be made. Firstly, the total net cash flow of £156 millions is close to twice the total investment, so there is a reasonable overall profit. However, the undiscounted payback time is about 5 years and 8 months (the total revenues for years 1–5 are still £10 millions short of covering the total costs). This is a long time before payback – 2 to 3 years would be preferable – so this project is beginning to look a little unexciting.

The NPV is negative at a discount factor of 10% (although still positive at 9%), and the NPV versus DF data give an IRR figure of 9.86% (by interpolation graphically between 9% and 10%).

This rate of return:

1. is higher than the interest rate being charged on the loan used to pay for the plant;
2. is higher than could probably be achieved from long-term investments on the money markets; but
3. matches the reality of bulk chemical production in western Europe; and
4. is probably below any likely figure for the hurdle rate of return for the company.

It follows that the project can be safely recommended (remember there are 5% contingencies in both plant cost and production cost calculations), but that there may well be better ways to invest £84 millions. A lot will then depend on any other reasons that there might be to want to build an acetone plant (defending a market, guaranteeing a raw material, *etc.*).

RECOMMENDED READING

General coverage of process engineering economics:
1. D. Allen, IChemE, 1991 (3rd Edition), "Economic Evaluation of Projects".
2. D. Brennan, IChemE, 1998, "Process Industry Economics".

Corporate finance:
3. B. Rothenberg & J. Newman, The Daily Telegraph/Kogan Page, 1993 (3rd Edition), "Understanding Company Accounts".
4. R.A. Brearly & S.C. Myers, McGraw-Hill, 1991 (4th Edition), "Principles of Corporate Finance"

Process equipment cost data (some of which also include advice on process plant economics):
5. M.S. Peters & K. Timmerhaus, McGraw-Hill, 1991 (4th Edition), "Plant Design and Economics for Chemical Engineers"
6. F.A. Holland & J.K. Wilkinson, McGraw-Hill, 1997 (7th Edition), "Process Economics" Section 9 of R.H. Perry, D.W. Green & J.O. Maloney, "Perry's Chemical Engineers' Handbook"
7. A.M. Gerrard (ed), IChemE, 2000 (4th Edition), "A Guide to Capital Cost Estimating"
8. J. Sweeting, IChemE, 1997 (1st Edition), "Project Cost Estimating – Principles and Practice"
9. R.K. Sinnott (ed), Butterworth-Heineman, 1999 (4th Edition), (Coulson and Richardson's) Chemical Engineering, Volume 6, of which Chapter 6 covers "Costing and Project Evaluation"

Cost data on other equipment (regularly republished):
10. Mott, Green & Wall (eds), Taylor & Francis, 2003 (34th Edition), "Spon's Mechanical & Electrical Services Price Book"
11. V.B. Johnson (eds), Butterworth-Heineman, 2005, "Laxton's Building Price Book"

Project management:
12. O.P. Kharbanda & E.A. Stallworthy, IChemE, 1985, "Effective Project Cost Control"

Occasional articles on costs and plant economics (monthly periodicals):
Chemical Engineering published by Chemical Week Publishing
Process Engineering published by Centaur Publishing

The Cost Engineer published by The Association of Cost Engineers
Oil and Gas Journal published by Penn Well Corporation

Regular reports on costs of chemicals and related materials (weekly periodicals):
European Chemical News published by Reed Business Press
Chemical Market Reporter published by Reed Business Press

CHAPTER 10

Hazard Studies and Risk Assessment[†]

ROBERT THORNTON

10.1 RESPONSIBILITIES OF DESIGNERS

It is the professional responsibility of all designers of process plant to consider the impact of their work on people, organisations and societies affected by their designs.

In its *Rules of Professional Conduct and Disciplinary Regulations Issue III* 7 December 2001, The Institution of Chemical Engineers defines these responsibilities for its members:

"Members when discharging their professional duties shall act with integrity, in the public interest, and to exercise all reasonable professional skill and care to:

(i) Prevent avoidable danger to health or safety.
(ii) Prevent avoidable adverse impact on the environment."

The responsibilities are not just confined to safety and environmental issues, though these are crucially important. Commercial, regulatory and management matters also play their part. Governments, society in general, designers' employers and clients, and, indeed, the designers themselves, all have legitimate interests in the safe and successful outcome of a project.

In this chapter we shall be looking at techniques which can be used by designers to help them recognise these issues and design plants and

---

[†] Much of the material in this chapter is taken from the Courtaulds' Design Manual. Courtaulds plc operated in the UK and Internationally until 1999.

processes which fulfil the expectations of the stakeholders. There are three steps in this:

- Identify the problems using Hazard Studies
- Assess the risk posed by each hazard to answer the question: how serious is it?
- Manage the risks so that they are eliminated or reduced to levels which are tolerable or acceptable to society.

One of the most powerful aspects of the system of hazard studies and risk assessments outlined in this chapter is the adaptability of the technique. It is possible to customise the procedure to cover many types of engineering project, and many aspects of engineering design. A large number of organisations use Hazard Studies. However, each one will have modified the technique to cover the particular needs, processes and operating environment of its business. The most important feature of this chapter is the ideas it contains, rather than laying down a rigid and prescriptive set of universally applicable rules.

The approach in this chapter is aimed primarily at the chemical process industries, but the ideas may be broadened and modified to cover many diverse situations.

10.2 DEFINITIONS

Although some of the words used above have an intuitive, if slightly "fuzzy" meaning, we need to be fairly precise about their definitions in the context of Hazard Studies and Risk Assessment.

A *Hazard* is an event or situation with the potential to cause harm. So it can include anything that may interfere with the successful outcome of the project. For example:

- Safety problems (fire, explosion, toxic release, release of kinetic energy, environmental challenges . . .).
- Commercial problems (cannot make a profit from the project).
- Marketing problems (nobody wants to buy the product or the project).
- Public relations problems (bad press, bad publicity).
- Government or regulatory problems (planning permission, patent infringement, process or product approval, illegal discharges . . .).

- Failure to deliver the design or project on time, within budget or within specifications.

A more conventional definition of hazard as a physical situation or condition with the potential to do harm to people, the environment or property is not inconsistent with this, provided the term "harm" is interpreted in a broad sense.

Although it is perhaps inevitable that designers and process engineers tend to concentrate on conventional safety issues, all the other matters are important and can be addressed by the techniques described in this chapter.

A *Risk* is a combination of the size of the hazard and the probability of its occurrence. So, for example, the hazard arising from a meteorite striking an oil refinery might be very large (fires, explosions and massive release of flammable or toxic chemicals), but the probability is very, very small. So the overall risk is also small, and almost certainly not worth worrying about. Conversely, the hazard arising from the accidental discharge from a relief valve might be relatively minor, but since the probability is quite high, we might judge the risk sufficiently large that we decide to take remedial action. We shall enlarge on these ideas later on, but do bear in mind that we are often looking at rather rare events, so just because a large hazard is rather unlikely does not mean that we can dismiss it without very careful thought.

The words "tolerable" and "acceptable" to society contain elements of judgement, and controversial judgement at that. Generally speaking, society will tolerate a risk if it is perceived that the benefits arising from some operation outweigh the potential disadvantages. This will usually involve some assurance that the risk is being managed or controlled in a responsible and adequate fashion. Most people will, for example, tolerate the risk imposed by a car journey, even though it is clearly a significant one. The benefits obtained, and the feeling of being "in control" outweigh the concern about a possible accident.

An acceptable risk is one which people instinctively consider to be so small that they never seriously worry about it. In northern latitudes, the risk arising from being struck by lightning might be an example – few people lie awake at night fretting about it.

The important point to realise is that these are not engineering judgements, but political and social ones. It is perfectly possible to quantify these risks and even gain an idea of the level of risk that society is prepared to tolerate or accept. However, these numbers vary very widely (and often without a clear or logical reason) according to the situation envisaged – and government or media campaigns can greatly affect perceptions overnight. This can be frustrating to engineers who look for some fixed and quantitative benchmark, but nevertheless in most societies they have to work within the bounds of public opinion, however irrational it may appear to them at times.

10.3   IDENTIFICATION OF HAZARDS

10.3.1   An Overview of the Hazard Study Framework

The first step in risk assessment in any project is to identify the potential hazards involved. This is done using Hazard Studies, which have been developed as a means of identifying and managing hazards in the design, building and early stages of operation of chemical plant. They are adaptable and may address both process and non-process hazards.

The Hazard Study approach is increasingly being used by major companies. Its particular attraction is that it encourages the integration of safety and plant design, and fits well with conventional project management. As well as addressing hazards, it ensures that members of the project team communicate with each other and work from a common set of data and assumptions. It also helps to set a timetable of key design activities.

In the project and capital authorisation procedures developed by one major company, there are six separate studies:

Hazard Study 1 aims to ensure that the understanding of the project, process and materials involved is sufficient to enable safety, health and environmental issues to be properly addressed. It also contributes to key policy decisions and establishes contacts, both internally and with external authorities. In essence it asks the question: "do you know what you're doing?" It is carried out early in the project when the scope is reasonably well understood, but there may still be alternative routes or processes or competing avenues for investment.

Hazard Study 2 helps to identify significant hazards which the detailed design then has to eliminate or otherwise address. The output is in effect a "shopping list" for the designer, covering aspects and features which must be built into the next stage of the design. It is asking the designer: "what ought I to be concerned about as I get into the fine detail of my work?" At this stage the process route has been decided, process flowsheets are available and mass and energy balances have been completed.

Hazard Study 3 is the same as the older "Hazard and Operability Study" (HAZOP). It is carried out on the firm, detailed engineering diagrams and is a thorough "line by line" investigation which seeks to identify issues missed or forgotten during the design process.

Hazard Study 4 is a check that the plant has been constructed to the intended design, all actions from previous studies have been carried out, operating and emergency instructions are satisfactory and the training of appropriate people has been executed and validated.

Hazard Study 5 ensures that the project, as implemented, meets Company and legal requirements.

Hazard Study 6 checks that all actions in Hazard Studies 1–5 have been completed and reviews them in the light of operational experience on the plant. This study is carried out after the plant or process has been running for a few months. As well as providing feedback to improve or modify the process and its operation, it forms a useful guidance for possible improvements in future designs.

In practice, Hazard Studies 1, 2 and 3 are used in a broadly similar pattern across the whole of the chemical process industries. However, the content and application of studies 4, 5 and 6 varies considerably between different companies and organisations, and indeed extra studies are often inserted. Since this chapter is aimed primarily at the *design* of process plant, the latter three studies will not be considered beyond this brief mention.

It has been found that well executed Studies 1 and 2 result in a plant design that is "right first time", and reduce the stress and amount of work needed at the HAZOP (Hazard Study 3) stage (see the "A brief history of Hazard Studies" box-out).

We will now look at the first three Hazard Studies in more detail.

A BRIEF HISTORY OF HAZARD STUDIES

A brief overview of the history of Hazard Studies, as experienced by one major UK chemical company, may be of help in setting this chapter in context.

Although the technique of Hazard Studies is a relatively new development in the process industries, the ideas behind it are far older, and date back to the beginnings of the industrial revolution. Until well into the second half of the 20th century, guardianship of safety in process design lay with experienced engineers who would peruse flowsheets and engineering diagrams and attempt to identify potential problems and safety issues.

As plants became larger and more complex, hazards became harder to spot and the consequences more severe. The old approach was not good enough. Major chemical companies like ICI, and workers such as Lawley and Kletz, pioneered the new procedure of Hazard and Operability Study (HAZOP study). This is dealt with in detail later in this chapter; it is a powerful way of identifying process hazards and operability problems at the design stage. The technique was highly successful and the ideas spread rapidly round the world.

HAZOP study remains one of the best ways of seeking out process hazards, and it is seated firmly at the heart of the Hazard Study framework.

Unfortunately, as a stand-alone technique, it is almost too successful and thorough. Inexperienced (and not so inexperienced) engineers would submit designs to HAZOP study, which would then identify tens or hundreds of shortcomings. These would have to be dealt with, and subsequent delays and administrative burden, not to mention the pressure on the morale of the people concerned, became unacceptable. So Hazard Studies 1 and 2 were developed as a way of formally identifying potential hazards and problem areas before any detailed work on the design is carried out. A well executed Hazard Study 2 will result in very few remedial actions appearing at the Hazard Study 3 (HAZOP) stage, as the designer will have already taken account of all important process "deviations".

10.3.2   Hazard Study 1

The output of Hazard Study 1 is a formal report with the following sections:

Title:
Date:
Contents: 1. Project and process
             2. Materials and hazards data
             3. Environmental aspects
             4. Health and toxicology
             5. Transport and siting
             6. Conformance with Company policies
             7. Consultation with external authorities
             8. Consent levels and Safety, Health and Environmental criteria
             9. Design guidelines and codes

(*continued*)

10. Organisational, human factors and emergency procedures
11. Sustainability
12. Further studies

The sections are discussed further below.

10.3.2.1 Section 1 – Project and Process. This starts with the project definition (objective, description, location, product and raw material specifications, quantity, quality, output requirements, throughput for equipment sizing purposes). Closely linked with this is a statement of the mode of operation (*e.g.* day time only or 24 hour operation, continuous or batch processing) and expected control philosophy (centralised computer or local, methods of data collection and storage, programmable controllers or "hard-wired").

Next is the process description (brief outline of basic chemistry, outline process flow diagram, batch sequencing if relevant, non-chemical process operations). An incident review is a key component, learning lessons and building on experiences culled from published information and data bases.

Although the need for these initial statements seems obvious and elementary, in practice it is surprising how many projects proceed a long, frustrating way before adequate definitions like these are produced.

10.3.2.2 Section 2 – Material Hazard Data. This section contains a list of all the materials to be found on the plant – raw materials, intermediates, by-products, effluents, emissions from other facilities close by, catalysts, support materials, services, principle materials of construction and materials encountered during construction and demolition.

There is, potentially, a great deal of information required here. Designers should be selective, but the section should include:

(i) important and relevant physical and chemical properties of the materials listed;
(ii) notes on significant chemical reactions between any of the materials listed;
(iii) notes on any material which is particularly toxic, reactive or has a severe environmental impact;

     (iv) estimates of the inventories. Most of the inventory of a plant will usually be in the storage areas (raw materials and product), though if there is a significant amount of a hazardous material present as a transitory intermediate, this should be noted;

     (v) comment on whether loss of containment could have off-site effects (some mathematical modelling of the dispersion of toxic releases, or the effects of fires and explosions may be required);

     (vi) means of storage and transport, and comments on segregation of materials if required;

     (vii) any special handling problems, including the cleaning up of spills and decontamination of equipment.

10.3.2.3   Section 3 – Environmental Aspects.   An environmental impact statement is needed. This will include a list or diagram showing all effluents and wastes (including packaging), and a description of how they will be disposed of, controlled and monitored. A check is also made on whether special mechanisms could result in loss of containment and environmental damage (such as earthquake, flood, wind, storm, fire and fire water run-off).

10.3.2.4   Section 4 – Health and Toxicology.   An occupational health statement is needed. Whilst some aspects of this are of a specialist nature, it should identify those materials which need particular attention to the control of emissions and leaks (minor escapes from joints, glands, seals *etc.*). Those materials where statutory exposure limits apply should be listed.

10.3.2.5   Section 5 – Transport and Siting.   The purpose of this section is to identify any risk caused by the siting of the plant and the movement of materials to, from and round the site. Existing plants and operations are important here, as is the land use around the proposed new plant (is it close to schools, hospitals, major roads, railways, airports and so on?).

    Attention needs to be paid to any issues arising from the chosen means of transport, such as the movement of raw materials and product through communities.

10.3.2.6   Section 6 – Conformance with Company Policies.   This section contains a copy of the Company policies on Health, Safety and the Environment, and lists the key objectives set under these policies. A description of how the proposed plant will conform to the policies and objectives, and the strategies used to ensure compliance are required.

10.3.2.7 Section 7 – Consultation with External Authorities. This is a list of government departments, statutory bodies, local authorities, citizens' groups and other interested parties which will need to be consulted prior to construction and during construction and operation of the plant. Alongside each item in the list is a named person responsible for the contact. The list should include a note describing why the contact should be made, and the sort of information that will be required.

10.3.2.8 Section 8 – Consent Levels and Health, Safety and Environmental Protection Criteria. This section lists the maximum levels for emissions to the environment, chemical concentrations in the working environment, noise, health effects and personal risk. These may be imposed by statute or the result of Company policies. These figures will have a direct effect on the design of various items of plant equipment.

10.3.2.9 Section 9 – Design Guidelines and Codes. Very little of a new chemical plant is designed for the first time – it is likely that previous generations of designers and engineers will have faced the same problems and after much thought have reached optimal solutions. This information is captured in various design standards and codes of practice. These are produced by national and international bodies, as well as government departments, trade associations and manufacturers. The purpose of this section is to list the standards, guidelines and codes of practice which need to be applied to the project. A comprehensive list is very long, but at this stage the designer should identify any key documents. Examples could include standards for vessel design, pipework, storage of flammable or toxic materials, storage of gases under pressure in liquid form, pressure relief systems, constructional standards, electrical standards, fire-fighting systems, guarding of machinery and so on.

10.3.2.10 Section 10 – Organisational, Human Factors and Emergency Requirements. This section is designed to highlight issues of staff availability and any training requirements. For example, thought should be given to the staffing strategy for the plant. Arrangements for the comfort and welfare of the staff need to be considered. Emergency procedures will need to be reviewed for currency and adequacy and to ensure that functionality and continuity are maintained during disruptive construction periods, always bearing in mind that extra staff and contractors will be on site during peak construction manning periods.

10.3.2.11 Section 11 – Sustainability. The resources of our planet are finite. Sustainable development – providing for human needs without compromising the ability of future generations to meet *their* needs – is one of the most significant issues facing society to-day.

The impact of industry on sustainability can be summarised as a "triple bottom line" covering three components:

- environmental responsibility;
- economic return (wealth creation);
- social development.

Professional engineering organisations are developing ideas in this field – for example see the publication from the Institution of Chemical Engineers *The Sustainability Metrics (Sustainable Development Progress Metrics – recommended for use in the process industries)* May 2002.

At this stage of the project design it may not be possible to evaluate the metrics quantitatively. However, relevant notes which summarise the strengths and weakness of the process from a sustainability viewpoint, and highlighting the areas which need to be addressed as the design develops, are needed. The site will need to be returned to its original state (or better!) after production has finished.

The answers to much of this will lie in the work already completed in Sections 2, 3, 4, 5, 8 and 10 of the Hazard Study 1 report, but the information is viewed in a different perspective.

10.3.2.12   Section 12 – Further Studies.   The need for further studies, if any, is noted here. A list of these studies and the responsibility for ensuring they happen is required.

10.3.3   Hazard Study 2

When a process flow diagram for a plant is available, but before detailed design commences, Hazard Study 2 may be carried out. This is a much more creative exercise than Hazard Study 1 (which is largely concerned with data collection). The purpose of this study is to look at the process area by area and identify the particular hazards and operating situations which the process engineer must take account of in his or her design.

10.3.3.1   Documentation.   The documentation required for Hazard Study 2 is:

- (i) a process flow diagram or block flowsheet (including non-process activities such as mechanical handling);
- (ii) a process description of each block, including control strategy;

(iii) a summary of moving objects (*e.g.* presses, palletisers, elevators, balers and traffic).

In some circumstances a Hazard Study 2 will include factors affecting the construction of the plant and its eventual demolition, as well as the demolition of any existing redundant structures.

10.3.3.2 Study Methodology. The study leader (who bears a lot of responsibility) divides the process into a series of blocks or sections. Each section is then subjected to a "brainstorming" exercise when the team ponders issues of (essentially) fire, explosion, toxic release and mechanical energy release, how they might arise and the control measures required – either to stop the event or cater for its consequences. The starting point for the examination of each section is a "prompt" from the table below.

The left-hand column of Table 1 suggests a possible hazardous scenario. The right-hand column suggests some of the factors which have to be present for the scenario to arise. Note that "acute exposure" means exposure to a substance over a short period of time (*e.g.* following a sudden release), whereas "chronic exposure" means exposure to low levels over a long period (such as small leaks and emissions). "Major financial factor" covers situations which might prove very costly if a given situation arises – for example, the poisoning of an expensive catalyst, or the downgrading of a particularly valuable product.

If the prompt reveals that a credible hazard exists, the design team should ask themselves questions, such as:

what exactly is the nature of the hazard?
how might the hazard be triggered?
how might it be prevented or controlled?
what effects will the hazard have on people, property or the environment if it occurs?
how might the effects of the hazard be mitigated?
are there any "knock-on" effects which the design needs to cater for?
. . . *etc.*

Clearly, this is a wide-ranging and open discussion. The study team should go far enough to ensure that the hazards and the likely means of control have been identified, but it is undesirable to go into detail, as this is the task for the designer. For example, if overheating is identified as a hazard, the team might suggest installing temperature monitoring, alarms and plant trips, but the details of the design of the system can be best handled by the designer(s).

Table 1 *Prompts for hazardous events and situations*

External fire	Fuel
	Release mechanism
	Source of ignition
Internal fire	Flammable mixture
	Source of ignition
Internal explosion	Flammable mixture
	Source of ignition
	Uncontrolled chemical reaction
Physical over/under pressure	
External explosion	Fuel
	Release mechanism
	Source of ignition
Acute exposure	Harmful agent
	Exposure mechanism
Chronic exposure	Harmful agent
	Exposure mechanism
Environmental damage	Gases
	Liquids
	Solids
	Release mechanism
Violent mechanical energy	Energy source
	Release mechanism
Noise	Sources
Anything else	Visual impact
	Major financial factor

The examination is repeated for each process block until the whole process has been covered.

10.3.3.3 Recording Hazard Study 2. Each organisation will have its own favourite way of recording the output from Hazard Study 2, but typically this will be in a tabular format under headings such as:

hazardous event or situation;
causes;
consequences;
protection mechanism or action required.

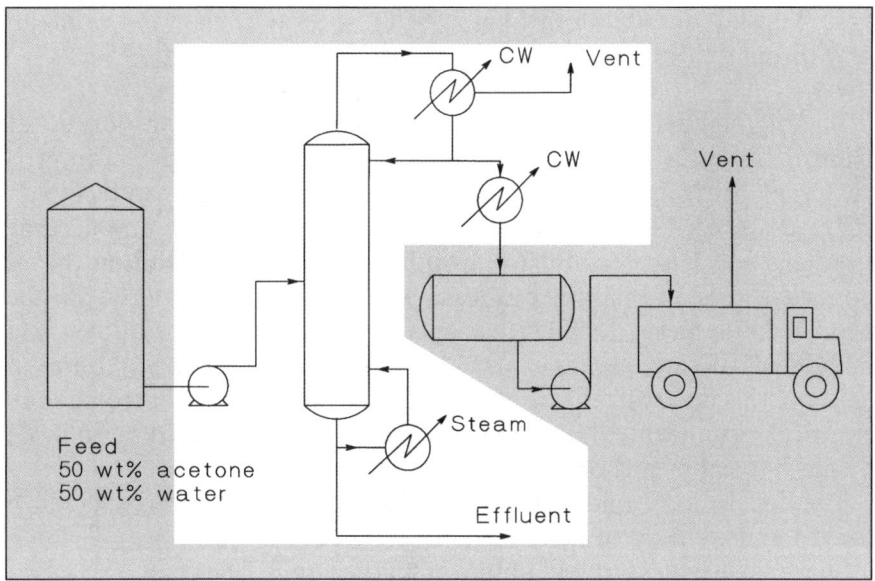

Figure 1 *Outline process flow diagram for the distillation of aqueous acetone*

The key requirement is that the output clearly and explicitly lists the features and safeguards that the process engineer needs to design into the plant so that it is both safe and readily operable.

10.3.3.4 Example 1. An example will make the methodology clearer. Figure 1 depicts an outline process flow diagram for a distillation plant recovering acetone (CH_3COCH_3) from a 50% by weight aqueous solution.

The designer has produced an outline process flow diagram, showing the stages and main equipment items, and is about to embark on a detailed design. The feedstock is pumped from a storage tank to a distillation column operating at atmospheric pressure. The column is equipped with a reboiler (steam heated) and a condenser (cooled with cooling water, CW). The bottom product, which is mostly water, runs to effluent. The top product (mostly acetone) is cooled and collected in a product tank, from which it is periodically transferred to a road tanker.

The relevant properties of acetone are:

- Vapour pressure 0.263 bar at 25°C
- Vapour pressure 0.133 bar at 10°C
- Boiling point at atmospheric pressure 59°C
- Lower flammable (explosive) limit in air 2.5% v/v; upper flammable limit 13% v/v

- Vapour in equilibrium with 50 wt.% acetone in water contains around 50% by volume of acetone.

The highlighted area represents a suitable "block" for a portion of Hazard Study 2. Looking at Table 1 we see that the first prompt is "external fire". Is this a credible scenario? The right-hand column suggests we need a fuel: that is certainly present (acetone and acetone solutions will burn readily). We also need a release mechanism: that is possible (pipes and flanges can leak, physical damage can occur to the plant from vehicles, fork lift trucks and maintenance operations, and pump seals or valve glands can leak). The third component is a source of ignition: this is very likely around any chemical plant (sparks from faulty electrical equipment, static electrical sparks, friction, sparks produced by mechanical means or hard surfaces striking).

The conclusion rapidly reached is that an external fire is a credible hazard and we need to design against it.

So we ask what exactly the nature of this hazard is. The plant operates at atmospheric pressure, so jet fires are unlikely and a pool of burning liquid is the most likely scenario. How might this be triggered? We may identify damage to plant as a possibility (we know we have road tankers in the vicinity) so we should design the layout of the plant with a view to keeping vulnerable areas away from moving vehicles. Pipes and vessels containing flammable liquids should be protected and mechanically strong.

Control of the hazard could well involve bunding the plant area so that any spills are contained. If the bunds slope to sumps or catch-pits, the surface area available for burning will be reduced. Other important control mechanisms will probably depend on the use of remotely operable shut-off valves on key pipelines (to stop the supply of fuel to any fire, and remotely operated because nobody will want to approach a fire to close valves manually).

How will a pool fire be fought? The designer may wish to consider the provision of fixed fire-fighting installations (such as foam monitors, hydrants, sprinklers and vessel cooling drenches) as well as making sure that there is adequate access for any fire-fighting team and appliances. This should also raise questions of how it is proposed to deal with contaminated fire-fighting water run-off, because draining to local rivers and water courses will not be acceptable.

The proximity of other plant, people and buildings should be examined, because they could all be affected by thermal radiation from a fire.

Instruments designed to detect and warn about leaks, acetone escapes and small fires before they become serious, and maybe even trigger automatic fire-fighting equipment, can also be installed.

Quite clearly, this discussion will raise many points for the designers to consider. Similarly, each of the other hazard prompts can be addressed. They will undoubtedly raise questions about the design of the venting system (toxic and flammable emissions), how to deal with a failure of the cooling water supply to the condenser, how to control and monitor the effluent discharge even under conditions of plant malfunction, instrument failure, loss of other services such as electrical supply and steam, human error, ease of safe maintenance and so on. The prompt "internal fire" may lead to a debate on the start-up of the system, when acetone vapour and air will be present initially.

After discussion on this block has been completed, other blocks are subject to a similar examination. The loading of the tanker will be interesting, and will raise issues of possible overfilling, containment of spillages, drive-away accidents (when the tanker drives off before proper disconnection of hoses) and failure of plant operating procedures, as well as many of the fire scenarios dealt with previously.

10.3.4 Hazard Study 3 (Hazard and Operability Study, HAZOP)

Hazard Study 3 (Hazard and Operability study, HAZOP) is carried out when the design of the process is finalised and detailed. The basic premise of a Hazard and Operability study is that hazards may occur on chemical plant when situations arise which the designer had not thought about. HAZOP sets out to identify these situations. There are many books and publications giving instruction, advice and background to these studies, and some organisations make these freely available (for example, the Royal Society of Chemistry Environmental Health and Safety Committee's *Notes on Hazard and Operability Studies 13 March 2001*).

HAZOP was, and still is, very directed towards chemical and process hazards (fire, explosion and toxic release). Although it often does identify mechanical, electrical and even civil engineering hazards, it is less strong in these areas, as the whole technique centres around the examination of "process" documents (flowsheets, engineering line diagrams, piping and instrumentation drawings and operating instructions).

A basic assumption is that where a set of process considerations has been identified by the process engineer, the resulting design has been carried out accurately and competently. HAZOP looks for situations unrecognised by the designer ("deviations") and examines whether there may be a hazardous implication. Thus, a HAZOP is NOT a place to carry out design work, a problem solving exercise, or a technique for bringing fresh minds to an old problem. The success of HAZOP depends on:

(i) the accuracy of the drawings and other data used for the study;
(ii) the technical skills and insights of the design team;
(iii) the ability of the team members to use the approach as an aid to their imagination in visualising process deviations, causes and consequences;
(iv) the ability of the team to maintain a sense of proportion, especially when assessing the seriousness of the hazards identified.

In practice, if the quality of the design or information available is deficient in some way, this will be quickly recognised during a HAZOP study, which will rapidly get "bogged down" with uncertainties and unanswered questions. This is a signal to go back and refine the design, so in this sense HAZOP study is a "fail safe" approach. For this reason, HAZOP studies are only really effective if they are undertaken on "final" chemical-engineering designs which are as good as the designer(s) can possibly make them.

10.3.5   Hazard Study 3 (HAZOP) Preliminaries

10.3.5.1   What the HAZOP Does.   The HAZOP process is there to identify those situations which the designer has forgotten or ignored and which could lead to a hazard. It often will be looking for complex and abstruse failures and situations and rare events. It should be borne in mind that even rare events can represent considerable risks (recalling that risk is a combination of the consequences of an event and its likelihood). Despite often bad publicity, the chemical industry is a very safe one and rare events need to be addressed.

10.3.5.2   When to Carry out a HAZOP.   There is no "right" answer to the question "When is the best time to conduct a HAZOP study?" although it should always be carried out on complete and firm flowsheets or Process & Instrument Diagrams (P&IDs). Provided Hazard Study 2 has been carefully executed, it is reasonable to apply for capital approval before the HAZOP (and after Hazard Study2). This minimises the pre-capital authorisation spend of any project. However, the danger is that poor initial design could lead to post-authorisation changes and cost increases.

10.3.5.3   Study Team Composition.   HAZOP study is a team effort. The team consists of the leader, a scribe and working team members. The ideal size seems to be 6–7 (including the leader and scribe). The minimum size is probably 5. Larger teams are unwieldy and costly in man-hours, though occasionally have to be tolerated.

It is important that the following expertise and functions are present on the team – of course, one person may carry out more than one function. For different types of process the list may be modified, or people can join the discussion when sections of interest to them are being examined.

Leader – who will direct and control the study. The leader should be trained and independent.
Scribe – who will record the discussion (see the separate section on recording).
Champion – a clearly identified person who will take responsibility for ensuring that actions are pursued and "closed out".
The designer (probably a process engineer).
Process engineer.
Electrical engineer.
Instrument engineer.
Chemist.
Someone with plant operating knowledge and expertise (may be a shift foreman or operator).
Business/technical representative of the customer or sponsor.

10.3.6 Information Needed by the Team

The study leader will specify the information that must be available to the team (and which will need to have been studied before the HAZOP meeting). The most important part is likely to be the complete P&ID (or flowsheet or Engineering Line Diagram, according to local terminology). It should show all vessels, pumps, process pipework, instruments, service connections and indicate floor levels or elevations.

A written process description is required.
An outline operating procedure is important if the process is continuous, and essential if it is a batch operation.
The information should be supplemented with isometric drawings, layouts, architects drawings, equipment data sheets and materials data sheets as necessary.

10.3.7 Carrying out a HAZOP

The methodology varies slightly according to whether the process is continuous or batch. In all cases it is the responsibility of the study leader to decide before the study how it will be conducted, the order in

which the study will proceed and the way in which the process will be
divided for examination. However, a degree of flexibility is always
desirable, and a good leader will be able to make changes to the
procedure "on the fly" – and also know when it may be desirable to
make these changes, and when not to.

10.3.7.1 Continuous Processes. The main drawing is examined pipeline by pipe-line. For each line the designer's intention is made clear and
written down. For example, the intention might be to convey a liquid
from vessel A to vessel B, at a flow rate of *x*, pressure of *y* and a
temperature of *z*. The composition is . . . and so forth. Since the process
is continuous, each part of the intention will be independent of time.

To each part of the intention a guide word is now applied. The guide
words are typically:

No or none (intention is not achieved);
More or too much (more than the intention takes place);
Less or not enough (less than the intention takes place);
Reverse (the reverse of the intention takes place);
As well as (something in addition to the intention);
Part of (the intention is only partially achieved);
Other than (something other than the intention);
Sooner (something happens before it is intended);
Later (something happens later than intended).

In many cases the application of the guide word will result in a credible
deviation from the designers intention – or will at least trigger ideas.

If a credible deviation is identified, potential causes and consequences
are discussed. From both *causes* and *consequences*, *hazards* may arise
which had not been foreseen by the designer. These hazards can be noted
and appropriate action taken.

10.3.7.2 Example 2. Consider Figure 2, which is a fragment taken
from the acetone distillation process described earlier in this chapter.
Here we are looking at the feed of acetone from the storage tank to the
distillation column. The designer has now been able to design this part
of the plant in detail.

If we look at the designer's intention for this pipe line it will read
something like:

"to convey 50 wt.% acetone solution in water from the Storage Tank T1
to the Distillation Column C1 using pumps P1a or P1b at a flow rate of

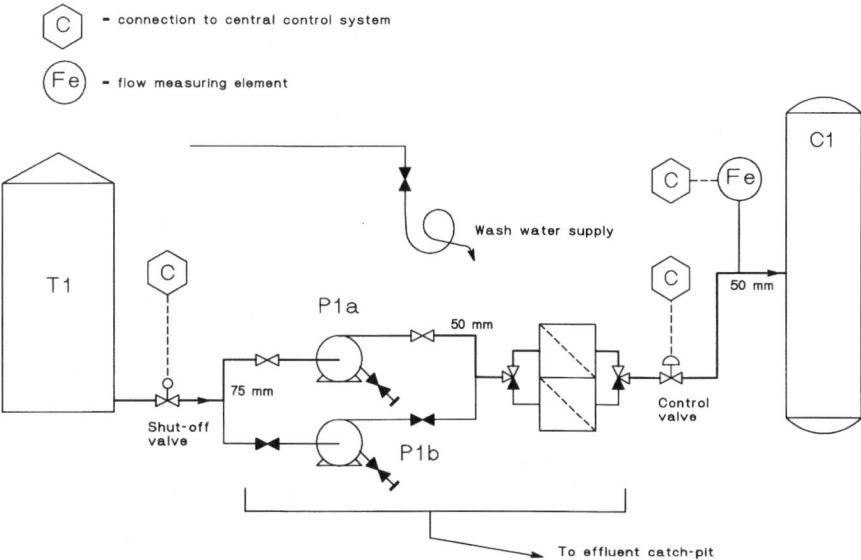

Figure 2 *Detailed P&I diagram of feed-pipe to acetone distillation column*

(say) 25 tonnes per hour and ambient temperature. The pressure at the inlet to the pipeline is just the static head of liquid in T1, which is at atmospheric pressure".

Take the word "flow" from the intention:

"No flow" is a credible deviation. Possible causes might be:

Storage tank T1 empty;
Pump P1a or P1b stopped;
Pump P1a or P1b faulty;
Both sets of block valves on the pumps closed;
Filters blocked;
Pipeline blocked;
High pressure in the distillation column;
Partial vacuum in the storage tank (because, say, a tank vent is blocked);
Spurious closure of the remotely operated shut-off valve;
Failure (to closed) of the flow control valve;
Faulty flow sensor or control system;

...and so on. The list may not be exhaustive.

Possible consequences of these may be:

Interruption to production, or product not to specification;

Damage to pumps from cavitation or dry running;

Overheating of liquid in pumps if there is no flow;

The need to maintain pumps or filters (can we do this safely? Can we isolate and drain these items safely?)

Sucking in of the walls of tank T1;

Unusually high pressure in column C1 (we need to investigate why there might be a high pressure here);

The need to investigate malfunctions of the instrumentation or control systems.

All these points might raise questions for the designer, such as "how do we know the situation is arising?" or "what do we do if it does arise?" Possible answers may encompass such recommendations as fitting level indicators and level alarms to T1, making sure T1 is sized correctly to enable continuous running of the plant under all normal eventualities, trips on the pumps, a re-examination of the venting or pressure relief systems of T1 and C1, provision of drainage points, ventilation and wash-down facilities around the filter and the pumps, facilities to test trip systems periodically without shutting down the plant.

After this is complete, the study team moves to the next deviation. Issues arising from application of the guide word "other than" to the parameter "flow" – which would mean that the flow is going to places other than intended (*e.g.* leaks) – could be interesting. Undoubtedly, questions would be asked about leakage from pump seals and valve glands. There are also issues concerning acetone escapes during the maintenance (removal and draining) of the pumps, and also during the cleaning of the duplex filter which would need to be addressed.

Some combinations of guide word and intention do not make sense (reverse composition?) and if no credible deviation can be generated the study moves on to the next guide word. However, the HAZOP study team should not be in too much of a hurry. Reverse composition might trigger thought of a phase reversal (*e.g.* droplets of water in oil instead of oil in water) and the consequence might be significant. Hazard generation is very much dependent on the skill and imagination of the study team.

In any case, it is worth having a few minutes "brainstorming" in the study looking at causes and consequences, before a more measured assessment of the significant scenarios.

10.3.7.3 Batch Processes. When batch processes are considered, the designer's intention and the conditions in a pipe may vary with time. This adds an extra dimension to the deviations, and "sooner" and "later" have particular significance.

The easiest way to approach such processes is for members of the study team to make themselves very familiar with the steps in the batch process so that it is clear that (for example) "no flow" might be the normal condition at step 1, but at step 5 it generates a series of hazardous causes and consequences.

In critical and complex batch operations it is sometimes helpful to mark up the study drawing to show the condition of the plant appropriate to a particular step in the process, and hazop this as a quasi-steady state. However, this is lengthy and tedious to apply as a general approach.

10.3.7.4 "Others". When all the inlet and outlet flows and pipes to a vessel have been examined, the guide word "other" is applied to the process and vessel. This is a catch-all because the pipe-line guide words do not generate all possible hazards. The prompt list (Table 2) lists some of these topics for further consideration.

10.3.7.5 Short Cut Approaches. In most normal chemical processes we are concerned with the same sets of intentions in pipe-lines (flow, pressure, temperature, composition, viscosity) and it is possible to come up with a ready made list of common deviations. This is easier than returning each time to first principles – but it should always be borne in mind that for unconventional processes this approach could miss important deviations.

The short cut approach yields a number of deviations such as:

---

no flow	too much pressure
too much flow	too little pressure
too little flow	too high temperature
reverse flow	too low temperature
	and so on.

These are captured in a prompt list illustrated in Table 2, along with some common "other" deviations. This is NOT a check list, and each deviation must be "brainstormed" or closely examined to search for ideas on causes, consequences and other possible situations arising.

10.3.7.6 Recording the Study. It is generally impracticable to record a whole HAZOP study verbatim. In most circumstances it suffices to

Table 2   *Typical process deviations for a chemical plant*

Parameter	Typical deviation (parameter plus guide word that makes sense)
Flow	No flow
	Less flow than intended
	More flow than intended
	Reverse flow
	Later flow
	Sooner flow
	Where else might flow be going?
Pressure	More pressure than intended
	Less pressure than intended
	Pressure gradient reversal
Temperature	Temperature higher than intended
	Temperature lower than intended
Level	Greater level
	Lower/less level
Viscosity	More viscous
	Less viscous
Composition	Missing component
	Extra component
	Different/wrong stream selected
	Contamination
	Extra phase present
Sequence	Step omitted
	Step shorter
	Step longer
	Wrong step/wrong order
Other	Start-up
	Shut down
	Relief system
	Power/service failure
	Corrosion/erosion
	Materials of construction
	Toxicity/asphyxia
	Maintenance (provision of isolation, etc.)
	Double valves (trapped liquids)
	Valve access
	Provision for testing instruments, trips, alarms
	Fire
	Static electricity
	Noise
	Thermal radiation
	Ionising radiation
	Sampling
	Spares
	Anything else?

record only those parts of the discussion which lead to hazards, or where the discussion is otherwise "interesting" (and therefore should be recorded).

The most successful way of recording seems to be one of making a brief note to demonstrate that a deviation has been considered, and completing a more detailed report describing the identified hazards. This report is usually in tabular form and has the headings:

Line identification and intention;
Guide word;
Deviation arising;
Cause(s);
Consequence(s);
Action(s) required (and by whom).

There are many HAZOP report shells which are sold to do this job, but they are easy to generate in a word processor or spread sheet. A typical page is shown in Figure 3. Make sure that hazards and actions are uniquely identified by a numbering system, so that they can be referenced by future reports and rapidly retrieved if required.

10.3.7.7 Closing out the HAZOP Study. Although this seems simple and obvious, experience has shown that "closing out" the study is far

Date		Scribe	Flowsheet issue: Flowsheet date:			Page	
Line identification and intention:							
Guide word	Deviation	Hazard no.	Causes	Consequences	Action no.	Action	By

Figure 3 *Typical report structure for a HAZOP study*

from easy. Not all the actions can be undertaken at one time, so that a list of completed and uncompleted actions and the responsibilities for them must be must be kept up to date and reviewed regularly. An action is not closed until it has been physically completed. Simply passing on an instruction to do it, or delegating an action, does not constitute closure. A check must be made that the action really has been completed.

10.4 ASSESSMENT OF RISK

Up to this point we have been exclusively concerned with *identifying* hazards. We have not been concerned with asking whether the hazards present any significant risk, or prioritising them in any way (recall that the risk is a combination of the severity of the hazard and the probability of its occurrence). For much of the time it will be intuitively obvious whether measures ought to be taken to eliminate or deal with an identified hazard. However, on occasion the decision might not be clear-cut and some form of risk assessment will be needed.

Risk assessment can be qualitative or quantitative. We shall deal with qualitative risk assessment first.

10.4.1 Qualitative Risk Assessment

In order to carry out a qualitative risk assessment, a way must be found to describe both the severity of a hazard and the probability of its occurrence. In each case, the exercise by the designers of considerable skill and judgment is needed, and the answers may well depend on local circumstances and the type of process or industry under consideration.

A "low severity" hazard might, for example, be one that is most likely to cause an interruption to production, or inconvenience to the work-force or people living near the plant. A medium level of severity could be a hazard likely to cause injuries to people. A high level of severity is one where the outcome of the hazard is probably one or more fatalities. A hazard may also be measured in financial terms from low severity (small financial implications) to high severity (crippling financial loss to an organisation).

The probability of a hazard may be described in qualitative terms as "rather unlikely", "unlikely" and "very unlikely". These terms have been selected to emphasise that we are normally dealing with uncommon events; if a hazardous event is thought to be more probable than "rather unlikely", appropriate steps to modify the design should be undertaken without further analysis.

Hazard Probability → Hazard Severity ↓	VERY UNLIKELY	UNLIKELY	RATHER UNLIKELY
LOW	LOW RISK	LOW RISK	MEDIUM RISK
MEDIUM	LOW RISK	MEDIUM RISK	HIGH RISK
HIGH	MEDIUM RISK	HIGH RISK	HIGH RISK

Figure 4 *Risk assessment matrix*

The collective experience of a design team can help in determining the probability of a hazardous event. As an example, if any member of the team has ever experienced a similar unusual event or set of conditions, the probability of a future occurrence might be classified as "rather unlikely". If a similar event is outside the direct experience of the team, but there are records or references in the literature to it, the probability could be classed as "unlikely". If the event is conceivably possible, but has never been recorded, the probability would be classed as "very unlikely".

It is worth commenting that the discussion involved in determining the severity and probability of a hazard can be as useful as the outcome, because the careful thought and analysis needed to reach any conclusion can raise the understanding of the hazard and its causes.

Once agreement has been reached on the severity and probability classifications, an idea of the risk can be obtained. A matrix approach is usually used to do this. An example is shown in Figure 4. An organisation may decide, as a matter of policy, to devote resources to addressing risks falling in the medium and high categories.

10.4.2 Quantitative Risk Assessment

Occasionally the qualitative and intuitive methods of risk assessment fail, and something better is required. For example, a designer may have taken steps to address some hazard, but be unsure whether these are sufficient. There is also the possibility of failure of equipment, control systems or operating procedures which may reduce margins of safety. In these circumstances, quantitative risk assessment may be considered. This is an attempt to put numbers to the risks so that we can judge them objectively.

In simple cases, such as the failure of a single component, breakdown of a simple system, or human error, the quantitative extension to risk assessment is not difficult. The first step is, of course, to quantify the severity of a hazard. In the case of fires, explosions or toxic releases, this is often done by calculating the number of fatalities likely to occur as a result. At first sight, such an approach may seem to suggest a morbid fascination with death – why not look at "near misses" or even injuries? The answer is that historical statistics are available for fatalities, and there is no argument about the degree of severity. Considering fatalities is the most objective way. There are other measurements we may be interested in. For financial concerns, the concept of cash loss following an accident is clear. For challenges to the environment, a quantitative measure of damage is much less clear, and at the time of writing, no widely accepted approach has been devised.

10.4.3   Calculation of Hazard Severity

The number of fatalities arising from any identified hazard will depend on several factors, such as the nature of the hazard, the number of people likely to be involved and whether there are any factors mitigating the effects of the hazard. There are many models of varying accuracy and complexity which are available to predict the effects of hazardous events, such as fires, explosions and toxic releases, on people and property. A discussion of them is beyond the scope of this chapter, but for further information the reader is directed to the appropriate chapters of the seminal work by FP Lees *Loss Prevention in the Process Industries* 2nd Edition (Butterworth Heinemann 1996). Designers should be aware that the effects of major accidents can be felt many kilometres off-site. It is often possible to take a simple view however – lesser and more common (but still serious) events, such as the rupture of a vessel, a small fire, or local release of a harmful material, will clearly have potentially fatal consequences to anyone close by. Fatalities arising from slips, trips, falls and contact with moving machinery are obvious and require no modelling.

10.4.4   Calculation of Frequency of Occurrence of Hazards

The next step is to calculate the frequency of occurrence of the hazardous event. Sometimes the issues can be complex. This is often the case where major hazards are concerned. For example, a large-scale release of toxic gases from a vessel under pressure may arise from a number of causes. For each of these causes a conscientious designer will have

Table 3 *Typical plant failure rates*

Event (service failures excluded)	Frequency, events per year
Control loop fails	0.5
Independent alarm or trip fails	0.2
Relief valve fails closed (fails to danger)	0.01
Relief valve lifts light (fails safe)	0.1
Bursting disk (rupture disk) fails	0.2
Pump fails	0.5
Fan fails	0.2
Compressor fails	9
Boiler fails	4
Heat exchanger fails (tube leaks, etc.)	0.1
Major vessel fails (inside design conditions)	10^{-5}
Major pipeline fails (per km)	10^{-4}

attempted to build-in several safeguards so that if one system fails, others will maintain plant integrity. The scenarios are complicated. In order to calculate the probability of such a release, the usual way is by constructing a "fault tree" which sets out in diagrammatic form all the faults which have to be present for the release to occur, calculates the probability of each and then combines these probabilities statistically.

Fortunately, in many cases, a hazard can be linked to a straightforward failure of a control system, malfunction of a plant trip, or a simple human error. In this case the problem reduces to obtaining an estimate for the frequencies of these occurrences. At first sight, obtaining credible data on failure rates of plant and people can seem a daunting task. However, we should be aware that we are usually looking only for "order of magnitude" estimates, and often reasonable estimates of failure rates can be gleaned from everyday experience or generic data.

Some figures used by one major company for well-maintained systems are given in Table 3 (source: Courtaulds plc).

10.4.5 The Concept of Testing Trips and Safety Devices

It is common on chemical plant to install safety devices such as trips and relief valves which protect the plant in the event of a malfunction of control systems or human error. Unfortunately, these devices can (and do) fail occasionally. The problem is that the failures cannot be seen until they are tested or until they are called upon to act (a plant may operate perfectly normally even though, say, a pressure relief valve is faulty, because under normal conditions the valve is never activated). It is thus necessary to test safety devices periodically to ensure they are functioning.

Given that we do tests in a routine fashion, what is the probability that a safety device will let us down because it has failed since the last test?

It may be shown that the hazard rate H arising from a failed safety device is given by the formula:

$$H = \frac{1}{2}ftD \tag{1}$$

where H = rate of occurrence of hazardous event (hazards per year), f = failure rate of the safety device (failures per year), D = frequency of the demand placed on the safety device (demands per year), t = time between successive tests of the safety device. The failure of the safety device is assumed to be a random event.

This simple formula only applies provided the product Dt is less than about 1 and $ft \ll 1$ (these conditions are usually satisfied, but should be checked!) For a much more detailed approach, including a discussion of the case where the approximation does not apply, the reader is referred to Chapter 13 of the book by F P Lees *Loss Prevention in the Process Industries* 2nd Edition (Butterworth Heinemann 1996).

In other words, if a safety device is called on to act, there is a probability equal to $\frac{1}{2}ft$ that it will fail to do so. The importance of this simple formula is the demonstration that the probability of a safety device proving ineffective is directly linked to the failure rate of the device and the time interval between tests.

It is very important to note that all this only applies to safety devices where the failure is hidden during normal plant operation. For, say, a normal control device where a failure would be immediately manifested by a malfunction of the plant, the hazard rate is simply the same as the failure rate. No amount of testing will help here.

10.4.6 Calculation of the Risk

The magnitude of the risk to people is normally taken as the Fatal Accident Rate (FAR). This is calculated by multiplying the size of the hazard (measured in fatalities per hazardous event) by the frequency of the hazardous event (measured in events per year). The FAR has units of fatalities per year.

10.4.7 Example 3

Figure 5 shows a (fictitious) arrangement for an autoclave. It is heated with high-pressure steam in a coil, and the temperature of the liquid

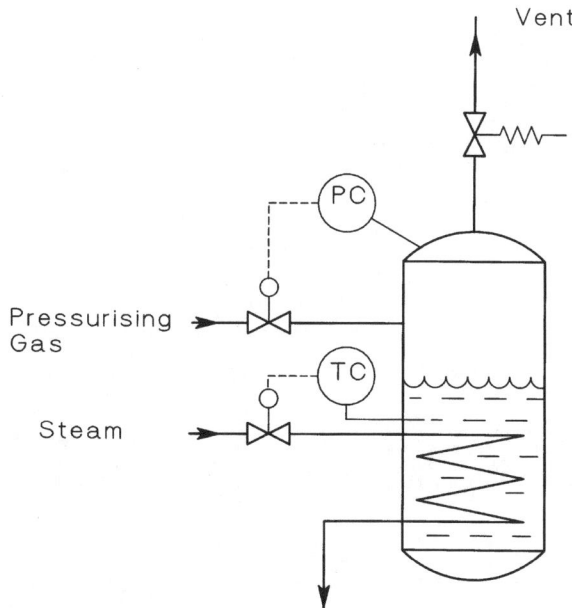

Figure 5 *Pressurised autoclave*

contents is maintained by a temperature controller operating on a modulating valve. In addition, a pressuring gas is applied to the head space through a control valve. In the event that either of the controllers should fail, a pressure relief valve is provided to vent the autoclave and prevent it being overpressurised. The relief valve is inspected and tested once per year. However, we note that the relief valve may not operate when called on, and this presents the potential hazard of autoclave overpressurisation.

For the purpose of the example, let us further suppose that someone is in the close vicinity of the vessel for 10% of the time, and that overpressurisation of the autoclave could lead to vessel rupture and a potential fatality to that person.

What is the magnitude of the risk (FAR) this arrangement presents?

The first step is to quantify the severity of the potential hazard. From the information we are given, we know that there is a 10% chance of a single fatality if vessel rupture occurs. In other words, there are 0.1 fatalities per hazardous event.

We next need to calculate the frequency of the hazardous event. A small fault tree, shown in Figure 6 will make this clear.

Using the frequencies suggested in Table 3, we can see that we might expect the temperature controller to fail about 0.5 times per year. The

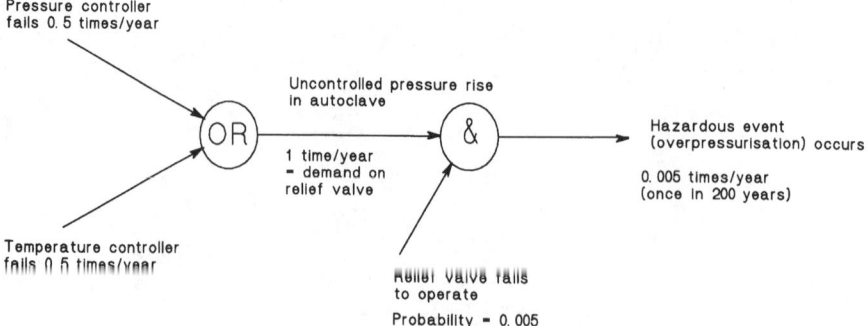

Figure 6 *Simple fault tree for pressurised autoclave*

pressure controller has a similar failure rate. The failure of either could lead to a high pressure in the autoclave, so the demand on the relief valve is once per year.

The time between regular inspection of the relief valve is 1 year, and the failure rate (to danger) is 0.01 failures per year. So the hazard rate arising from this demand on the relief valve is given by Equation (1) above to yield

$$H = \frac{1}{2}ftD = 0.5 \times 0.01 \times 1 \times 1 = 0.005 \text{ hazardous events per year}$$

The magnitude of the risk arising from this arrangement is thus:

$$\text{FAR} = \text{hazard frequency} \frac{\text{events}}{\text{year}} \times \text{hazard severity} \frac{\text{fatalities}}{\text{event}}$$

$$= 0.005 \times 0.1$$

$$= 0.0005 \frac{\text{fatalities}}{\text{year}} \text{ or } 5 \times 10^{-4} \frac{\text{fatalities}}{\text{year}}$$

10.4.8 The Numerical Perspective

The calculation of numerical values of risk is of little use unless some attempt is made to place the numbers in some sort of perspective. The following argument was developed originally by ICI and is further discussed by F P Lees in his book *Loss Prevention in the Process Industries* 2nd Edition (Butterworth Heinemann 1996).

In the UK the historical FAR in the chemical industry during the latter part of the 20th century was about 3.5 fatalities per 10^8 hours worked. Of these, about half were chemical process-related accidents (the remainder were falls, vehicle accidents, *etc.*). So, for continuously operated chemical plant (8760 hours per year) the fatal process accident rate was about $0.5 \times 3.5 \times 8760/10^8$ per year, or about 0.00015 per year.

Now, for a "typical" plant there may be around five possible potentially fatal hazardous events that can be identified. Clearly, this is a considerable generalisation, but we are only interested in order of magnitude estimates here. This means that, for any identified hazard, the historical average FAR is about 0.00003 per year, or 3×10^{-5} fatalities per year for the individuals most at risk.

This figure can be used as a first benchmark when deciding what to do about identified risks. If we look at the FAR for the individual most at risk calculated in Example 3 above, we see that it is about 5×10^{-4} fatalities per year. This is an order of magnitude higher than the industrial average, and would indicate that further levels of protection on the autoclave would be necessary.

Note that an FAR of 3×10^{-5} fatalities per year for an identified risk does not imply a level of tolerability or acceptability – it is simply the average value for the chemical industry in recent years. It is probably not far from a value which might be considered tolerable for on-site workers in the conventional chemical industries – but the figure is subject to major changes through shifts in public opinion and political pressure, and depends on the nature of the industry.

For people not working on-site, but simply living near chemical plant, the boundaries of tolerability and acceptability are set much lower. In the UK the FAR arising from lightning strikes is about 1×10^{-7} fatalities per year for each individual. It is unlikely that any man-made risk exceeding this would be acceptable off-site.

At first sight it may seem that the accuracy of this approach is severely limited by the rather crude quality of the failure-rate data available. However, it should be borne in mind that this is very much an "order of magnitude" exercise. In practice, far greater errors occur because of failure to *identify* significant hazards in the first place, thus leaving whole branches off the fault tree. Other large errors crop up because of logical mistakes in constructing fault trees. These can include problems in recognising "common mode" failures (where a single breakdown can interfere with multiple safety systems, so they are not independent), and statistical mistakes. A common example of the latter is the temptation to multiply failure rates of parallel systems together in order to get a combined failure rate. This is completely *wrong*, as an examination of the units on each number will immediately reveal.

This brief excursion for quantitative risk assessment is intended only to illustrate what can be done in simple cases. For a detailed discussion of this controversial topic, it is essential that designers consult more advanced texts and up-to-date sources. Recent statistics from the UK (Health and Safety Commission (UK) *Statistics for fatal injuries 2004/5*) suggest that the FAR fell in the late 1990's to around 2 per 10^8 hours worked in UK manufacturing industry as a whole and has remained roughly constant since.

10.5   CONTROL OF CHEMICAL PLANT

It is a common misconception among process designers that safety devices built into chemical plant can be expected to work at all times – at least if they are well maintained. Quite clearly, the discussion in this chapter has demonstrated that this is not the case. Designers should always be aware that their safety and protection systems can, and will, fail at some time. So an element of redundancy in such systems is nearly always required.

It follows that it is good practice to separate chemical plant safety systems from control systems, so that the number of components common to both is minimised. Whilst it is quite possible to specify that control systems should raise alarms and trips when measured variables move out of bounds, a likely reason for this is that a part of the control system failed in the first place. So the control system cannot be relied upon to raise alarms reliably.

Increasingly, the control of chemical plant is being undertaken by advanced, programmable electronic equipment (see chapter 7). This brings its own set of problems. The equipment may fail electrically (though on the whole electronic systems are very reliable), but in addition there may be programming errors in the system. These are extremely hard to discover, and can lurk for years before becoming apparent when the system is presented with a set of inputs that the programmer had not thought about. Consequently, conservative engineers tend to specify that safety related applications should be mechanical or "hard wired" and not programmable. This is not so much because such systems are more reliable than programmable ones – it is just that when they go wrong it is usually in a simpler and more easily understood fashion. However, for cost reasons there is often pressure to use programmable systems in safety related applications. In this case, the programming must be carried out under strictly controlled conditions, following established guidelines such as those laid down by the Institution of Electrical Engineers.

10.6 SUMMARY

Process plant designers have a professional responsibility to take all reasonable steps to ensure that the equipment they design does not harm people or the environment. To this end, they are required to identify hazards that may arise from their activities, and then assess the risks that arise from those hazards.

Hazard identification takes place throughout all the stages of plant design, and a staged system of Hazard Studies may be used to facilitate this.

The risk arising from a hazard is a combination of the severity of the hazard and the likelihood of its occurrence. Once a hazard has been identified, the risk posed by it should be assessed. This assessment can be intuitive (most "obvious" hazards are dealt with this way), formal (but qualitative) or quantitative.

Based on the risk assessment, a decision can be made whether to accept the risk, tolerate it or modify the design and/or operating procedures to minimise it.

REFERENCES

1. *The Sustainability Metrics (Sustainable Development Progress Metrics) – recommended for use in the process industries*, The Institution of Chemical Engineers, Rugby, 2002.
2. *Notes on Hazard and Operability Studies*, The Royal Society of Chemistry, London, 2001.
3. F. Crawley, M. Preston and B. Tyler, *HAZOP – Guide to Best Practice*, The Institution of Chemical Engineers, Rugby, 2000.
4. F. Crawley and B. Tyler, *Hazard Identification Methods*, The Institution of Chemical Engineers, Rugby, 2003.
5. F.P. Lees, *Loss Prevention in the Process Industries*, 2nd edn, Butterworth Heinemann, Oxford, 1996.
6. T.A. Kletz, *Learning from Accidents in Industry*, Butterworth, Oxford, 1988.
7. T.A. Kletz, *An Engineer's View of Human Error*, The Institution of Chemical Engineers, 2nd edn, Rugby, 1991.

Subject Index